# Soil Science and Management

# Soil Science and Management

Edited by **Brian Bechdal**

R CALLISTO REFERENCE

New York

Published by Callisto Reference,
106 Park Avenue, Suite 200,
New York, NY 10016, USA
www.callistoreference.com

**Soil Science and Management**
Edited by Brian Bechdal

International Standard Book Number: 978-1-63239-567-2 (Hardback)

Printed in the United States of America.

# Contents

# Preface

Soil science is a scientific discipline that focuses on the study of soil as a natural resource that is abundant on the surface of the Earth, including the study of soil formation, its classification and mapping as well as the chemical, physical, biological, and fertility properties of soils along with the relation of these properties to the use and management of soils. Soil science tends to draw from various areas of specialization such as pedology, microbiology, physics, edaphology and chemistry. The work done in this field can be said to be very much restricted and directed by the problems and challenges facing our civilization's need to sustain the land that supports it. Soil scientists usually specialize in soil physics, soil chemistry, pedology, soil microbiology and applied soil science in related disciplines. Soil scientists have raised concerns about the need to preserve soil in today's world with a growing population, possible future water crisis, land degradation and increasing per capita food consumption. New opportunities and avenues in soil research are driven by a need to understand soil in terms of greenhouse gases, climate change and carbon sequestration. Scientists are researching as well as exploring the diversity and dynamics of this resource and therefore continuing to yield fresh discoveries and insights.

I wish to thank all the contributing authors who took out time to share their researches with us. I am grateful to those who put their hard work, effort and expertise into these researches as well as those who were supportive in this endeavour.

**Editor**

# The Effect of Minjingu Phosphate Rock and Triple Superphosphate on Soil Phosphorus Fractions and Maize Yield in Western Kenya

**Robert Orangi Nyambati[1] and Peter Asbon Opala[2]**

[1] Kenya Forestry Research Institute, Maseno Regional Centre, P.O. Box 5199, Kisumu, Kenya
[2] Department of Soil Science, Maseno University, P.O. Box, Private bag, Maseno, Kenya

Correspondence should be addressed to Robert Orangi Nyambati; nyambatir@yahoo.com

Academic Editors: J. A. Entry and W. Peijnenburg

We tested the effects of triple superphosphate (TSP) and Minjingu phosphate rock (MPR), when applied at phosphorus (P) rates of 50 or 250 kg P ha$^{-1}$ in a factorial combination with urea or Tithonia diversifolia green manure as nitrogen sources, on P availability and maize yields for two seasons at Nyabeda and Khwisero in Kenya. Phosphorus availability was determined by the Olsen method or sequential fractionation. There was no significant difference in Olsen P as influenced by TSP and MPR at 50 kg P ha$^{-1}$ irrespective of the N source at both sites in both seasons. However, at 250 kg P ha$^{-1}$, TSP gave significantly higher Olsen P than MPR. The labile P fractions generally followed the same trend as the Olsen P. Maize yields increased with increasing amount of P applied. Generally, there was no significant difference between TSP and MPR on maize yields irrespective of the N source. The Olsen-P, Resin-P, and sodium bicarbonate inorganic P correlated well with maize yields when TSP was used but the correlations between these P tests and maize yields for MPR were not consistent and therefore their use on soils treated with MPR should be exercised with caution.

## 1. Introduction

More than 80% of the soils in the densely populated highlands of western Kenya are inherently low in P and this seriously limits the productivity of maize which is the staple food crop in the area [1]. Although judicious application of inorganic P fertilizers is recognized as the most effective method for alleviating P deficiencies, their high cost, inaccessibility, and erratic and unprofitable crop responses limit their use, particularly on smallholder farms [2]. Much research on soil fertility management in western Kenya has therefore been devoted to testing nutrient inputs that are thought to be inexpensive, locally available, and sustainable, as alternatives to conventional fertilizers. The use of organic materials (OMs) and phosphate rocks (PR) has in particular received considerable research attention in recent years [3–5]. This has been given impetus by a paradigm shift in soil fertility management towards the integrated soil fertility management (ISFM). In this strategy, the use of organic and inorganic

nutrient sources in combination is advocated on smallholder farms [6] with the OMs being utilized mainly as sources of N, because of their low P content, while inorganic fertilizers are used to supply P. In addition, to increase the range of OMs that smallholder farmers can use in integrated soil fertility management (ISFM), some nontraditional sources of nutrients such as agroforestry shrubs are being tested with Tithonia diversifolia green manure (tithonia) showing great promise [7]. The combination of OMs such as tithonia with inorganic P fertilizers is particularly attractive since it has been demonstrated to reduce P sorption and increase P availability in these P fixing soils [8].

Phosphorus when applied to the soil undergoes a series of reactions which transforms it to varying forms of availability to the plant. Determination of the available P usually employs extracting solutions which dissolve and remove specific forms of soil P from the solid phase. Currently the Olsen method is used to estimate inorganic P availability in acid soils of western Kenya but there are concerns that it may

TABLE 1: Some selected chemical and physical characteristics of the soils at Nyabeda and Khwisero.

| Parameter | Nyabeda | Khwisero |
|---|---|---|
| pH (1 : 2.5 soil : water) | 5.40 | 5.20 |
| Exchangeable acidity (cmol$_c$ k g$^{-1}$) | 0.10 | 1.00 |
| Organic carbon % | 1.58 | 1.75 |
| Exchangeable Ca (cmol$_c$ kg$^{-1}$) | 4.65 | 3.53 |
| Exchangeable Mg (cmol$_c$ kg$^{-1}$) | 1.90 | 1.25 |
| Exchangeable K (cmol$_c$ kg$^{-1}$) | 0.09 | 0.07 |
| Olsen extractable P (mg kg$^{-1}$) | 3.9 | 4.4 |
| P sorption at P conc. of 0.2 mg P L$^{-1}$ | 316 | 375 |
| Clay (%) | 55 | 35 |
| Sand (%) | 25 | 45 |
| Silt (%) | 20 | 20 |
| Textural class | Clay loam | Sandy clay loam |

not accurately predict available P because it was originally developed for alkaline soils. In addition, when used with phosphate rock (PR), it may not extract undissolved PR [9], which provides a significant, if not the main portion of P, that is directly available to plants. Sequential chemical extraction procedures, on the other hand, are used to separate extractable soil P into different inorganic and organic fractions [10]. The underlying assumption in these approaches is that the readily available soil P is removed first with mild extractants, while the less available or plant unavailable P can only be extracted with stronger acids and alkalis. However, according to [11], methods of determination of available P, in an agronomic context, never measure the quantity of P available to the crop but measure a fraction of soil P that is somewhat related to that portion of soil P that is plant available. A knowledge of the relationships is useful for a better understanding of the capacity of P fractions to sustain an adequate supply of P to crops [12]. The objective of this study was therefore to (i) compare effect of two P sources, Triple superphosphate (TSP) and Minjingu phosphate rock (MPR) when applied with either urea or tithonia green manure on soil phosphorus fractions and maize yields, and (ii) determine the relationship between the P fractions and maize yields in acid soils of western Kenya.

## 2. Materials and Methods

*2.1. Site and Soil Descriptions.* A field experiment was conducted for two consecutive seasons from March 2007 to February 2008 at two sites in western Kenya, Nyabeda (34°15'E, 020'N) which is at an altitude of 1330 masl and Khwisero (34°30'E, 07'N) which is at 1430 masl. Both sites receive a mean annual rainfall of 1800 mm. The sites were chosen on the basis of contrasting soil characteristics (Table 1) with the soils at Khwisero being Acrisols while at Nyabeda they are Ferralsols.

The experiment was laid out as a randomized complete block design with four replications with plot sizes measuring 6 m by 6 m. The treatments consisted of two P sources, TSP (20% P) and MPR (13% P), each applied at the rate

of 50 kg P ha$^{-1}$ or 250 kg P ha$^{-1}$. The two P rates represent two different strategies of P application, a large one-time application (250 kg P ha$^{-1}$) that is expected to provide a strong residual effect for several cropping seasons of maize and seasonal applications of 50 kg P ha$^{-1}$. These two P sources were combined with either urea or tithonia, which were applied seasonally to provide N at a rate of 60 kg ha$^{-1}$. In addition, a control treatment with no P input but with urea applied to supply 60 kg N ha$^{-1}$ and a tithonia alone treatment (tithonia applied at 1.82 t ha$^{-1}$ to provide 60 kg N ha$^{-1}$) were included. Initial characterization of tithonia (consisting of leaves and tender stems) showed that, on average, it had 3.3% N, 0.30% P, and 4.0% K. At the application rate of 60 kg N ha$^{-1}$, tithonia therefore also provided 6 kg of P. To ensure that the total P rates among treatments with urea and tithonia as N sources did not differ, TSP or MPR in the treatments with tithonia was therefore applied to supply only the difference required to provide the appropriate P rate. Potassium was blanket applied to all treatments at 60 kg ha$^{-1}$. The intention was to ensure that the other macronutrients were not limiting to plant growth while studying the P effects. The nutrient inputs were evenly spread within the appropriate experimental plots and incorporated to a depth of 0–0.15 m at the time of planting in each of the seasons, apart from MPR or TSP at the P rate of 250 kg ha$^{-1}$, which were applied in the first season only. However, only half of the urea was applied at planting and the rest was applied 5 weeks later. The experiment was run for two cropping seasons (March to August 2007 and September 2007 to February 2008). In both seasons, maize (Hybrid 511) was planted at a spacing of 0.75 m by 0.25 m and grown using the recommended agronomic practices of the area and its grain yield determined at the end of each season.

*2.2. Soil Sampling and Analyses.* In each season, soil samples (0–0.15 m) were collected in each plot at the ninth week after planting of the maize crop. Each soil sample was a composite collected from nine sampling points per plot. The soils were air-dried and prepared for analyses using standard procedures. The available soil P was determined using two methods, (i) the conventional Olsen method [13] in which available P was determined by shaking 2.5 g of air-dried soil with 50 mL of 0.5 M NaHCO$_3$ (pH 8.5) for 30 minutes and the inorganic P in the extract determined colorimetrically, and (ii) sequential fractionation by the method of [10] in which the following P fractions were determined: Resin extractable inorganic P (Resin-P$_i$), sodium bicarbonate extractable inorganic and organic P (NaHCO$_3$-P$_i$ and NaHCO$_3$-P$_o$), sodium hydroxide extractable inorganic and organic P (NaOH-P$_i$ and NaOH-P$_o$), and dilute hydrochloric acid extractable P (HCl-P$_i$).

*2.3. Mathematical Calculations and Data Analyses.* The relative agronomic effectiveness (RAE) of MPR compared to TSP was calculated as

$$\text{RAE} = \frac{(Y_{\text{MPR}} - Y_{\text{control}})}{(Y_{\text{TSP}} - Y_{\text{control}})} \times 100, \quad (1)$$

The Effect of Minjingu Phosphate Rock and Triple Superphosphate on Soil Phosphorus Fractions and Maize Yield in
Western Kenya

3

TABLE 2: Olsen extractable soil P ($mg\,kg^{-1}$) at Nyabeda and Khwisero.

| Treatment | Nyabeda | | Khwisero | |
|---|---|---|---|---|
| | Season 1 | Season 2 | Season 1 | Season 2 |
| (1) Control | 3.8 | 4.4 | 3.6 | 5.8 |
| (2) Tithonia ($6\,kg\,P\,ha^{-1}$) | 4.0 | 4.9 | 4.3 | 6.1 |
| (2) TSP + urea ($50\,kg\,P\,ha^{-1}$) | 4.8 | 5.4 | 5.7 | 7.7 |
| (3) TSP + tithonia ($50\,kg\,P\,ha^{-1}$) | 5.2 | 5.4 | 6.8 | 7.5 |
| (4) MPR + urea ($50\,kg\,P\,ha^{-1}$) | 3.9 | 5.0 | 4.2 | 7.1 |
| (5) MPR + tithonia ($50\,kg\,P\,ha^{-1}$) | 4.8 | 5.6 | 4.8 | 6.9 |
| (6) TSP + urea ($250\,kg\,P\,ha^{-1}$) | 21 | 12 | 22 | 24 |
| (7) TSP + tithonia ($250\,kg\,P\,ha^{-1}$) | 18 | 14 | 20 | 28 |
| (8) MPR + urea ($250\,kg\,P\,ha^{-1}$) | 7.7 | 9.0 | 9.4 | 16 |
| (9) MPR + tithonia ($250\,kg\,P\,ha^{-1}$) | 8.0 | 11 | 10 | 16 |
| s.e.d | 2.3 | 0.9 | 1.7 | 1.9 |
| CV. | 11 | 8 | 13 | 10 |

TSP: triple superphosphate; MPR: Minjingu phosphate rock; s.e.d.: standard error of difference between means; CV: coefficient of variance.

where $Y_{MPR}$ is maize grain yield from MPR, $Y_{TSP}$ is maize grain yield from TSP, and $Y_{control}$ is maize grain yield from control (0 P).

The Genstat statistical package [14] was used to conduct analysis of variance (ANOVA) to determine the effect of treatments on the soil P fractions and maize grain yields. The standard error of difference between means (s.e.d.) was used to compare the treatment means. Mention of statistical significance refers to $P < 0.05$ unless otherwise stated. The relationships between relevant parameters were determined by using correlation analysis.

# 3. Results

## 3.1. Olsen P.
The Olsen P generally increased with increasing rates of P application (Table 2). There was no significant difference in Olsen P as influenced by TSP and MPR at the P rate of $50\,kg\,ha^{-1}$ irrespective of whether they were applied in combination with urea or tithonia at both sites in both seasons. However at P rate of $250\,kg\,ha^{-1}$, TSP gave significantly higher Olsen P values than MPR with same N source at both sites in both seasons. For the same P source, there was no significant effect of N source, that is, tithonia and urea on Olsen P in the first season at both sites. In the second season, however, combining TSP or MPR with tithonia generally gave higher Olsen P values than their respective combinations with urea. Between the sites, the Olsen P values were generally higher at Khwisero than Nyabeda for comparable treatments with the differences being more marked in the second season.

## 3.2. Soil P Fractions

### 3.2.1. Labile P (Resin-P and Sodium Bicarbonate Inorganic and Organic P).
Both the resin-P and $NaHCO_3$-$P_i$ were low and, on average, only 4% and 2% of the resin-P and $NaHCO_3$-$P_i$ fractions, respectively, of the sequentially extracted P were recovered in these fractions (Tables 3, 4, 5, and 6). The addition of the P sources significantly increased the resin-P and $NaHCO_3$-$P_i$ values above the control at the higher P rates ($250\,Kg\,ha^{-1}$) but not the lower rates ($50\,Kg\,ha^{-1}$) in both seasons at both sites. At the higher P rate, there was no significant difference in resin-$P_i$ between TSP + urea and TSP + tithonia in the first season but, in the second season, TSP + tithonia was better than TSP + urea at both sites. But the difference between MPR combined with urea or tithonia was not significant at both sites in both seasons. The $NaHCO_3$-$P_i$ followed the same trend as that of resin-P although its values were lower. In general, TSP had higher resin-P and $NaHCO_3$-$P_i$ than MPR at both sites in both seasons at similar P application rates. Bicarbonate extractable $P_o$ was higher than the resin-P and $NaHCO_3$-$P_i$ in all the treatments at both sites at the two sampling times and represented an average of 11% (range 8–15%) of the sum of the sequentially extracted P fractions and an average of 59% of the labile P fraction (Tables 4–6). However, the $NaHCO_3$-$P_o$ was unaffected by treatments at both sites.

### 3.2.2. Moderately Labile P ($NaOH$-$P_i$ and $NaOH$-$P_o$) and Nonlabile P ($HCl$-$P_i$).
The moderately labile fraction constituted 80% of the sequentially extracted P that was recovered. There was no significant treatment effect on the moderately labile fraction at the lower P rates at both sites in both seasons. However, at the high P rate, TSP applied in combination with urea gave higher $NaOH$-$P_i$ than its combination with tithonia at Nyabeda in the first season (Table 3). The TSP treatments also gave higher $NaOH$-$P_i$ than similar MPR treatments at the higher rate at Nyabeda in the first season. However, the difference between MPR applied in combination with urea and tithonia was not significant. A similar trend was observed at Khwisero in this season. In the second season, there was an increase in $NaOH$-$P_i$ above the control for the lower P rate of $50\,kg\,ha^{-1}$ but this was not significant. A similar trend as in the first season whereby the TSP treatments gave higher $NaOH$-$P_i$ than the MPR treatments at the higher P rate was observed at both sites although statistical significance was not

TABLE 3: Sequentially extractable soil P fractions ($mg\ kg^{-1}$) at Nyabeda in the first season.

| Treatment | Resin-$P_i$ | NaHCO$_3$-$P_i$ | NaOH-$P_i$ | HCl-$P_i$ | NaHCO$_3$-$P_o$ | NaOH-$P_o$ | Sum of fractions |
|---|---|---|---|---|---|---|---|
| (1) Control | 8.5 | 5.2 | 81 | 9.2 | 47 | 265 | 416 |
| (2) Tithonia (6 kg P ha$^{-1}$) | 8.5 | 5.6 | 75 | 9.2 | 44 | 274 | 416 |
| (3) TSP + urea (50 kg P ha$^{-1}$) | 9.5 | 6.9 | 85 | 9.7 | 47 | 291 | 449 |
| (4) TSP + tithonia (50 kg P ha$^{-1}$) | 9.6 | 8.6 | 77 | 9.7 | 44 | 268 | 417 |
| (5) MPR + urea (50 kg P ha$^{-1}$) | 8.4 | 6.2 | 76 | 8.2 | 43 | 285 | 427 |
| (6) MPR + tithonia (50 kg P ha$^{-1}$) | 9.8 | 7.4 | 84 | 11 | 45 | 275 | 432 |
| (7) TSP + urea (250 kg P ha$^{-1}$) | 28 | 27 | 143 | 9.6 | 46 | 281 | 535 |
| (8) TSP + tithonia (250 kg P ha$^{-1}$) | 29 | 27 | 125 | 10 | 47 | 299 | 537 |
| (9) MPR + urea (250 kg P ha$^{-1}$) | 27 | 14 | 99 | 11 | 43 | 288 | 483 |
| (10) MPR + tithonia (250 kg P ha$^{-1}$) | 26 | 16 | 103 | 12 | 44 | 268 | 469 |
| s.e.d. | 5.1 | 2.1 | 10 | 0.63 | NS | NS | 18 |
| CV | 11 | 8 | 13 | 12 | 10 | 11 | 7 |

TSP: triple superphosphate; MPR: Minjingu phosphate rock; s.e.d.: standard error of difference between means; CV: coefficient of variance.

TABLE 4: Sequentially extractable soil P fractions ($mg\ kg^{-1}$) at Khwisero in the first season.

| | Resin-$P_i$ | NaHCO$_3$-$P_i$ | NaOH-$P_i$ | HCl-$P_i$ | NaHCO$_3$-$P_o$ | NaOH-$P_o$ | Sum of fractions |
|---|---|---|---|---|---|---|---|
| (1) Control | 7.4 | 7.4 | 72 | 9.4 | 57 | 322 | 475 |
| (2) Tithonia | 8.4 | 7.7 | 72 | 10 | 56 | 398 | 522 |
| (3) TSP + urea (50 kg P ha$^{-1}$) | 11 | 9.2 | 72 | 10 | 52 | 327 | 481 |
| (4) TSP + tithonia (50 kg P ha$^{-1}$) | 15 | 11 | 84 | 10 | 57 | 329 | 506 |
| (5) MPR + urea (50 kg P ha$^{-1}$) | 11 | 8.8 | 71 | 11 | 52 | 328 | 482 |
| (6) MPR + tithonia (50 kg P ha$^{-1}$) | 12 | 7.9 | 71 | 11 | 52 | 327 | 481 |
| (7) TSP + urea (250 kg P ha$^{-1}$) | 43 | 25 | 135 | 12 | 57 | 356 | 628 |
| (8) TSP + tithonia (250 kg P ha$^{-1}$) | 50 | 29 | 116 | 11 | 60 | 354 | 620 |
| (9) MPR + urea (250 kg P ha$^{-1}$) | 39 | 16 | 109 | 21 | 58 | 345 | 588 |
| (10) MPR + tithonia (250 kg P ha$^{-1}$) | 33 | 16 | 101 | 21 | 54 | 337 | 571 |
| s.e.d. | 6.6 | 2.8 | 11 | 1.5 | NS | NS | 22 |
| CV | 8 | 12 | 11 | 12 | 10 | 13 | 10 |

TSP: triple superphosphate; MPR: Minjingu phosphate rock; s.e.d.: standard error of difference between means; CV: coefficient of variance.

always attained. The NaOH-$P_o$ fraction was on average 60% of the sum of fractions (range 46–72%) and constituted the largest of the P fractions in the studied soils but the difference in this P fraction among the treatments was not significant. The HCl-$P_i$ fraction was small and accounted for only 2% of the sum of all fractions. All treatments with MPR at the high P rate of 250 kg ha$^{-1}$ gave significantly higher HCl-$P_i$ values than the control at Khwisero in both seasons while at Nyabeda only MPR applied in combination with tithonia at 250 kg P ha$^{-1}$ had higher HCl-$P_i$ values than the control in both seasons.

### 3.3. Maize Grain Yields.

Maize strongly responded to application of P at both sites with yields generally increasing with increasing amount of P applied (Table 7). Even tithonia applied alone (6 kg P ha$^{-1}$), more than doubled yields compared to the control at both sites in the two seasons. At the same P rate, there was no significant difference between TSP and MPR as P sources irrespective of whether tithonia or urea was used as N sources at Nyabeda in both seasons. At Khwisero, TSP was superior to MPR when tithonia was used as the N source, at similar P rates, in the first season. There were however no significant differences between MPR and

TSP, at same P rate, when urea was used at this site in the first season. In the second season, no significant differences were observed between TSP and MPR at similar P rates at Khwisero irrespective of the N source used. Among the N sources, tithonia and urea were equally effective at Nyabeda irrespective of P source or rate in the second season but, in the first season, TSP combined with urea at 250 kg ha$^{-1}$ was superior to tithonia at similar P rate. At Khwisero, tithonia was generally superior to urea in the second season. The RAE for MPR at Nyabeda ranged from 55% to 113% (mean 83%) while at Khwisero the RAE was higher with a range of 64% to 171% (mean 112%). Generally combining MPR with tithonia gave a higher RAE at Nyabeda in the first season but at Khwisero combining it with urea gave higher RAE. In the second season however, combining MPR with tithonia consistently gave higher RAE than its combination with urea at both sites.

### 3.3.1. Relationships between Phosphorus Fractions with Maize Grain Yields.

Olsen P correlated well with maize yields at Khwisero in both seasons while at Nyabeda the correlations were also significant except when MPR was used with urea or tithonia in the second season (Table 8). There were also

The Effect of Minjingu Phosphate Rock and Triple Superphosphate on Soil Phosphorus Fractions and Maize Yield in Western Kenya

5

TABLE 5: Sequentially extractable soil P fractions (mg kg$^{-1}$) at Nyabeda in the second season.

| Treatment | Resin-$P_i$ | NaHCO$_3$-$P_i$ | NaOH-$P_i$ | HCl-$P_i$ | NaHCO$_3$-$P_o$ | NaOH-$P_o$ | Sum of fractions |
|---|---|---|---|---|---|---|---|
| (1) Control | 5.4 | 9.6 | 77 | 9 | 44 | 308 | 453 |
| (2) Tithonia | 7.1 | 9.2 | 79 | 10 | 45 | 327 | 477 |
| (3) TSP + urea (50 kg P ha$^{-1}$) | 9.8 | 11 | 91 | 11 | 46 | 324 | 493 |
| (4) TSP + tithonia (50 kg P ha$^{-1}$) | 8.1 | 10 | 84 | 9 | 44 | 317 | 472 |
| (5) MPR + urea (50 kg P ha$^{-1}$) | 9.0 | 10 | 85 | 10 | 41 | 322 | 477 |
| (6) MPR + tithonia (50 kg P ha$^{-1}$) | 9.6 | 10 | 93 | 10 | 43 | 324 | 490 |
| (7) TSP + urea (250 kg P ha$^{-1}$) | 33 | 21 | 143 | 11 | 44 | 333 | 585 |
| (8) TSP + tithonia (250 kg P ha$^{-1}$) | 40 | 23 | 135 | 10 | 44 | 331 | 582 |
| (9) MPR + urea (250 kg P ha$^{-1}$) | 20 | 17 | 116 | 11 | 43 | 313 | 520 |
| (10) MPR + tithonia (250 kg P ha$^{-1}$) | 28 | 19 | 127 | 12 | 45 | 320 | 551 |
| s.e.d. | 3.2 | 1.7 | 7.1 | 1.4 | NS | NS | 13 |
| CV | 9 | 10 | 11 | 11 | 8 | 13 | 10 |

TSP: triple superphosphate; MPR: Minjingu phosphate rock; s.e.d.: standard error of difference between means; CV: coefficient of variance.

TABLE 6: Sequentially extractable soil P fractions (mg kg$^{-1}$) at Khwisero in the season.

| Treatment | Resin-$P_i$ | NaHCO$_3$-$P_i$ | NaOH-$P_i$ | HCl-$P_i$ | NaHCO$_3$-$P_o$ | NaOH-$P_o$ | Sum of fractions |
|---|---|---|---|---|---|---|---|
| (1) Control | 13 | 10 | 85 | 10 | 59 | 297 | 474 |
| (2) Tithonia | 15 | 13 | 85 | 11 | 72 | 284 | 480 |
| (3) TSP + urea (50 kg P ha$^{-1}$) | 15 | 12 | 106 | 10 | 63 | 295 | 501 |
| (4) TSP + tithonia (50 kg P ha$^{-1}$) | 17 | 12 | 103 | 10 | 58 | 298 | 498 |
| (5) MPR + urea (50 kg P ha$^{-1}$) | 17 | 10 | 101 | 10 | 54 | 307 | 499 |
| (6) MPR + tithonia (50 kg P ha$^{-1}$) | 20 | 15 | 93 | 13 | 60 | 283 | 484 |
| (7) TSP + urea (250 kg P ha$^{-1}$) | 46 | 30 | 158 | 13 | 76 | 288 | 611 |
| (8) TSP + tithonia (250 kg P ha$^{-1}$) | 59 | 38 | 165 | 12 | 78 | 300 | 652 |
| (9) MPR + urea (250 kg P ha$^{-1}$) | 46 | 23 | 134 | 20 | 62 | 293 | 578 |
| (10) MPR + tithonia (250 kg P ha$^{-1}$) | 44 | 24 | 145 | 24 | 67 | 323 | 641 |
| s.e.d. | 6.5 | 3.2 | 13 | 2.1 | NS | NS | 25 |
| CV | 8 | 10 | 12 | 10 | 8 | 11 | 9 |

TSP: triple superphosphate; MPR: Minjingu phosphate rock; s.e.d.: standard error of difference between means; CV: coefficient of variance.

significant correlations between Resin-$P_i$ and maize yield in both seasons at both sites except when MPR was used in combination with tithonia in the both seasons at Nyabeda and in the second season at Khwisero. NaHCO$_3$-$P_i$ followed a similar trend to that of Resin-$P_i$. The relationship of the grain yield with organic P fractions (NaHCO$_3$-$P_o$ and NaOH-$P_o$) and the nonlabile fraction (HCl-$P_i$) was generally weak and not significant.

## 4. Discussion

The higher amounts of Olsen P, resin-$P_i$, and NaHCO$_3$-$P_i$ fractions from the application of TSP compared to MPR are ascribed to the higher solubility of TSP compared to MPR whose dissolution is usually slow [15]. The increase in resin-$P_i$ concentration, even when the relatively insoluble MPR was applied, is attributed to the low pH of these soils (5.4 at Nyabeda and 5.2 at Khwisero) which are lower than the upper pH limit of 6.0 for PR dissolution [16]. The soils at these sites are therefore suitable for P replenishment using MPR. The higher magnitude of the increase in available P

concentrations in the soil as influenced by addition of MPR at Khwisero than that at Nyabeda is likely due to a higher rate of dissolution of MPR in the Khwisero soil than in the Nyabeda soil, as the former was more acidic than the latter.

The NaOH-$P_i$ concentrations significantly increased with the increase in the application rates of TSP and MPR at both sites, but the increase was higher for the TSP treatments than for the MPR treatments. TSP is very soluble and it might therefore have quickly been fixed and then transformed mainly into the NaOH-$P_i$ pool in these high P fixing soils. The NaOH-$P_i$ fraction is a sink for soluble P sources and is the less readily available P that is associated with Al and Fe oxides that dominate Ferralsols and Acrisols such as those in this study. Similar findings were reported by [17] who found that the rate of increase in NaOH-$P_i$ concentration in a high P fixing Acrisols was higher when the soils were treated with TSP compared with the time when the soils were treated with a PR. The failure of the lower P rates to significantly increase P availability is attributed to P-fixation and uptake by plants. At the higher P rates, most of the P-fixation sites are quenched and the P buffer capacity exceeded

TABLE 7: Effect of treatments on maize grain yield (t ha$^{-1}$) at Nyabeda and Khwisero.

| Treatment | Nyabeda | | Khwisero | |
|---|---|---|---|---|
| | Season | | | |
| | 1 | 2 | 1 | 2 |
| (1) Control | 0.95 | 0.43 | 1.13 | 0.10 |
| (2) Tithonia | 2.21 | 0.76 | 2.87 | 0.57 |
| (3) TSP + urea (50 kg P ha$^{-1}$) | 3.47 | 1.58 | 3.81 | 0.94 |
| (4) TSP + tithonia (50 kg P ha$^{-1}$) | 3.69 | 1.35 | 4.34 | 1.59 |
| (5) MPR + urea (50 kg P ha$^{-1}$) | 3.22 | 1.63 | 3.74 | 1.11 |
| (6) MPR + tithonia (50 kg P ha$^{-1}$) | 3.12 | 1.47 | 3.19 | 1.46 |
| (7) TSP + urea (250 kg P ha$^{-1}$) | 6.49 | 2.95 | 5.00 | 1.71 |
| (8) TSP + tithonia (250 kg P ha$^{-1}$) | 5.05 | 2.50 | 4.34 | 1.59 |
| (9) MPR + urea (250 kg P ha$^{-1}$) | 4.03 | 2.28 | 5.34 | 2.02 |
| (10) MPR + tithonia (250 kg P ha$^{-1}$) | 4.07 | 2.00 | 4.97 | 2.66 |
| s.e.d. | 0.50 | 0.37 | 0.61 | 0.25 |
| CV | 15 | 18 | 12 | 17 |

TSP: triple superphosphate; MPR: Minjingu phosphate rock; s.e.d.: standard error of difference between means; CV: coefficient of variance.

and therefore the higher the amounts of P detected in soil solution [18]. The absence of major changes in the HCl-$P_i$ fraction, other than the MPR treatments at the higher rates, is to be expected because the soils in this study were very acidic and this favours the formation of Al and Fe $P_i$ compounds associated with the NaOH-$P_i$ fraction, over the insoluble Ca-P compounds such as hydroxyapatite which dominate the HCl-$P_i$ fraction [19]. The higher HCl-$P_i$ in the MPR treatments compared to TSP is an indication of its lower solubility since it represents the undissolved PR. This could however become available over time for plant use as indicated by our results which show that soils treated with MPR at a rate of 250 kg P ha$^{-1}$ maintained a significantly higher level of the available P fractions (resin-$P_i$ and NaHCO$_3$-$P_i$) and maize yields than the control in the second season even though the MPR was applied only in the first season. One-time application of high rates of fertilizer P inputs has been termed as the high-input strategy [20] and this increases the soil P capital that serves as a major sink for added P and gradually releases plant available P for several years. Our study was however terminated after only two seasons and so a firm conclusion on the ability of the 250 kg P ha$^{-1}$ to sustain yields compared to the seasonal application of 50 kg P ha$^{-1}$ cannot be made because five seasons would have been required to have the P rates of the two strategies (one-time application of 250 kg P ha$^{-1}$ versus seasonal application of 50 kg P ha$^{-1}$) equal. Soil organic P fractions were largely unaffected by application of the inorganic P sources and even tithonia green manure. This is consistent with other research using sequential extraction procedures on tropical soils [21]. Organic P is sensitive to microbial activity and has a fast turnover rate and hence is difficult to increase [22].

Without P application, but with urea as N source (control), maize grain yield averaged <1.0 t ha$^{-1}$. Such low maize grain yields are typical in this area, which has highly P deficient soils [20]. At both sites, the soils that received no P fertilizer had an Olsen P values of <6 mg kg$^{-1}$, which is below the critical P concentration of 10 mg kg$^{-1}$ for maize [23]. The response to P was therefore expected. On the basis of soil P levels alone, the crop at Khwisero would be expected to perform better than at Nyabeda, which was the case in the first season. In the second season however, the poor rainfall could have limited the plants ability to utilize the available P optimally and thus Nyabeda with higher clay content is likely to have retained more water than the sandy soil at Khwisero. This could partially explain the poorer performance of maize at Khwisero compared to Nyabeda in this season. The variations in maize yields observed between the two seasons are attributed mainly to the differences in rainfall. In the first season, the rainfall was generally high (>1000 mm) and hence higher maize yields were obtained than in the second season when the rainfall was low (<500 mm) and poorly distributed. The similar yields obtained by tithonia as a source of N compared to urea suggest that tithonia can provide N as effectively as urea. This is attributed to tithonia's fast decomposition rate, hence its ability to mineralize N during the growing season [24]. Tithonia can therefore be used as a substitute for urea in integrated soil fertility management systems. The RAE of MPR in our study was generally high and is consistent with other studies in the area [4, 25] that have concluded that MPR is a reactive PR and is therefore suitable for direct application. The better performance of MPR at Khwisero than Nyabeda is consistent with the available P being higher at Khwisero due to higher dissolution of MPR occasioned by the lower pH at this site as earlier discussed.

There were good correlations obtained between the Olsen P, Resin-$P_i$ and NaHCO$_3$-$P_i$, and maize grain yields when TSP was used as the P source indicating that these fractions can be used to predict P availability to maize in these soils when TSP is used. But given that sequential fractionation is not easily amenable to routine analyses, Olsen P is an adequate test for evaluating plant available P from TSP in these soils. The correlations between these tests and grain yields for MPR treatments were not however consistent and therefore their use on soils treated with MPR especially when combined with tithonia should be exercised with caution. Although plants can only take up available P, other fractions of P, such as NaOH-$P_i$, are also depleted due to crop growth [18]. This fraction of P, though not readily available to plants, may replenish available P when depleted. This may explain some of the significant correlations obtained with the NaOH-$P_i$. Though organic fractions were far greater than the available inorganic P, they did not correlate with the maize yields. This is because $P_i$ is released into solution after mineralization of $P_o$ and could only contribute to maize uptake if availability coincided with the crops growing period [26], which is not usually easy to achieve. The organic P fractions are however an important source of P in low input systems that are common on smallholder farms [27] and application of organic materials that are likely to enhance the soil organic matter content should therefore be encouraged on smallholder farms.

The Effect of Minjingu Phosphate Rock and Triple Superphosphate on Soil Phosphorus Fractions and Maize Yield in Western Kenya

7

TABLE 8: Relationships between phosphorus fractions with maize grain yields at Nyabeda and Khwisero.

| Treatment | P fraction | Nyabeda | | Khwisero | |
|---|---|---|---|---|---|
| | | Season 1 | Season 2 | Season 1 | Season 2 |
| | | $R^2$ | | | |
| TSP + Urea[a] | Resin-$P_i$ | $0.74^{***}$ | $0.70^{***}$ | $0.54^{**}$ | $0.40^*$ |
| | NaHCO$_3$-$P_i$ | $0.77^{***}$ | $0.50^{**}$ | $0.62^{**}$ | $0.43^*$ |
| | NaHCO$_3$-$P_o$ | ns | ns | ns | ns |
| | NaOH-$P_i$ | ns | $0.60^{**}$ | $0.51^{**}$ | $0.37^*$ |
| | NaOH-$P_o$ | ns | ns | ns | ns |
| | HCl-$P_i$ | ns | ns | ns | ns |
| | Olsen P | $0.66^{***}$ | $0.45^*$ | $0.54^{**}$ | $0.40^*$ |
| TSP + Tithonia[a] | Resin-$P_i$ | $0.53^{**}$ | $0.43^*$ | $0.48^*$ | $0.43^*$ |
| | NaHCO$_3$-$P_i$ | $0.44^*$ | $0.39^*$ | $0.45^*$ | $0.44^*$ |
| | NaHCO$_3$-$P_o$ | ns | ns | ns | ns |
| | NaOH-$P_i$ | $0.40^*$ | $0.41^*$ | $0.44^*$ | 0.29 |
| | NaOH-$P_o$ | ns | ns | ns | ns |
| | HCl-$P_i$ | ns | ns | ns | ns |
| | Olsen P | $0.59^{**}$ | $0.40^*$ | $0.55^{**}$ | $0.50^{**}$ |
| MPR + Urea[a] | Resin-$P_i$ | $0.49^*$ | $0.53^{**}$ | $0.74^{***}$ | $0.58^{**}$ |
| | NaHCO$_3$-$P_i$ | ns | $0.39^*$ | $0.70^{***}$ | $0.58^{**}$ |
| | NaHCO$_3$-$P_o$ | ns | ns | ns | ns |
| | NaOH-$P_i$ | $0.33^*$ | $0.51^{**}$ | $0.51^{**}$ | $0.49^*$ |
| | NaOH-$P_o$ | ns | ns | ns | ns |
| | HCl-$P_i$ | ns | ns | ns | ns |
| | Olsen P | $0.50^{**}$ | ns | $0.76^{***}$ | $0.53^{**}$ |
| MPR+ Tithonia[a] | Resin-$P_i$ | ns | ns | $0.64^{**}$ | ns |
| | NaHCO$_3$-$P_i$ | ns | ns | $0.67^{**}$ | ns |
| | NaHCO$_3$-$P_o$ | ns | ns | ns | ns |
| | NaOH-$P_i$ | $0.39^*$ | ns | $0.36^*$ | ns |
| | NaOH-$P_o$ | ns | ns | ns | ns |
| | HCl-$P_i$ | ns | ns | ns | ns |
| | Olsen P | $0.40^*$ | ns | $0.68^{***}$ | $0.48^*$ |

[a]P fractions for P rates of 50 and 250 kg ha$^{-1}$ have been included in the analysis. $^*$, $^{**}$, and $^{***}$ significance at 0.05, 0.01, and 0.001 probability levels, respectively; ns: not significant.

## 5. Conclusion

Phosphorus availability was influenced by P rate and source. There was no significant difference in Olsen P as influenced by TSP and MPR at the P rate of 50 kg ha$^{-1}$ irrespective of whether they were applied in combination with urea or tithonia at both sites in both seasons. However, at P rate of 250 kg ha$^{-1}$, TSP gave significantly higher Olsen P values than MPR. The labile P fractions generally followed the same trend as the Olsen P. Only the MPR treatments significantly increased the NaOH inorganic P above the control with no P input. The organic P fractions were not affected by treatments. Maize strongly responded to application of P at both sites with yields generally increasing with increasing amount of P applied. Generally, at the same P rate, there was no significant difference between TSP and MPR as P sources irrespective of whether they were combined with tithonia or urea. The relative agronomic effectiveness of MPR was high (83% at Nyabeda and 112% at Khwisero) suggesting that MPR is suitable for direct application at these sites. Between the N sources, tithonia was as equally effective or in some cases better than urea in increasing maize yields. There were good correlations obtained between the Olsen P, Resin-P and NaHCO$_3$ inorganic P, and maize grain yields when TSP was used indicating that these fractions can be used to predict P availability to maize in these soils when TSP is the P source. But given that sequential fractionation is laborious and therefore not easily amenable to routine soil analyses, the Olsen P is an adequate test for evaluating plant available P from TSP in these soils. The correlations between these P tests and grain yields for MPR were however not consistent and therefore their use on soils treated with MPR especially when combined with tithonia should be exercised with caution.

## Conflict of Interests

The authors declare that there is no conflict of interests regarding the publication of this paper.

## Acknowledgments

The authors are grateful to Robin Chacha, Samuel Kirui, and Augustine Wandabwa who assisted with soil analyses, V. Oeba for helping in statistical analyses, and Tom Ochinga for managing the field experiments. Financial support for this study was provided by KEFRI.

## References

[1] B. Jama and P. van Straaten, "Potential of East African phosphate rock deposits in integrated nutrient management strategies," *Annals of Brazilian Academy of Sciences*, vol. 78, no. 4, pp. 781–790, 2006.

[2] S. M. Nandwa and M. A. Bekunda, "Research on nutrient flows and balances in East and Southern Africa: state-of-the-art," *Agriculture, Ecosystems and Environment*, vol. 71, no. 1–3, pp. 5–18, 1998.

[3] P. A. Opala, J. R. Okalebo, and C. Othieno, "Comparison of effects of phosphorus sources on soil acidity, available phosphorus and maize yields at two sites in western Kenya," *Archives of Agronomy and Soil Science*, vol. 59, no. 3, 2013.

[4] S. T. Ikerra, E. Semu, and J. P. Mrema, "Combining Tithonia diversifolia and minjingu phosphate rock for improvement of P availability and maize grain yields on a chromic acrisol in Morogoro, Tanzania," *Nutrient Cycling in Agroecosystems*, vol. 76, no. 2-3, pp. 249–260, 2006.

[5] M. W. Waigwa, C. O. Othieno, and J. R. Okalebo, "Phosphorus availability as affected by the application of phosphate rock combined with organic materials to acid soils in western Kenya," *Experimental Agriculture*, vol. 39, no. 4, pp. 395–407, 2003.

[6] N. Sanginga and P. L. Woomer, *Integrated Soil Fertility Management in Africa: Principles, Practices and Development Process*, Tropical Soil Biology and Fertility Institute of the International Centre for Tropical Agriculture, Nairobi, Kenya, 2009.

[7] B. Jama, C. A. Palm, R. J. Buresh et al., "*Tithonia diversifolia* as a green manure for soil fertility improvement in western Kenya: a review," *Agroforestry Systems*, vol. 49, no. 2, pp. 201–221, 2000.

[8] G. Nziguheba, C. A. Palm, R. J. Buresh, and P. C. Smithson, "Soil phosphorus fractions and adsorption as affected by organic and inorganic sources," *Plant and Soil*, vol. 198, no. 2, pp. 159–168, 1998.

[9] A. A. Rivaie, P. Loganathan, J. D. Graham, R. W. Tillman, and T. W. Payn, "Effect of phosphate rock and triple superphosphate on soil phosphorus fractions and their plant-availability and downward movement in two volcanic ash soils under Pinus radiata plantations in New Zealand," *Nutrient Cycling in Agroecosystems*, vol. 82, no. 1, pp. 75–88, 2008.

[10] M. J. Hedley, J. W. B. Stewart, and B. S. Chauhan, "Changes in inorganic and organic soil phosphorus fractions induced by cultivation practices and by laboratory incubations," *Soil Science Society of America Journal*, vol. 46, no. 5, pp. 970–976, 1982.

[11] H. Tiessen and J. O. Moir, "Characterization of available P by sequential extraction," in *Soil Sampling and Methods of Analysis*, M. R. Carter, Ed., pp. 75–86, Lewis Publications, Boca Raton, Fla, USA, 1993.

[12] A. Samadi, "Contribution of inorganic phosphorus fractions to plant nutrition in alkaline-calcareous soils," *Journal of Agricultural Science and Technology*, vol. 8, pp. 77–89, 2006.

[13] S. R. Olsen, C. V. Cole, F. S. Watanabe, and L. A. Dean, "Estimation of available phosphorus in soils by extraction with sodium bicarbonate," U.S.D.A. Circular 939, 1954.

[14] GENSTAT, "The GenStat Teaching Edition," GenStat Release 7. 22 TE, VSN International, 2010.

[15] S. H. Chien, W. R. Clayton, and G. M. McClellan, "Kinetic dissolution of phosphate rocks in soils," *Soil Science Society of America Journal*, vol. 44, pp. 260–264, 1980.

[16] R. E. White, "Recent developments in the use of phosphate fertilizer on New Zealand pastures," *Journal of the Australian Institute of Agricultural Science*, vol. 2, pp. 25–32, 1989.

[17] A. K. N. Zoysa, P. Loganathan, and M. J. Hedley, "Comparison of the agronomic effectiveness of a phosphate rock and triple superphosphate as phosphate fertilisers for tea (*Camellia sinensis* L.) on a strongly acidic Ultisol," *Nutrient Cycling in Agroecosystems*, vol. 59, no. 2, pp. 95–105, 2001.

[18] R. J. Buresh, P. C. Smithson, and D. T. Hellums, "Building soil phosphorus capital in Africa," in *Replenishing Soil Fertility in Africa*, R. J. Buresh, P. A. Sanchez, and F. Calhoun, Eds., pp. 114–149, Soils Science Society of America special publication, Madison, Wis, USA, 1997.

[19] F. Iyamuremye and R. P. Dick, "Organic amendments and phosphorus sorption by soils," *Advances in Agronomy*, vol. 56, pp. 139–185, 1996.

[20] P. A. Sanchez, K. D. Shepherd, M. J. Soule et al., "Soil fertility replenishment in Africa: an investment in natural resource capital," in *Replenishing Soil Fertility in Africa*, R. J. Buresh, P. A. Sanchez, and F. Calhoun, Eds., pp. 1–46, Soils Science Society of America special publication, Madison, Wis, USA, 1997.

[21] A. Paniagua, M. J. Mazzarino, D. Kass, L. Szott, and C. Fernandez, "Soil phosphorus fractions under five tropical agroecosystems on a volcanic soil," *Australian Journal of Soil Research*, vol. 33, no. 2, pp. 311–320, 1995.

[22] H. Tiessen, J. W. B. Stewart, and C. V. Cole, "Pathways of phosphorus transformations in soils of differing pedogenesis," *Soil Science Society of America Journal*, vol. 48, no. 4, pp. 853–858, 1984.

[23] J. R. Okalebo, K. W. Gathua, and P. L. Woomer, *Laboratory Methods of Soil and Plant Analysis. A Working Manual*, TSBF-CIAT, SACRED Africa, KARI, SSEA, Nairobi, Kenya, 2nd edition, 2002.

[24] C. N. Gachengo, C. A. Palm, B. Jama, and C. Othieno, "Tithonia and senna green manures and inorganic fertilizers as phosphorus sources for maize in Western Kenya," *Agroforestry Systems*, vol. 44, no. 1, pp. 21–36, 1998.

[25] P. K. Mutuo, P. C. Smithson, R. J. Buresh, and R. J. Okalebo, "Comparison of phosphate rock and triple superphosphate on a phosphorus- deficient Kenyan soil," *Communications in Soil Science and Plant Analysis*, vol. 30, no. 7-8, pp. 1091–1103, 1999.

[26] A. Abunyewa, E. K. Asiedu, and Y. Ahenkorah, "Fertilizer phosphorus fractions and their availability to maize on different land forms of vertisols in the coastal savanna zone of Ghana," *West African Journal of Applied Ecology*, vol. 5, pp. 63–73, 2004.

[27] P. A. Sánchez and J. G. Salinas, "Low-input technology for managing oxisols and ultisols in tropical America," *Advances in Agronomy*, vol. 34, pp. 279–406, 1981.

# Organochlorine Pesticide Residues in Sediments and Waters from Cocoa Producing Areas of Ondo State, Southwestern Nigeria

Aderonke Adetutu Okoya,[1] Aderemi Okunola Ogunfowokan,[2] Olabode Idowu Asubiojo,[2] and Nelson Torto[3]

[1] Institute of Ecology and Environmental Studies, Obafemi Awolowo University, Ile-Ife, Nigeria
[2] Department of Chemistry, Obafemi Awolowo University, Ile-Ife, Nigeria
[3] Department of Chemistry, Rhodes University, Grahamstown 6140, South Africa

Correspondence should be addressed to Aderonke Adetutu Okoya; ronkeokoya@yahoo.com

Academic Editors: D. Lin, C. Martius, and D. van Tuinen

This study investigated levels of organochlorine pesticide (OCP) residues in water and sediment samples from eleven rivers serving as drinking water sources and receiving runoff from nearby cocoa plantations in Ondo State, Nigeria. Twenty-two composite samples of surface water and sediments (0–3 cm) were collected randomly using grab technique and replicated thrice per season. The efficiency of the two techniques [supercritical fluid extraction (SFE) and liquid/liquid extraction (LLE)] was evaluated with percentage analyte recoveries $98.17 \pm 0.03$ to $134.72 \pm 0.02$ for SFE and $84.82 \pm 3.32$ to $1102.83 \pm 3.17$ for LLE. Determination of OCPs by gas chromatography with electron capture detection gave higher concentrations for sediments compared to the equivalent water samples. The commonly occurring pesticide residues in the sediments were (range, $\mu$g g$^{-1}$) cis-chlordane 0.03–6.99; $\alpha$-endosulfan 0.03–6.99; p,p$'$-DDE 0.08–19.04; and dieldrin 0.01–7.62; in the sediments and dieldrin (not detected-1.51 $\mu$g L$^{-1}$) in water samples, during the dry season. OCP levels were significantly ($P < 0.05$) higher in dry season than wet season among the rivers. The study concluded that most of the rivers in cocoa growing areas were contaminated with OCPs associated with agricultural activities.

## 1. Introduction

In spite of the benefits (especially with respect to food production and health management) derived from the use of pesticides, the environmental consequences of the widespread use, handling, and disposal methods of pesticides are of great concern [1, 2]. It had been reported that the constraints during acquisition [3] and application of pesticides in the humid forest zones could lead to the risky handling of obsolete pesticides which represents a threat to health and environment. This is because their toxicity has led to the deterioration of human health [4] especially with respect to cancer, neurological damage, and abnormal immune system including the foetus and foetus reproductive system [5–8] as well as fish kills, honey bee poisonings, and the contamination of livestock products [9].

The most commonly used pesticides are the organochlorine pesticides and they are considered to be responsible for the various environmental consequences. The largest regional example of pesticide contamination and human health is perhaps that of the Aral sea-region in Asia [10]. Human health effects of pesticides are caused by inhalation and ingestion through skin contact, handling of pesticide products, breathing of dust or spray and pesticides consumed on/in food, water and aquatic organisms. In view of their toxicity to some plants and insects and persistence in the

TABLE 1: Percentage recoveries, retention times, and response factors of the pesticide standard mixture.

| Standards | % Recovery | | Retention time (min) | Response factor |
|---|---|---|---|---|
| | LLE | SFE | | |
| HCB | 97.79 ± 1.13 | 103.14 ± 0.05 | 8.31 | 0.41 |
| αBHC | 90.27 ± 8.34 | 99.59 ± 0.03 | 8.58 | 0.48 |
| βBHC | 93.30 ± 2.09 | 99.57 ± 0.01 | 9.74 | 0.56 |
| γBHC | 98.58 ± 3.15 | 130.41 ± 16.01 | 9.52 | 0.54 |
| Heptachlor | 99.46 ± 1.22 | 99.91 ± 0.02 | 10.52 | 0.96 |
| Aldrin | 96.61 ± 1.48 | 99.85 ± 0.02 | 11.51 | 1.07 |
| Trans-Chlordane | 94.02 ± 1.79 | 134.72 ± 0.02 | 14.07 | 1.10 |
| Cis-Chlordane | 96.01 ± 1.96 | 100.29 ± 0.01 | 14.59 | 0.60 |
| α-endosulfan | 96.01 ± 1.85 | 100.29 ± 0.01 | 14.59 | 0.60 |
| p,p′-DDE | 94.48 ± 1.11 | 99.86 ± 0.02 | 15.41 | 0.68 |
| Dieldrin | 87.93 ± 2.33 | 98.17 ± 0.03 | 15.83 | 1.03 |
| o,p′-DDD | 102.83 ± 3.17 | 124.95 ± 0.02 | 16.38 | 0.81 |
| Endrin | 91.40 ± 3.53 | 107.90 ± 0.04 | 17.16 | 1.08 |
| p,p′-DDD | 91.68 ± 1.54 | 115.10 ± 0.03 | 17.72 | 0.80 |
| β-endosulfan | 84.82 ± 3.32 | 115.35 ± 0.01 | 18.12 | 1.05 |
| p,p′-DDT | 90.79 ± 4.56 | 115.89 ± 0.02 | 19.09 | 0.86 |
| Methoxychlor (I. S.) | | | 22.96 | — |

environment, many synthetic organochlorine compounds have found extensive use as pesticides [1, 11]. Pesticides usage is indeed responsible for the current ability of the developed countries to produce and harvest large amount of food crops on relatively small amount of land with a relatively small input of human labour.

One major problem of cocoa growing is presented by diseases and pests [12–14]. Large scale spraying of pesticides against these diseases has been employed by most farmers. A lot of effort has been channeled by environmental protection agencies and organizations in developed countries towards regulation of organochlorine pesticides use in order to prevent their concentrations from exceeding permissible levels, particularly in our food supply. Pesticides contamination was also reported in areas where citric crops are predominant by Pitarch et al. [15]. Five watersheds relevant to the sustainability of the area were also monitored [15–17] for pesticides in Salmonid-bearing streams under the auspices of Washington State Departments of Ecology and Agriculture. Residues of organochlorine pesticides were investigated in the water and surface sediments from the lower reaches of the Yangtze River to evaluate their pollution and potential risks. The study reported that there was no obvious trend of declining DDT concentrations in the sediments from the river [18].

The need for pesticide regulation and enforcement of such with respect to pest management, safeguarding users and consumers' health, and the protection of the environment is a task that must be achieved. In Nigeria, apart from the national guidelines and standards for industrial effluents, gaseous emission and hazardous wastes [19], the National Agency for Food and Drugs Administration and Control (NAFDAC) also made the Pesticide Registration Regulation, 1996 under Decree 15 of the Federal Republic of Nigeria, 1993.

The law made application for the registration of pesticides compulsory with prescribed guidelines. The regulation specified that "No pesticide shall be manufactured, formulated, imported, advertised, sold or distributed in Nigeria unless it has been registered in accordance with the provisions of these Regulations" [20]. In spite of the laudable provisions, lack of enforcement of the law by the regulatory agency still makes the containment of the risk associated with the use of pesticides an elusive task in Nigeria. The marketing of pesticides in Nigeria is very much unorganized and lacks proper legislative control. This has made it difficult to determine the various market sizes, types, and shares of pesticides in use. Hence there is no dependable official statistics on the type and amount of pesticides imported into the country [21]. However, environmental pollution control is just beginning to receive the desired attention in Nigeria. An important component of this is in the reliable information on the levels of key pollutants in different media and settings in the country. Akinnifesi et al. [22] investigated the physicchemical characteristics of soils in the main cocoa producing area of Ondo state and found that fungicide residues caused a significant increase in soil acidity, organic matter, and copper concentrations. The general occurrence, persistence, and consequences in the environment of OCPs and the fact that it is inevitable, that persistent organic pollutants (POPs) contaminated sites will continue to represent an environmental issue for contemporary and future generations to address [23] make it important to determine their levels in some areas of likely predominance.

Ondo State is a major cocoa producing area in Nigeria. In addition to supplying much-needed raw materials for some local industries, cocoa export is an important source of foreign exchange earnings for the country. In Nigeria, there

Organochlorine Pesticide Residues in Sediments and Waters from Cocoa Producing Areas of Ondo State,
Southwestern Nigeria

11

TABLE 2: Sampling sites and their geographical position.

| S/N | Sample | GPS location of sampling point | Local name of study unit |
|---|---|---|---|
| 1 | $R_1S_1$ | 07°10′27.9″N 004°51′54.22′E | Agoo river at Ile-Oluji |
| 2 | $R_2S_2$ | 07°18′36.6″N 005°39′55.3″E | Ose river at Ose |
| 3 | $R_3S_3$ | 07°16′27.1″N 005°9′56.9″E | Ala river at Akure |
| 4 | $R_4S_4$ | 07°6′39.8″N 004°49′26.7″E | Luwa river at Ondo |
| 5 | $R_5S_5$ | 07°15′35.8″N 005°22′46.4′E | Ogbese river at Ogbese |
| 6 | $R_6S_6$ | 07°10′17.4″N 004°43′5.5′E | Oni river at Ifetedo/Oke-Igbo |
| 7 | $R_7S_7$ | 07°13′56″N 005°3′54.1E | Aponmu river at Aponmu |
| 8 | $R_8S_8$ | 07°13′56″N 004°15′00″E | Osun river at Osogbo |
| 9 | $R_9S_9$ | 07°13′56″N 004°30′00″E | Opa river at Ile-Ife |
| 10 | $R_{10}S_{10}$ | 07°24′10.9″N 005°00′49.5″E | Owena-Osun river at Owena-Ijesa |
| 11 | $R_{11}S_{11}$ | 07°11′52.2″N 005°01′14.6″E | Owena-Ondo river in Ondo |

seems to be paucity of data on the monitoring of pesticide residues in the country. Cocoa farmers in Nigeria have a long history of pesticide usage on their farms. Cocoa, being a plantation crop, had been subjected to large volume of insecticides annually since 1957 especially for the control of the brown cacao *mirid, Sahlbergella singularis* Haglund [20, 21]. Hence, this study is designed to provide information on the levels of OCPs in the sediments and surface water from rivers that flow through the main cocoa-producing areas of Ondo State of Nigeria, where the cocoa farmers have employed pesticide spraying operations on their cocoa farms.

## 2. Materials and Methods

### 2.1. Sample Collection, Preservation, Preparation, and Storage

#### 2.1.1. Sampling Sites.
Sediment and water samples of rivers in some cocoa producing areas of Ondo State, Nigeria were collected as shown in Figure 1. These rivers included Oluwa, Owena-Osun, Owena-Ondo, Ose, Ogbese, Ala, Agoo, Aponmu, Oni, Opa and Osun. Opa, and Osun rivers were sampled where there was no cocoa plantation to serve as controls.

#### 2.1.2. Water samples.
Grab sampling technique was used to collect six core surface water samples randomly which was homogenized to form a composite sample (2.5 L) per river. Each was replicated thrice per season to give a total of 66 samples. Concentrated sulphuric acid (5.0 mL) was added to each of the samples immediately after the collection to prevent microbial degradation of samples. The samples were kept cool during transportation to the laboratory and then stored at 4°C in a refrigerator, until analysed.

#### 2.1.3. Sediment Samples.
Sediment samples were collected from the 0–3 cm depth from the same site as water samples, wrapped up in aluminium foil and then put in a polyethylene bag. Samples were kept cool during transportation to the laboratory. At the laboratory, they were freeze-dried prior to sample preparation and analysis. Sediment samples were later thawed and air-dried at ambient temperature. The composite dried sediment samples were processed through 2.0 mm stainless steel sieve. The less than 63 $\mu$m soil samples were prepared using the 63 $\mu$m stainless steel sieve prior to analysis.

### 2.2. Extraction of OCPs from the Sediment and Surface Water Samples.
The sample cell was packed with some glass-wool, after which 3 g sediment samples fortified with pesticides standards in the concentration range of 1–50 ppm and with 500 $\mu$L modifier (methanol/acetone mixture ratio 2 : 3) spiked onto the sediment was introduced and glass-wool was added to fill the cell completely. The cell was pressurized to 300 bar at 60°C with SC-CO$_2$ (density = 0.872 g/mL).

The pressure was maintained for 20 min (static extraction) and dynamic extraction was carried out for another 30 min. The extract was collected into a glass tube containing 5 mL acetone and then concentrated to about 2 mL on a vacuum rotary evaporator.

TABLE 3: Mean concentration ($\mu$g/g) of OCPs in sediment samples of some rivers in Ondo State during the wet season.

| Analytes | Agoo | Ose | Ala | Oluwa | Ogbese | Oni | Aponmu | Osun | Opa | Owena-Osun | Owena-Ondo |
|---|---|---|---|---|---|---|---|---|---|---|---|
| HCB | ND | ND | 0.04[b] ± 0.03 | 5.55[a] ± 0.01 | 0.02[b] ± 0.01 | 0.16[b] ± 0.07 | 0.06[b] ± 0.05 | 0.11[b] ± 0.07 | 0.13[b] ± 0.02 | ND | NS |
| α-BHC | ND | ND | ND | 0.99[a] ± 0.03 | 0.04[b] ± 0.02 | ND | 0.17[b] ± 0.13 | 0.25[b] ± 0.03 | ND | ND | NS |
| β-BHC | ND | ND | ND | 7.25[a] ± 0.01 | ND | 0.29[b] ± 0.13 | 1.62[b] ± 0.77 | 0.18 ± 0.05[b] | ND | ND | NS |
| γ-BHC | ND | ND | ND | 5.28[a] ± 0.01 | 0.04[b] ± 0.01 | 0.39[b] ± 0.17 | 0.79[b] ± 0.06 | 0.15[b] ± 0.01 | 0.64[b] ± 0.53 | ND | NS |
| Heptachlor | ND | ND | ND | 1.16[a] ± 0.01 | ND | ND | 0.58[b,a] ± 0.04 | **0.01[b]** ± 0.01 | 1.09[a] ± 0.06 | ND | NS |
| Aldrin | ND | ND | ND | 2.76[a] ± 0.02 | ND | ND | ND | ND | ND | ND | NS |
| Trans-chlordane | 1.74[b] ± 0.02 | ND | **0.01[b]** ± 0.01 | 4.58[a] ± 0.03 | ND | ND | 0.49[b] ± 0.27 | 0.23[b] ± 0.12 | ND | 0.58[b] ± 0.01 | NS |
| Cis-chlordane | ND | ND | 0.02[b] ± 0.01 | ND | 0.36[b] ± 0.17 | 0.02[b] ± 0.01 | ND | 0.03[b] ± 0.02 | 2.99[a] ± 0.61 | ND | NS |
| α-endosulfan | ND | ND | 0.02[b] ± 0.01 | ND | 0.36[b] ± 0.07 | 0.02[b] ± 0.01 | ND | 0.03 ± 0.02[b] | 2.12[a] ± 0.19 | ND | NS |
| β-endosulfan | 0.14[b] ± 0.07 | ND | 0.06[b] ± 0.02 | 1.50[a] ± 0.01 | 0.02[c] ± 0.01 | ND | 0.88[b] ± 0.02 | 0.04[c] ± 0.02 | ND | ND | NS |
| p,p′-DDE | ND | ND | 0.06[b] ± 0.01 | 7.91[a] ± 0.04 | 1.51[b] ± 0.02 | 0.37[b] ± 0.08 | 5.01[a] ± 1.51 | 0.13[b] ± 0.15 | 0.35[b] ± 0.31 | ND | NS |
| o,p′-DDD | 0.02[b] ± 0.01 | ND | ND | 5.18[a] ± 0.03 | ND | ND | 0.09[b] ± 0.02 | **0.01[b]** ± 0.01 | ND | ND | NS |
| p,p′-DDD | 1.31[b] ± 0.08 | ND | **0.01[b]** ± 0.01 | 10.28[a] ± 0.01 | ND | ND | 1.48[b] ± 0.17 | 0.13[b] ± 0.11 | ND | ND | NS |
| p,p′-DDT | 17.4[a] ± 0.01 | 0.11[b] ± 0.00 | **0.01[b]** ± 0.01 | 1.15[b] ± 0.15 | ND | 0.02[b] ± 0.02 | 0.18[b] ± 0.06 | 0.43[b] ± 0.01 | ND | ND | NS |
| Dieldrin | 4.56[b] ± 0.02 | 0.57[c] ± 0.06 | 3.03[c,b] ± 0.03 | 8.82[a] ± 0.04 | 0.99[c,b] ± 0.53 | 0.12[c] ± 0.06 | 0.46[c] ± 0.03 | 0.61[c] ± 0.11 | 3.43[c,b] ± 0.01 | 0.07[c] ± 0.03 | NS |
| Endrin | ND | ND | **0.01[b]** ± 0.01 | 4.43[a] ± 0.04 | ND | ND | 0.32[b] ± 0.01 | 0.12[b] ± 0.01 | ND | ND | NS |

* Data in the same column followed by the same alphabets are not significantly different at $\alpha = 0.05$ using the new Duncan Multiple Range Test, NS: no sample, ND: not detected.

Organochlorine Pesticide Residues in Sediments and Waters from Cocoa Producing Areas of Ondo State, Southwestern Nigeria

13

TABLE 4: Mean concentration ($\mu$g/g) of OCPs in sediment samples of some rivers in Ondo State during the dry season.

| Analytes | Agoo | Ose | Ala | Oluwa | Ogbese | Oni | Aponmu | Osun | Opa | Owena-Osun | Owena-Ondo |
|---|---|---|---|---|---|---|---|---|---|---|---|
| HCB | 1.89$^b$ ± 0.05 | ND | 0.51$^b$ ± 0.06 | 25.18$^a$ ± 3.05 | 6.05$^b$ ± 0.02 | 0.33$^b$ ± 0.12 | 4.69$^b$ ± 1.53 | NS | NS | ND | 0.04$^b$ ± 0.03 |
| $\alpha$-BHC | 5.03$^{b,a}$ ± 0.02 | ND | 0.95$^c$ ± 0.08 | 0.01$^c$ ± 0.01 | 2.32$^{b,c}$ ± 1.53 | 0.18$^c$ ± 0.03 | 8.07$^a$ ± 3.00 | NS | NS | ND | 0.05$^c$ ± 0.01 |
| $\beta$-BHC | 4.60$^b$ ± 0.22 | 0.51$^b$ ± 0.77 | 3.26$^b$ ± 0.05 | 0.01$^b$ ± 0.01 | 0.10$^b$ ± 0.06 | 1.31$^b$ ± 0.06 | 10.91$^a$ ± 6.66 | NS | NS | ND | 0.14$^b$ ± 0.11 |
| $\gamma$-BHC | 4.05$^b$ ± 0.01 | ND | 0.22$^b$ ± 0.04 | 0.01$^b$ ± 0.01 | 9.08$^a$ ± 0.02 | 0.93$^b$ ± 0.08 | 3.89$^b$ ± 0.04 | NS | NS | ND | 0.05$^b$ ± 0.01 |
| Heptachlor | ND$^b$ | ND | ND | ND | ND | ND | ND | NS | NS | ND | ND |
| Aldrin | 6.55$^a$ ± 0.02 | ND | 0.03$^b$ ± 0.01 | 0.01$^a$ ± 0.01 | 0.05$^b$ ± 0.03 | 0.80$^b$ ± 0.37 | 1.39$^b$ ± 0.05 | NS | NS | ND | 0.10$^b$ ± 0.07 |
| Trans-chlordane | 2.64$^b$ ± 0.01 | ND | 3.80$^b$ ± 0.03 | 81.32$^a$ ± 7.06 | 0.26$^b$ ± 0.06 | 0.47$^b$ ± 0.11 | 2.88$^b$ ± 1.53 | NS | NS | 0.58$^b$ ± 0.08 | 0.39$^b$ ± 0.06 |
| Cis-chlordane | 3.44$^b$ ± 0.02 | 1.99$^{c,b}$ ± 0.05 | 0.20$^c$ ± 0.07 | 1.06$^{c,b}$ ± 0.03 | 6.99$^a$ ± 0.05 | 2.28$^{c,b}$ ± 1.03 | 0.23$^c$ ± 0.15 | NS | NS | 1.03$^{c,b}$ ± 1.00 | 0.03$^c$ ± 0.01 |
| $\alpha$-endosulfan | 3.44$^b$ ± 0.02 | 1.99$^{c,b}$ ± 0.01 | 0.20$^c$ ± 0.08 | 1.06$^{c,b}$ ± 0.03 | 6.99$^a$ ± 0.02 | 2.28$^{c,b}$ ± 1.53 | 0.23$^c$ ± 0.15 | NS | NS | 1.03$^{c,b}$ ± 0.71 | 0.03$^c$ ± 0.01 |
| $\beta$-endosulfan | 1.36$^b$ ± 0.01 | ND | 5.84$^a$ ± 1.00 | 7.04$^a$ ± 0.02 | 1.98$^b$ ± 0.01 | 0.80$^b$ ± 0.69 | 0.09$^b$ ± 0.07 | NS | NS | 0.01$^b$ ± 0.01 | 0.08$^b$ ± 0.02 |
| P,p'-DDE | 17.89$^a$ ± 0.03 | 7.75$^b$ ± 0.66 | 3.26$^b$ ± 0.05 | 2.79$^b$ ± 0.04 | 1.21$^b$ ± 1.15 | 3.37$^b$ ± 3.08 | 19.04$^a$ ± 1.25 | NS | NS | 0.08$^b$ ± 0.03 | 0.19$^b$ ± 0.15 |
| o,p'-DDD | 3.12$^b$ ± 1.09 | ND | ND | 11.15$^a$ ± 1.36 | 0.09$^b$ ± 0.03 | 0.25$^b$ ± 0.09 | 9.54$^a$ ± 4.10 | NS | NS | 0.01$^b$ ± 0.01 | 0.09$^b$ ± 0.07 |
| P,p'-DDD | 2.05$^b$ ± 0.01 | ND | 7.05$^b$ ± 0.13 | 57.40$^a$ ± 6.76 | 0.06$^b$ ± 0.05 | 1.54$^b$ ± 1.36 | 0.74$^b$ ± 0.16 | NS | NS | 0.01$^b$ ± 0.01 | 0.08$^b$ ± 0.07 |
| Dieldrin | 2.38$^{c,b}$ ± 0.01 | 2.39$^{c,b}$ ± 0.01 | 7.62$^a$ ± 5.72 | 0.01$^d$ ± 0.01 | 0.13$^d$ ± 0.05 | 3.13$^b$ ± 1.53 | 0.18$^d$ ± 0.03 | NS | NS | 0.65$^{c,d}$ ± 0.53 | 0.05$^d$ ± 0.03 |
| Endrin | 11.4$^b$ ± 0.05 | ND | 0.22$^c$ ± 0.16 | 21.28$^a$ ± 3.17 | 0.11$^c$ ± 0.03 | 0.39$^c$ ± 0.12 | 0.54$^c$ ± 0.15 | NS | NS | ND | 0.02$^c$ ± 0.01 |

*Data in the same column followed by the same alphabets are not significantly different at $\alpha$ = 0.05 using the new Duncan Multiple Range Test, NS: no sample, ND: not detected.

TABLE 5: Mean concentration (μg/L) and seasonal variation of organochlorine compounds in some rivers in Ondo State.

| Analytes | Agoo-Ile-Oluji Dry | Wet | Ose Dry | Wet | Ala Dry | Wet | Oluwa Dry | Wet | Ogbese Dry | Wet | Oni Dry | Wet | Aponmu Dry | Wet | Osun Dry | Wet | Opa Dry | Wet | Owena-Osun Dry | Wet | Owena-Ondo Dry | Wet |
|---|---|---|---|---|---|---|---|---|---|---|---|---|---|---|---|---|---|---|---|---|---|---|
| HCB | ND | ND | ND | ND | ND | ND | ND | ND | ND | ND | ND | ND | ND | ND | ND | ND | ND | ND | ND | ND | ND | ND |
| α-BHC | ND | ND | ND | ND | ND | ND | ND | ND | ND | ND | ND | ND | ND | ND | ND | ND | ND | ND | ND | ND | ND | ND |
| β-BHC | ND | ND | ND | ND | ND | ND | ND | ND | ND | ND | ND | ND | ND | ND | ND | ND | ND | ND | ND | ND | ND | ND |
| γ-BHC | ND | ND | ND | ND | ND | ND | ND | ND | ND | ND | ND | ND | ND | ND | ND | ND | ND | ND | ND | ND | ND | ND |
| Heptachlor | ND | ND | ND | ND | ND | ND | ND | ND | ND | ND | ND | ND | ND | ND | ND | ND | ND | ND | ND | ND | 0.01[a] ± 0.01 | ND |
| Aldrin | ND | ND | ND | ND | ND | ND | ND | ND | ND | ND | ND | ND | ND | ND | ND | ND | ND | ND | ND | ND | ND | ND |
| Trans-chlordane | ND | ND | ND | ND | ND | ND | ND | ND | ND | ND | ND | ND | ND | ND | ND | ND | ND | ND | ND | ND | ND | ND |
| Cis-chlordane | 1.65[a] ± 0.01 | ND | ND | ND | ND | ND | 0.13[b] ± 0.09 | ND | ND | ND | ND | ND | 0.34[a] ± 0.03 | ND | ND | ND | ND | ND | ND | ND | ND | ND |
| α-endosulfan | 1.65[a] ± 0.01 | ND | ND | ND | ND | ND | 0.13[b] ± 0.02 | ND | ND | ND | ND | ND | 0.01[c] ± 0.01 | ND | ND | ND | ND | ND | ND | ND | ND | ND |
| β-endosulfan | ND | ND | ND | ND | ND | ND | ND | ND | ND | ND | ND | ND | 0.01[c] ± 0.01 | ND | ND | ND | ND | ND | ND | ND | ND | ND |
| p,p'-DDE | ND | ND | ND | ND | ND | ND | 0.11[a] ± 0.03 | ND | ND | ND | ND | ND | ND | ND | ND | ND | ND | ND | ND | ND | ND | ND |
| o,p'-DDD | ND | ND | ND | ND | ND | ND | 0.01[a] ± 0.01 | ND | ND | ND | ND | ND | ND | ND | ND | ND | ND | ND | ND | ND | ND | ND |
| p,p'-DDD | ND | ND | ND | ND | ND | ND | ND | ND | ND | ND | ND | ND | ND | ND | ND | ND | ND | ND | ND | ND | ND | ND |
| p,p'-DDT | ND | ND | ND | ND | ND | ND | 0.02[a] ± 0.01 | ND | ND | ND | ND | ND | ND | ND | ND | ND | ND | ND | ND | ND | ND | ND |
| Dieldrin | ND | ND | ND | ND | 1.07[b] ± 0.06 | ND | 1.51[a] ± 0.05 | ND | ND | ND | ND | ND | ND | ND | ND | 0.25[a,b] ± 0.04 | ND | ND | ND | ND | ND | ND |
| Endrin | ND | ND | ND | ND | ND | ND | ND | ND | ND | ND | ND | ND | ND | ND | ND | ND | ND | ND | ND | ND | ND | ND |

Data in the same column followed by the same alphabets are not significantly different at α = 0.05 using the new Duncan Multiple Range Test. ND: not detected.

Organochlorine Pesticide Residues in Sediments and Waters from Cocoa Producing Areas of Ondo State, Southwestern Nigeria

15

TABLE 6: Results of the correlation tests for OCPs in sediment samples during the wet season.

| OCPs/PCBs | HCB | α-BHC | β-BHC | γ-BHC | HEPT ACHLOR | ALDRIN | TRANS-CHLORDANE | CIS-CHLORDANE | α-ENDO SULFAN | β-ENDO SULFAN | p,p'-DDE | o,p'-DDD | p,p'-DDD | p,p'-DDT | DIELDRIN | ENDRIN |
|---|---|---|---|---|---|---|---|---|---|---|---|---|---|---|---|---|
| HCB | 1.00 | **0.96** | **0.98** | **0.99** | **0.68** | **0.99** | **0.92** | -0.06 | -0.04 | **0.86** | **0.83** | **0.99** | **0.98** | 0.01 | **0.81** | **0.99** |
| α-BHC | | 1.00 | **0.97** | **0.97** | **0.69** | **0.96** | **0.90** | -0.09 | -0.07 | **0.90** | **0.88** | **0.96** | **0.96** | -0.00 | **0.76** | **0.97** |
| β-BHC | | | 1.00 | **0.99** | **0.72** | **0.98** | **0.92** | -0.09 | -0.07 | **0.94** | **0.92** | **0.98** | **0.99** | 0.00 | **0.78** | **0.99** |
| γ-BHC | | | | 1.00 | **0.78** | **0.98** | **0.91** | 0.02 | 0.04 | **0.91** | **0.89** | **0.98** | **0.98** | -0.00 | **0.81** | **0.99** |
| HEPTACHLOR | | | | | 1.00 | **0.67** | **0.63** | **0.53** | **0.59** | **0.74** | **0.75** | **0.68** | **0.70** | -0.04 | **0.70** | **0.69** |
| ALDRIN | | | | | | 1.00 | **0.92** | -0.07 | -0.06 | **0.86** | **0.83** | **0.99** | **0.98** | 0.02 | **0.81** | **0.99** |
| TRANS-CHLORDANE | | | | | | | 1.00 | -0.11 | -0.09 | **0.86** | **0.80** | **0.93** | **0.96** | 0.37 | **0.89** | **0.93** |
| CIS-CHLORDANE | | | | | | | | 1.00 | **0.92** | -0.11 | -0.06 | -0.07 | -0.09 | -0.08 | 0.17 | -0.08 |
| α-ENDOSULFAN | | | | | | | | | 1.00 | -0.09 | -0.03 | -0.06 | -0.07 | -0.06 | 0.21 | -0.06 |
| β-ENDOSULFAN | | | | | | | | | | 1.00 | **0.98** | **0.86** | **0.92** | 0.07 | **0.71** | **0.89** |
| p,p'-DDE | | | | | | | | | | | 1.00 | **0.84** | **0.88** | -0.02 | **0.66** | **0.86** |
| o,p'-DDD | | | | | | | | | | | | 1.00 | **0.98** | 0.02 | **0.81** | **0.99** |
| p,p'-DDD | | | | | | | | | | | | | 1.00 | 0.14 | **0.84** | **0.99** |
| p,p'-DDT | | | | | | | | | | | | | | 1.00 | 0.43 | 0.01 |
| DIELDRIN | | | | | | | | | | | | | | | 1.00 | **0.80** |
| ENDRIN | | | | | | | | | | | | | | | | 1.00 |

*Bold $r$ values are significant at $P < 0.0001$.

TABLE 7: Results of the correlation tests for OCPs in sediment samples during the dry season.

| | HCB | α-BHC | β-BHC | γ-BHC | HEPTA CHLOR | ALDRIN | TRANS-CHLORDANE | CIS-CHLORDANE | α-ENDO SULFAN | β-ENDO SULFAN | p,p'-DDE | o,p'-DDD | p,p'-DDD | p,p'-DDT | DIELDRIN | ENDRIN |
|---|---|---|---|---|---|---|---|---|---|---|---|---|---|---|---|---|
| HCB | 1.00 | 0.13 | 0.08 | 0.19 | 0.06 | 0.01 | **0.96** | 0.23 | 0.23 | **0.75** | 0.15 | **0.82** | **0.95** | 0.07 | -0.14 | **0.86** |
| α-BHC | | 1.00 | **0.93** | **0.64** | **0.92** | **0.63** | -0.05 | 0.31 | 0.31 | 0.04 | **0.90** | **0.57** | -0.07 | **0.63** | 0.07 | 0.16 |
| β-BHC | | | 1.00 | 0.39 | **0.90** | **0.50** | -0.05 | 0.06 | 0.06 | 0.09 | 0.88 | **0.58** | -0.05 | 0.51 | 0.24 | 0.09 |
| γ-BHC | | | | 1.00 | 0.46 | 0.39 | -0.08 | **0.86** | **0.86** | 0.11 | 0.45 | 0.21 | -0.09 | 0.37 | 0.27 | -0.02 |
| HEPTACHLOR | | | | | 1.00 | 0.81 | -0.05 | 0.21 | 0.21 | 0.13 | **0.94** | **0.48** | -0.05 | **0.83** | 0.30 | 0.26 |
| ALDRIN | | | | | | 1.00 | -0.04 | 0.34 | 0.34 | 0.34 | **0.76** | 0.24 | -0.05 | **0.98** | 0.21 | 0.41 |
| TRANS-CHLORDANE | | | | | | | 1.00 | 0.03 | 0.03 | **0.76** | 0.04 | **0.75** | **0.99** | 0.03 | -0.09 | **0.89** |
| CIS-CHLORDANE | | | | | | | | 1.00 | 1.00 | 0.21 | 0.27 | 0.04 | 0.03 | 0.31 | 0.15 | 0.19 |
| α-ENDOSULFAN | | | | | | | | | 1.00 | 0.21 | 0.27 | 0.04 | 0.03 | 0.31 | 0.15 | 0.19 |
| β-ENDOSULFAN | | | | | | | | | | 1.00 | 0.09 | **0.52** | **0.81** | 0.14 | 0.44 | 0.69 |
| p,p'-DDE | | | | | | | | | | | 1.00 | **0.57** | 0.03 | **0.76** | 0.25 | 0.32 |
| o,p'-DDD | | | | | | | | | | | | 1.00 | **0.73** | 0.28 | -0.13 | **0.73** |
| p,p'-DDD | | | | | | | | | | | | | 1.00 | 0.03 | -0.02 | **0.88** |
| p,p'-DDT | | | | | | | | | | | | | | 1.00 | 0.26 | 0.47 |
| DIELDRIN | | | | | | | | | | | | | | | 1.00 | -0.01 |
| ENDRIN | | | | | | | | | | | | | | | | 1.00 |

* Bold r values are significant at P < 0.0001.

Organochlorine Pesticide Residues in Sediments and Waters from Cocoa Producing Areas of Ondo State, Southwestern Nigeria

17

The extraction chamber was depressurized according to the equipment manual. The reduced extract was taken for GC-ECD analysis. Triplicate analyses of the sediment samples from each study site were carried out. The same procedure above was used to extract OCPs from raw sediment samples.

However, standard liquid/liquid extraction method was used to isolate the OCPs from both raw and spiked surface water samples as suggested by Fatoki and Awofolu [24]. The reduced extracts were taken for GC-ECD analyses. Triplicate analyses of the water samples from each study site were done.

*2.3. Gas Chromatographic Analysis of Extracts from Sediment and Water Samples.* One microlitre each of processed sample for GC analysis was injected in turns into the GC-ECD system XL PerkinElmer used in a split less mode and equipped with a $^{63}$Ni electron capture detector, column: Zebron ZB 35 fused silica capillary column 30 cm × 0.25 mm × 0.25 $\mu$m (film thickness) for analyses. The injector and detector temperatures were maintained at 250°C and 300°C, respectively. The oven temperature was initially maintained at 50°C (hold 1 min), ramped to 200°C at 40°C/min (hold 2 min), ramped to 240°C at 4°C/min (hold 1 min), and finally ramped to 270°C at 4°C/min (hold 5 min). The carrier gas was 99.999% nitrogen gas. The carrier gas flow rate was 14 psi for optimum performance. The extraction efficiencies at the different temperature were determined by comparison of the peak areas sample extract with those of the pesticide standard mixture. The response factors were determined according to standard method [25].

The result of the GC-ECD determination of the OC standard mixture and the calculated response factors is presented in Table 1. All the 22 compounds were well resolved and eluted within a reasonable time of less than 30 minutes under the optimized gas chromatograph-electron capture detector (GC-ECD) conditions. The identities of the OCPs in sample extracts were confirmed by spiking and comparing their retention times with those of standards and concentrations were determined by computer calculation making use of both the response factors of the OCPs and the internal standard. The sample clean-up techniques achieved high analyte recoveries of 98.17 ± 0.03 to 134.72 ± 0.02 for SFE and 84.82 ± 3.32 to 102.83 ± 3.17 for LLE with RSD of less than 6% in both cases (Table 1).

*2.4. Statistical Analysis of the Data.* The various data obtained were subjected to New Duncan Multiple Range and Pearson Correlation tests.

## 3. Results and Discussion

The geographical locations of the sampling sites are presented in Table 2. The results of various organochlorine pesticide (OCP) residues in the sediment are as presented in Tables 3 and 4 on seasonal basis. Low concentrations of OCPs were observed for samples taken during the wet season relative to those for the dry season. This is expected because of the dilution at the former season and the fact that the transport and dispersion of pollutants in the aquatic environment is controlled by advection (mass movement) and mixing or diffusion [26]. In the sediment samples, all the analytes except heptachlor, which was not detected (<0.02 $\mu$g/g detection limit), were found at appreciably higher concentration with the following ranges ($\mu$g/g): HCB (ND—25.18 ± 3.05); $\alpha$ – BHC (ND—8.07 ± 3.00); $\beta$-BHC (ND—10.91 ± 6.66); $\gamma$-BHC (ND—9.08 ± 0.02); aldrin (ND—6.55 ± 0.02); Trans-chlordane (ND—81.32 ± 7.06); Cis-chlordane (0.03 ± 0.01–6.99 ± 0.05); $\alpha$-endosulphan (0.03 ± 0.01–6.99 ± 0.02); $\beta$-endosulphan (ND—7.04 ± 0.02); p,p'-DDE (0.08 ± 0.03–19.04 ± 1.25); o,p'-DDD (ND—11.15 ± 1.36); p,p'-DDD (ND—57.40 ± 6.76); dieldrin (0.01 ± 0.01–7.62 ± 5.72) and endrin (ND—21.28 ± 3.17) in almost all the rivers in the dry season (Table 4) than the range ($\mu$g/g) in the wet season: HCB (ND—5.55 ± 0.01); $\alpha$-BHC (ND—0.99 ± 0.03); $\beta$-BHC (ND—7.25 ± 0.01); $\gamma$-BHC (ND—5.28 ± 0.01); heptachlor (ND—1.16 ± 0.01); aldrin (ND—2.76 ± 0.02); Trans-chlordane (ND—4.58 ± 0.03); Cis-chlordane (ND—2.99 ± 0.61); $\alpha$-endosulphan (ND—2.12 ± 0.19); $\beta$-endosulphan (ND—1.50 ± 0.01); p, p'-DDE (ND—7.91 ± 0.04); o,p'-DDD (ND—5.18 ± 0.03); p,p'-DDD (ND—10.28 ± 0.01); dieldrin (0.07 ± 0.03–8.82 ± 0.04), and endrin (ND—4.43 ± 0.04) (Table 3). The high concentration of p,p'-DDT (17.4 ± 0.01) $\mu$g/g in sediments from Agoo River compared to the other DDT metabolites suggests an indication of recent usage of DDT (probably with different trade name) in the study area, more so that at the time of sampling Agoo River, a farmer was spraying a cocoa plantation adjacent to the river. Some of the pesticides detected in the sediments, such as chlordane, heptachlor, DDT, DDE, and endosulfan are known to have endocrine and estrogenic disrupting properties [27], which may greatly impact on the biodiversity of the aquatic ecosystem. The presence of DDT (Table 3) and some of its degradation residues in the matrix can be attributed to their wide usage before their banning [28]. Since they are persistent enough and degrade slowly and easily accumulate in the soil, the transportation of these pesticides both sorbed onto solids and dissolved by the surface water down to the water sources is expected [29]. The persistent half-life of DDT in aquatic environments has been suggested to be approximately 5 years [30], 10–20 years (estimated from studies) in bivalves [31]. As various DDT metabolites persist for a long time in the environment, their gradual degradation occurs under aerobic conditions as DDE and as DDD [32].

Table 5 shows the results of the determination of organochlorine pesticide (OCP) residues in water samples from various rivers on seasonal basis. It has been observed that the concentrations of analytes were very low in the water samples (Table 5) compared to their concentrations in sediment samples (Tables 3 and 4) in both dry and wet season. These results prove that these compounds are not hydrophilic and tend to accumulate in sediment and subsequently in fatty tissue of organisms [33, 34]. Levels of OCPs in the surface water samples from all the rivers in the study area were in trace concentration (<0.01 $\mu$g/L detection limit) for the wet season analyses. However, some elevated levels (Table 5) of 0.34 $\mu$g/L of trans-chlordane in Aponmu River, 1.51, 0.11, 0.13 and 0.13 $\mu$g/L of Dieldrin, p,p'-DDE, cis-chlordane and $\alpha$-endosulphan, respectively, in Oluwa River, elevated levels

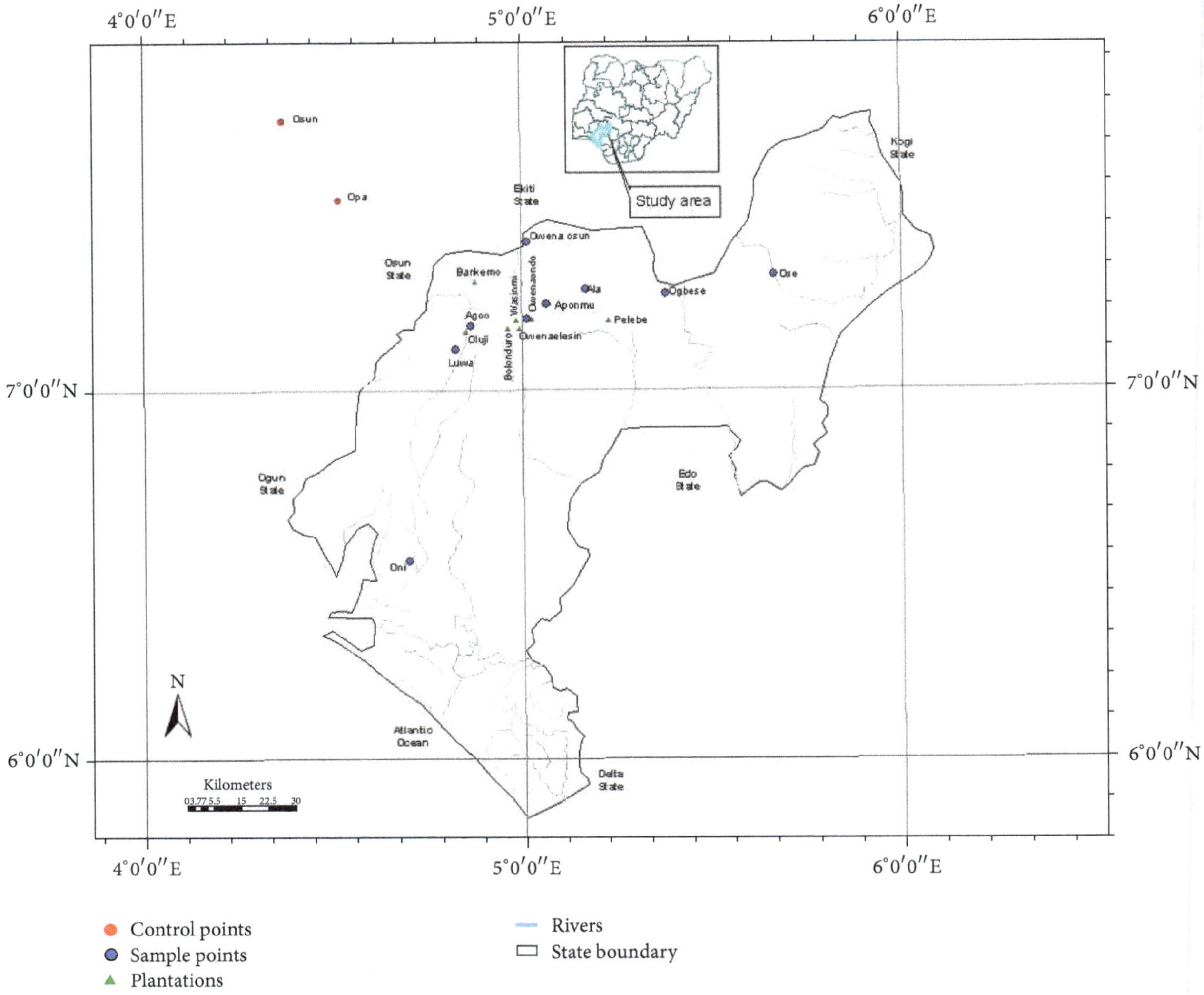

FIGURE 1: Map of the study area showing the geographical locations of sampling points.

of 1.65 and 1.65 μg/L of cis-chlordane and α-endosulfan, respectively, in Agoo River gave cause for concern. On the basis of the percentage of the analyte present in each river and the various concentration of each analyte detected, Oluwa River appeared to be the most contaminated. Other compounds were not detected because these compounds are hydrophobic and tend to accumulate in fatty tissue of organisms and sediment.

The result for dry season analyses also indicated that the concentration of most of the analytes is at trace concentration except for trans-chloride (0.34 μg/L), cis-chlordane, and α-endosulphan (0.01 μg/L each), which are detected in Aponmu river and those detected in Oluwa River: cis-chlordane and α-endosulphan (0.13 μg/L), p,p′-DDE (0.11 μg/L), o,p′-DDD (0.01 μg/L), p,p′-DDT (0.02 μg/L), and dieldrin (1.51 μg/L). The concentration of cis-chlordane and α-endosulphan in Agoo River at Ile-Oluji was 1.65 μg/L. The amount of aldrin in Owena river at Owena-Ondo was 0.01 μg/L.

It was not surprising to detect DDT in water samples of Oluwa River in Ondo State. DDE, which is the most often, recognized metabolite of the DDT, is a very less degradable compound. Another major reason of the presence of DDD in the river, despite the fact that its production is completely banned in the world, is due to the fact that DDD was produced and sold by another name "Rothane" for several years [28]. The results showed that the concentrations of most of the organochlorine compounds analyzed were below the maximum acceptable concentration of 0.1 μg/L value set by the European Union (EU) for the protection of aquatic environment.

3.1. Statistical Analysis. There is no significant difference in the mean concentrations of most of the analytes detected in water and sediment in the results of Duncan multiple range tests (Tables 3, 4, and 5). The correlation tests (Table 6) carried out for sediment for wet season analysis revealed that most of the organochlorine compounds correlate positively ($0.71 \leq r \leq 1.00$) with each other at probability level of 0.0001 while o, p′-DDD, p,p′-DDD were negatively correlated at probability range of 0.03 to 0.89. However dieldrin had very low but

Organochlorine Pesticide Residues in Sediments and Waters from Cocoa Producing Areas of Ondo State, Southwestern Nigeria

19

positive correlation within the probability range of 0.0158 to 0.2371. There were strong correlations (Table 7) between hexachlorobenzene (HCB) and trans-chlordane ($r = 0.96$), $\beta$-endosulphan ($r = 0.75$), o,p-DDD ($r = 0.82$), p,p,-DDD ($r = 0.95$), endrin ($r = 0.86$).

## 4. Conclusions

Supercritical fluid extraction (SFE) and liquid/liquid extraction (LLE) have been successfully employed for sample clean-up of sediment and water samples, respectively, in the determination of OCPs in the matrices. The concentrations of contaminants in water samples were very low as compared to concentrations in sediment samples. These results prove that these compounds are hydrophobic and tend to accumulate in sediments and also in fatty tissue of organisms [29]. On the basis of the percentage of the analyte present in each river and the various concentration of each analyte, Oluwa river is the most contaminated. The high concentration of these contaminants is of great concern, especially when the increasing accumulation potential of these compounds in the food chain is considered.

Due to lack of a similar survey for same study area, results could not be compared to determine past situation of the contamination and to estimate time trend. Moreover due to absence of such kind of previous study, numerical decrease factors could not be determined for the region. However, the present study can serve as reference data in the future, if routine monitoring of our environment is embarked upon by all concerned.

## Acknowledgments

The fellowship granted to A. A. Okoya by the Southern and Eastern Africa Network of Analytical Chemists (SEANAC) body which was utilized at the Department of Chemistry, University of Botswana, Gaborone, Botswana is highly acknowledged. The authors are also grateful to the authorities of the Obafemi Awolowo University for releasing her to utilize the grant. This study, however, forms part of the Ph.D. thesis of A. A. Okoya The comments and useful suggestions of the reviewers have considerably improved the quality of the paper.

## References

[1] B. Colin, *Environmental Chemistry*, W.H. Freeman and Company, 1999.

[2] G. Anitescu and L. L. Tavlarides, "Supercritical extraction of contaminants from soils and sediments," *Journal of Supercritical Fluids*, vol. 38, no. 2, pp. 167–180, 2006.

[3] A. Eqani, R. N. Malik, A. Alamdar, and H. Faheem, "Status of organochlorine contaminants in the different environmental compartments of Pakistan: a review on occurrence and levels," *Bulletin of Environmental Contamination and Toxicology*, vol. 88, no. 3, pp. 303–310, 2012.

[4] L. J. Bao, K. A. Maruya, S. A. Snyder, and E. Y. Zeng, "China's water pollution by persistent organic pollutants," *Environmental Pollution*, vol. 163, pp. 100–108, 2012.

[5] B. Hileman, "Environmental estrogens linked to reproductive abnormalities, cancer," *Chemical and Engineering News*, vol. 72, no. 5, pp. 19–23, 1994.

[6] D. L. Davis and H. L. Bradlow, "Can environmental estrogens cause breast cancer?" *Scientific American*, vol. 273, no. 4, pp. 166–172, 1995.

[7] W. R. Kelce, C. R. Stone, S. C. Laws, L. E. Gray, J. A. Kemppainen, and E. M. Wilson, "Persistent DDT metabolite p,p'-DDE is a potent androgen receptor antagonist," *Nature*, vol. 375, no. 6532, pp. 581–585, 1995.

[8] P. Cocco, A. Blair, P. Congia et al., "Proportional mortality of dichloro-diphenyl-trichloroethane (DDT) workers: a preliminary report," *Archives of Environmental Health*, vol. 52, no. 4, pp. 299–303, 1997.

[9] D. Pimentel, D. Andow, R. Dyson-Hudson et al., "Environmental and social costs of pesticides: a preliminary assessment," *Oikos*, vol. 34, no. 2, pp. 126–140, 1980.

[10] UNEP, *The Aral Sea: Diagnostic Study for the Development of an Action Plan for the Conservation of the Aral Sea*, Nairobi, Kenya, 1993.

[11] S. A. M. A. S. Eqani, R. N. Malik, and A. Mohammad, "The level and distribution of selected organochlorine pesticides in sediments from River Chenab, Pakistan," *Environmental Geochemistry and Health*, vol. 33, no. 1, pp. 33–47, 2011.

[12] M. Wessel, "Review of the use of fertilizer on cacao in Western Nigeria," Annual Report 79, CRIN, Ibadan, Nigeria, 1966.

[13] L. K. Opeke, *Tropical Tree Crops*, John Wiley & Sons, 1st edition, 1982.

[14] N. E. Egbe, E. A. Ayodele, and C. R. Obatolu, *Soil and Nutrition of Cocoa, Coffee, Kola, Cashew and Tea. Progress in Tree Research*, CRIN, Ibadan, Nigeria, 2nd edition, 1989.

[15] E. Pitarch, C. Medina, T. Portolés, F. J. López, and F. Hernández, "Determination of priority organic micro-pollutants in water by gas chromatography coupled to triple quadrupole mass spectrometry," *Analytica Chimica Acta*, vol. 583, no. 2, pp. 246–258, 2007.

[16] A. Johnson and J. Cowles, "Quality assurance project plan: Washington state surface water monitoring program for pesticides in salmonid habitat: a study for the Washington state," Department of Agriculture Conducted by the Washington State Department of Ecology. Publication Number 03-03-104, http://www.ecy.wa.gov/biblio/0303104.html, 2003.

[17] C. Burke and P. Anderson, "Addendum to the quality assurance project plan for the surface water monitoring program for pesticides in salmonid-bearing streams, addition of the skagit-samish watersheds and extension of program through June 2009," Washington State. Department of Ecology, Olympia, Wash, http://www.ecy.wa.gov/biblio/0303104add.html, 2006.

[18] Z. Tang, Q. Huang, Y. Yang, X. Zhu, and H. Fu, "Organochlorine pesticides in the lower reaches of Yangtze River: occurrence, ecological risk and temporal trends," *Ecotoxicology and Environmental Safety*, vol. 87, pp. 89–97, 2013.

[19] Federal Environmental Protection Agency (FEPA), *Guidelines and Standards for Environmental Pollution Control in Nigeria*, 1991.

[20] Federal Republic of Nigeria Official Gazette, Federal Government Press, Lagos, Nigeria. 83, 303–307, 1996.

[21] P. N. Ikemefuna, "Pesticide importation and consumption in Nigeria," in *Proceedings of the Pesticide Training Workshop on Environmentally Safe and Effective Control of Pests*, UNIFECS/FEPA, Lagos, Nigeria, 1998.

[22] T. A. Akinnifesi, O. I. Asubiojo, and A. A. Amusan, "Effects of
fungicide residues on the physico-chemical characteristics of
soils of a major cocoa-producing area of Nigeria," *Science of the
Total Environment*, vol. 366, no. 2-3, pp. 876–879, 2006.

[23] R. Weber, C. Gaus, M. Tysklind et al., "Dioxin- and POP-
contaminated sites—contemporary and future relevance and
challenges: overview on background, aims and scope of the
series," *Environmental Science and Pollution Research*, vol. 15, no.
5, pp. 363–393, 2008.

[24] O. S. Fatoki and O. R. Awofolu, "Levels of organochlorine
pesticide residues in marine-, surface-, ground- and drinking
waters from the eastern cape province of South Africa," *Journal
of Environmental Science and Health B*, vol. 39, no. 1, pp. 101–114,
2004.

[25] R. O. Awofolu and O. S. Fatoki, "Persistent organochlorine
pesticide residues in freshwater systems and sediments from the
Eastern Cape, South Africa," *Water SA*, vol. 29, no. 3, pp. 323–
330, 2003.

[26] C. N. Hewitt and R. M. Harrison, "Chapter 1," in *Understanding
our Environment*, R. E. Hester, Ed., Royal Society of Chemistry,
London, UK, 1st edition, 1986.

[27] A. M. Soto, K. L. Chung, and C. Sonnenschein, "The pesticides
endosulfan, toxaphene, and dieldrin have estrogenic effects on
human estrogen-sensitive cells," *Environmental Health Perspectives*, vol. 102, no. 4, pp. 380–383, 1994.

[28] L. P. Van Dyk, I. H. Wiese, and J. E. Mullen, "Management and
determination of pesticide residues in South Africa," *Residue
Reviews*, vol. 82, pp. 37–124, 1982.

[29] A. Aydin and T. Yurdun, "Residues of organochlorine pesticides
in water sources of istanbul," *Water, Air, and Soil Pollution*, vol.
111, no. 1-4, pp. 385–398, 1999.

[30] J. P. Villeneuve, P. P. Carvalho, S. W. Fowler, and C. Cattini, "Levels and trends of PCBs, chlorinated pesticides and petroleum
hydrocarbons in mussels from the NW Mediterranean coast:
comparison of concentrations in 1973/1974 and 1988/1989,"
*Science of the Total Environment*, vol. 237-238, pp. 57–65, 1999.

[31] J. L. Sericano, T. L. Wade, E. L. Atlas, and J. M. Brooks, "Historical perspective on the environmental bioavailability of DDT
and its derivatives to Gulf of Mexico oysters," *Environmental
Science and Technology*, vol. 24, no. 10, pp. 1541–1548, 1990.

[32] D. W. Klumpp, H. Huasheng, C. Humphrey, W. Xinhong, and S.
Codi, "Toxic contaminants and their biological effects in coastal
waters of Xiamen, China. I. Organic pollutants in mussel and
fish tissues," *Marine Pollution Bulletin*, vol. 44, no. 8, pp. 752–
760, 2002.

[33] Washinton State Pest Monitoring Programme, *Pesticides and
PCBs in Marine Mussels 1995*, Washington State Department of
Ecology, 1996.

[34] S. Y. A. Chau and B. K. Afghan, *Analysis of Pesticides in Water*,
Ontario, Canada, 1982.

# Carbon and Water in Upper Montane Soils and Their Influences on Vegetation in Southern Brazil

**M. B. Scheer,[1,2] G. R. Curcio,[2] and C. V. Roderjan[3]**

[1] *Research and Development Assistance (APD/DMA), SANEPAR, Curitiba, PR, Brazil*
[2] *Brazilian Agricultural Research Corporation (Embrapa Florestas), Colombo, PR, Brazil*
[3] *Federal University of Paraná (UFPR), Curitiba, PR, Brazil*

Correspondence should be addressed to M. B. Scheer; mauriciobs@sanepar.com.br

Academic Editors: P. Conte and W. Ding

Considering the many environmental functions of the upper montane soils, the aims of this study were (1) to verify if the soils of upper montane forests and grasslands of Caratuva Peak (1850 m a.s.l.) have similar characteristics to those found in other highlands in southern and southeastern Brazil; (2) to reinforce the geomorphological and pedological factors that impose the establishment of each type of vegetation in these highlands; and (3) to estimate potential soil carbon stocks and potential soil water retention. Folic and haplic histosols were found in the grasslands, and dystrophic regosols were found in the forests. The soils were dystrophic, extremely acidic, and saturated with Al and total organic carbon. In contrast to the grasslands, the upper montane forests were prevalent in valleys and subjected to morphogenetic processes resulting in soils that contained thicker mineral horizons. The grasslands occupied ridges and divergent convex ramps, and the pedogenetic processes in these regions promoted thicker histic horizons. The potential water retention capacities were high and strongly related to the high porosities of histic horizons associated with the gleyic horizons. In particularly, the carbon stocks were two- to three-fold higher than those found in soil ecosystems from the same latitude but lower altitude.

## 1. Introduction

Soil organic matter, which accumulates to a depth of 1 meter, weighs approximately $1500 \times 10^{15}$ g (Pg = petagrams) and constitutes the majority of continental organic carbon [1, 2]. Considering the soil surrounding native vegetation, Brazil stocks 5% of the global carbon reservoir [1], indicating that a single country can significantly influence global carbon fluxes [3]. According to published studies [4–7], stock levels of below ground carbon increase concomitantly with higher altitudes. In southern Brazil, carbon stock levels of isolated soils can be two to three-fold greater in high altitudes compared to low altitudes from the same latitude [7].

Since the upper montane soil horizons include histic and humic horizons, they play an important role in the immobilization of carbon and regulation of hydric fluxes [8–10]. The vast accumulation of organic matter in these environments originates from a lack of decomposition by microbes owing to unfavorable conditions, such as low temperature, high humidity, low respiration, lack of available nutrients, and high exchangeable Al concentrations [9–11]. Moreover, the organic matter is highly humified and forms relatively stable organometallic complexes, predominantly with Al and Fe that serve as poor substrates for decomposition by microorganisms [9, 11–13].

The humic substances contribute to the sustainability of these ecosystems by increasing their nutrient contents and increasing their capacity for cation exchange and water retention. In turn, the ecosystem recycles carbon matter through above- and below-ground litterfall of developing vegetation [9, 10]. During rainy periods, the upper montane soils minimize erosion by becoming waterlogged and draining slowly; thus, regulating the hydric fluxes to the headwaters of the river basins [10, 14, 15]. In dryer periods, upper montane soils collect water from horizontal precipitation (water present in clouds) that commonly occurs in upper montane forests

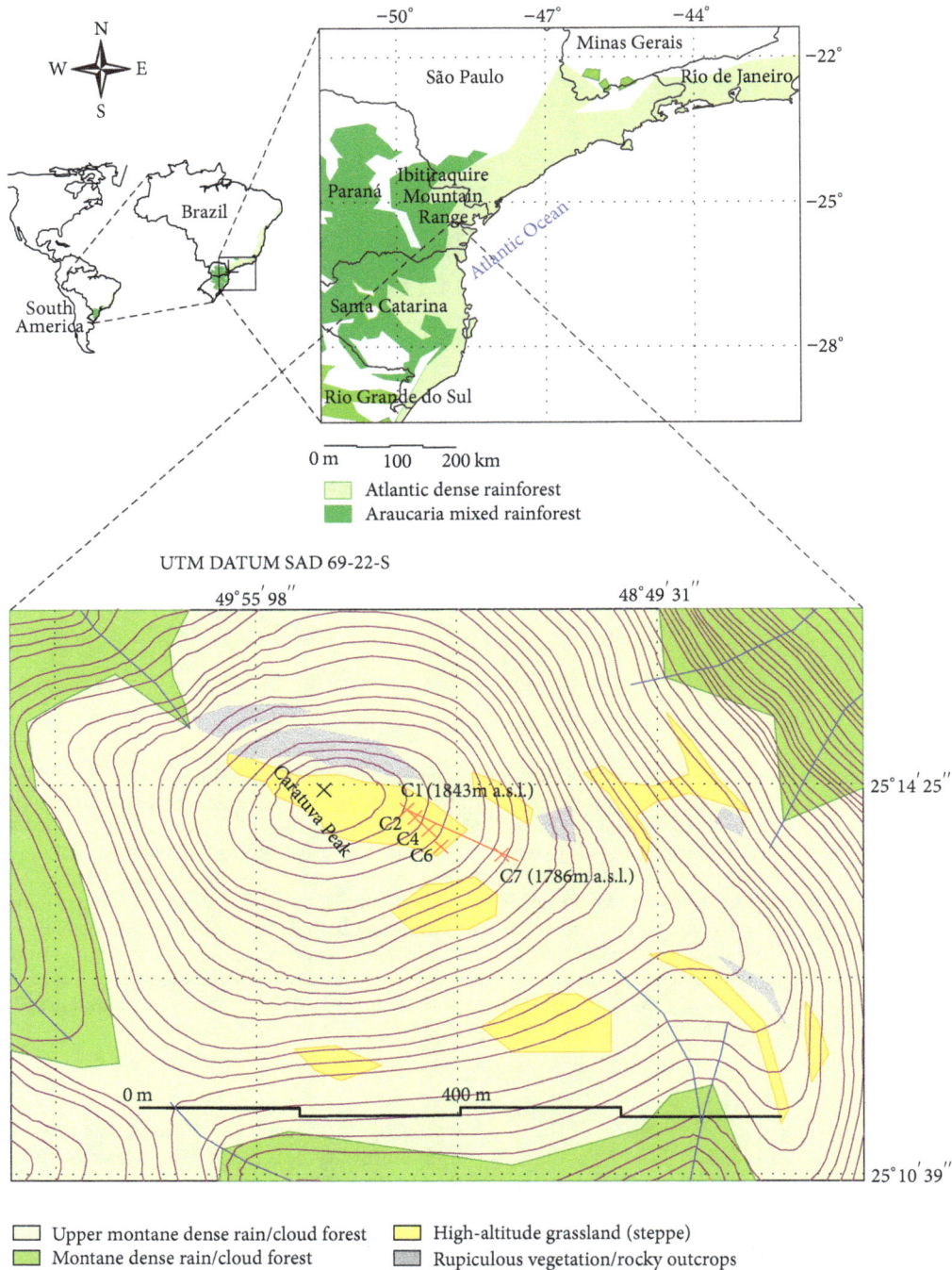

FIGURE 1: Sampling sequence in the Caratuva Peak, in the Ibitiraquire Mountain Range, southern Brazil.

[14, 15]. This water can be retained by vegetation and soil or forwarded to the headwaters immediately downstream, thereby avoiding major losses.

Soil from the mountaintop regions of many upper montane environments has been developing for a few thousand years (some of them starting from the middle and late Holocene period). In this period, significant climate change promoted the invasion of forests on grasslands. Since then,

these two types of vegetation have coexisted in a restricted landscape and are controlled by geomorphological features [7] forming the unique environment endemic to upper montane regions today.

Estimates of soil carbon stocks at regional, national, and global scales are important for the understanding of changes in carbon flux [16, 17]. To improve our capacity for providing, promoting, and executing actions that mitigate the consequences of the global climate change related to the greenhouse gases, we need to acknowledge the importance of carbon stocks in different soil types and study the factors that influence their distribution [3, 16, 18]. Considering that a significant amount of organic matter is stocked in the surface of soil (up to 30 cm deep), there is a great risk of carbon dioxide liberation to the atmosphere when vegetation is burned, cut, or converted to support agriculture or livestock [17]. The loss of natural and soil ecosystems can have a significant economic impact [14]. Thus, the need to understand soil organic matter dynamics in different areas in the world is great [19].

Upper montane soils are still understudied [3, 6, 9, 20, 21], and their importance is not yet fully appreciated [10]. In southern Brazil, some studies have characterized soils in upper montane rain/cloud forests [7, 22–28], but information about high altitude grassland soils is scarce [7, 27, 28]. According to previous studies conducted in the state of Paraná, southern Brazil, soils from grasslands and forests of upper montane ecosystems are undeveloped, extremely acidic, highly concentrated with carbon and have low base saturation and high exchangeable Al saturation. These observations are consistent with those reported in other upper montane regions of southeastern Brazil, which include different lithologies and vegetation types [9, 10, 29] and upper montane cloud forests of Aparados da Serra Geral, in the state of Santa Catarina, also in southern Brazil [26]. Another important environmental function described for these soils is their high capacity for water retention (in some cases more than $500 \, L \, m^2$), which results from high total porosities of their histic and gleyic horizons [7].

Considering that studies about upper montane soils are very scarce, the aims of this study were (1) to verify if the soils of upper montane forests and grasslands of Caratuva Peak (1850 m a.s.l.) have similar characteristics to those found in other highlands of the mountain ranges in southern and southeastern Brazil, which vary in altitude; (2) to reinforce the geomorphological and pedological factors that impose the establishment of each type of vegetation in these highlands; and (3) to estimate potential soil carbon stocks and potential soil water retention.

## 2. Material and Methods

*2.1. Study Area.* The study area is located in Caratuva Peak (transect between 1840 and 1770 m a.s.l.). This is the second highest mountain of southern Brazil, located in the Ibitiraquire Mountain Range, a subrange of Serra do Mar (Sea Mountain Range) with coordinates 25°14′S and 48°49′W (Figure 1). The Caratuva Peak is within the Paraná Peak State Park, that spans parts of the municipalities of Antonina, Morretes, and Campina Grande do Sul.

Like most of the mountains in the sea mountain range of Paraná, their upper portions are formed by alkali granites embedded in high-grade metamorphic terrains with contacts clearly failing in keeping with their late to posttectonic origin [30]. According to Mineropar, intrusive, igneous rocks from Ibitiraquire Mountain Range (Graciosa granite) originated

nearly 550 million years ago in the late pre-Cambrian to Cambrian period. It was cut by faults in the NE-SW axis by the Brasiliano (or Pan African) cycle (events of the end of the late Proterozoic) and the Ponta Grossa Arch, a major structure in the NW-SE basin of the Paraná River [31].

The upper parts of Ibitiraquire Mountain Range are covered with natural vegetation such as dense rain/cloud forests and grasslands (steppes). In the studied slope, the high-altitude grasslands are located near the summit, which lacks stony and rocky elements and is dominated by hydric divergence. The upper montane cloud forests occupy the crests downslope of the upper montane region and associated valleys with strong, undulating, and mountainous reliefs dominated by hydric convergence (Figure 2).

The Atlantic rainforest covers the greatest extension of Ibitiraquire Mountain Range, reaching the upper montane formation, with the highest part of the slopes reaching higher than 1830 m a.s.l. The majority of these forests are well preserved due to their isolation and difficult accessibility, despite some sites having been damaged by fires, such as other slopes of Caratuva Peak in 2007. In some slopes, located in the west, on the occidental side of the sea mountain range (at about 1300 m a.s.l.), there is an ecotonal zone between the Atlantic rainforest (Montane dense rainforest) and the Montane Araucaria forest. The high-altitude grasslands appear in the uppermost parts of the mountains, beginning from 1750 m a.s.l. in Caratuva Peak and from 1350 m a.s.l. in some other mountains. They are in excellent condition with no signs of human interference with the exception of mountain trails.

The sampling points were in a topographical sequence beginning near the sharp ridge of Caratuva Peak (1850 m a.s.l) covered by high-altitude grasslands and dominated by *Chusquea pinifolia* (Nees) Ness (Poaceae; caratuva) with about 0.5 m height on histosols (according to World Reference Base for Soil Resources: FAO-WRB) in divergent, convex ramps. After an abrupt alteration in slope, the landscape changes to concave, converging ramps, with colluvionar stability that provide a Cg (gleyic) layer and a grassland/shrubland physiognomy to histosols (2.2 m height), which are also dominated by the caratuva. After another abrupt break of the slope (1790 m a.s.l.), the studied ramp gives rise to the upper montane cloud forest situated on mineral soils with a histic layer (Figures 1 and 2).

As previously commented, in the studied high-altitude grasslands, *C. pinifolia* is the dominant species, comprising at least 25% of the vegetation. In a survey of this physiognomy covering many mountains of the Ibitiraquire Mountain Range, 85 vascular plant species were found [32, 33]. Additionally, *Croton mullerianus* L. R. Lima (Euphorbiaceae), *Rhynchospora exaltata* Kunth (Cyperaceae), *Lagenocarpus triquetrus* (Boeck.) Kuntze (Cyperaceae), and *Machaerina austrobrasiliensis* M. T. Strong are among the most important species. Tree species with extensive influence on the structure of the neighboring upper montane forests, of about 5 m in height, include *Myrsine altomontana* M.F. Freitas and Kin.-Gouv. (Myrsinaceae), *Handroanthus catarinensis* A. Gentry (Bignoniaceae), and *Symplocos corymboclados* Brand (Symplocaceae). These species also occur on the

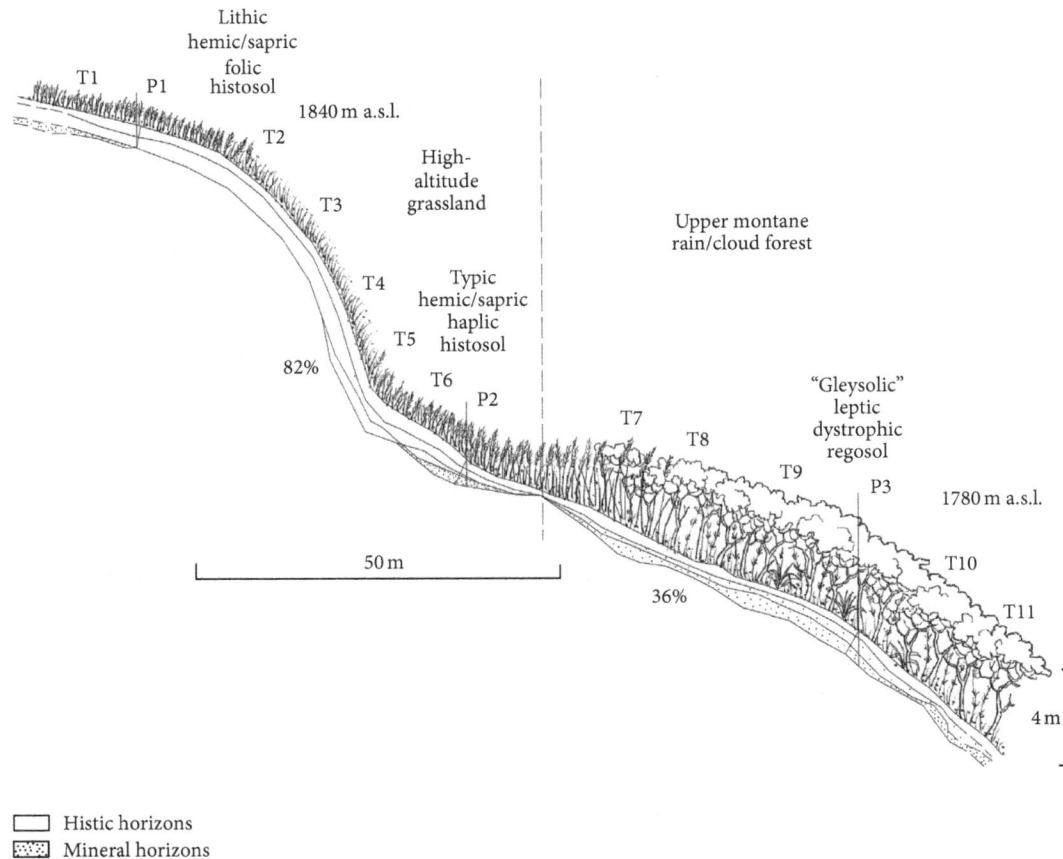

FIGURE 2: Toposequence indicating the sampling points (auger = T and profiles = P) in the Caratuva Peak, of the Ibitiraquire Mountain Range in southern Brazil.

grasslands, although sparsely, with low coverage, and in a shrub form, approximately 0.5 m in height, on the mountaintop grasslands with an average of 2.2 m height in the grasslands immediately downslope of the convergent ramps with higher colluvionar stability.

The change in vegetation from grassland to forest is abrupt, with a small ecotonal area with a shrubland physiognomy containing species of both formations in a range of 2 to 10 meters. According to [34, 35], *Ilex microdontia* Reisseck (Aquifoliaceae), *Siphoneugena reitzii* D. Legrand (Myrtaceae), *Myrceugenia seriatoramosa* (Kiaersk) D. Legrand and Kausel, and *Drimys angustifolia* Miers. (Winteraceae) contribute by 57.5% of the phytosociological importance (PAP ≥ 10 cm) of the sample sites in the Ibitiraquire Mountain Range, including the Caratuva Peak. The estimated basal area was 40.3 m$^2$ ha$^{-1}$, and the tree density was 4600 ind ha$^{-1}$.

The climate of upper portions of the sea mountain range of Paraná is classified as Cfb (subtropical), according to the Köppen classification system. It is predominantly wet, and the average temperature of the coldest month is between −3°C and 18°C, while the average temperature of the warmest month is below 22°C [22]. Roderjan and Grodski [36] observed an absolute minimum temperature of −5°C, an average annual temperature of 13.4°C, and an absolute maximum temperature of 30°C at 1385 m a.s.l. in

an upper montane forest at Anhangava Peak, approximately 24 km southwest of the Caratuva Peak. Based on the mean temperature variation of 0.56°C at each 100 m elevation, a mean temperature of 10.8°C can be estimated for the ridge of the Caratuva Peak. The precipitation in the sea mountain range is well distributed throughout the year and varies greatly depending on the local topography. Measurements in the coastal region exceed 2000 mm per year, and on the slopes of the mountains these values can reach 3500 mm [37].

*2.2. Sample Collection and Data Analysis.* The cores were carried out along a transect (toposequence), spanning high altitude grasslands and upper montane forests (Figures 1 and 2). After the soil survey, three profiles were analyzed, in which two of them represented the histosols of the grasslands and the regosols of the forests. Complementary superficial (0–20 cm) and subsuperficial (20–40 cm) samples were collected with a hand auger. In these profiles, morphological descriptions and definition of the diagnostic horizons and samplings were done following Santos et al. [38], in order to classify the soils according to the Brazilian Soil Classification System [39]. The Von Post scale [39] was used to classify the stage of decomposition of organic matter in the field.

Physicochemical analyses were based on air-dried soil, and granulometric analysis was carried out according to the densitometer method [40]. Due to the limitation of the laboratorial procedure for histic material, only the clay fractions for the majority of the samples were obtained. The chemical and physicohydric analyses were performed according to published reports [40]. The potential water retention capacities were estimated by the volumetric humidity and calculated by the subtraction of the aeration porosity from the total porosity [41]. Data were reported by units of $L\,m^{-2}$ and $m^3\,ha^{-1}$ according to the thicknesses of the horizons.

The total organic carbon (TOC) analyzed by dry weight method was done in Stable Isotope Laboratory of the Center for Nuclear Energy in Agriculture, Brazil (Table 1). Soil carbon stocks ($kg\,C\,m^{-2}$) were calculated according to Batjes [1] and Yimer et al. [17]. First, the TOC ($g\,kg^{-1}$) of each horizon was multiplied by the soil density ($kg\,m^{-3}$) and by its thickness (in m). The density of course particles ($\geq 2$ mm), generally subtracted from the soil density to use only the fraction $\leq 2$ mm [13, 19] was a variable not necessary due to its negligible values. Data were reported by units of $Mg\,C\,ha^{-1}$. The stocks of C of all points of the transect were based on their thicknesses and on the correspondent concentrations and densities of the nearest profiles.

The Ibitiraquire Mountain Range has an estimated 790 ha of high-altitude grasslands [44]. The soil C stocks were estimated for these ecosystems as an approach to predict their magnitude. The soil C stocks for the upper montane ecosystems, including both the grasslands and forests of Sea Mountain Range in Paraná were estimated based on the areas calculated by PRÓ-ATLÂNTICA [45]. However, we emphasize that in this study only the potential soil C stocks were estimated. Further estimations on scales of greater detail and consideration of additional types of soils present in other portions of the mountains are needed.

To compare chemical properties among the main horizons of 0 to 20 cm and 20 to 40 cm and the soil C stocks of the studied grasslands and forests, Kruskal Wallis tests ($P < 0,05$) were performed. Spearman correlation coefficients ($R_s$) were calculated to verify the behavior of the chemical parameters along profile depths.

## 3. Results and Discussion

*3.1. High-Altitude Grassland Soils in Caratuva Peak.* A predominance of histosols in the high-altitude grasslands in Caratuva Peak was observed. The histic horizons (mainly "O", humid colors 10YR 3/3 and N2/) presented with high concentrations of TOC ($400\,g\,kg^{-1}$). Profile 1 (nearer to the ridge) was classified as "lithic hemic/sapric folic histosol" and profile 2 (about 100 m downslope) as "typic/hemic sapric haplic histosol". The first profile was comprised of seasonal hydromorphic soil (O and O2/H horizons) and its location in a divergent convex slope, was more drained than profile 2 (Table 1 and Figure 3(a)). In the haplic histosol (less drained than the upslope folic histosols), three subhorizons were identified including two histic (O and H) and one mineral (Cg). The mineral subhorizon was not considered to be a terric histosol because it had a medium texture and was found to be less than 30 cm thick in the 1 m depth of soil surveyed.

This reducing environment was influenced by its position within a divergent concave slope (near the end of the ramp) combined with the low saturated permeability of the deepest mineral horizons, which delay the run off of the system (Table 3). Both soils had a prevalence of hemic herbaceous materials in topsoil.

The characteristics of these extremely acidic soils are similar to those found in other high-altitude grasslands of the Sea Mountain Range in southern Brazil [7, 27, 28] and those studied in soils associated with rocky outcrops in Mantiqueira Mountain Range in southeastern Brazil [9, 10, 46, 47]. The Al, $m\%$ (aluminum saturation) and cation exchange capacity (CEC) of the histic horizons increased with depth ($R_s = 1$; $P < 0.01$; Tables 1 and 2). This was due to the mineral fraction (mainly clay) in the system. The bases of the histic horizons ($S$ = sum of bases) were probably drained in relation to depth of values of three- to sixfold lower than on the surface (Tables 1 and 2). The P concentrations were considered to be in the low to medium range [48], and the base saturation (BS%), very low in topsoil, decreased proportionally with depth ($R_s = -1$; $P < 0, 01$; Tables 1 and 2).

*3.2. Upper Montane Rain/Cloud Forest in Caratuva Peak.* In the third part of the toposequence downslope, the soil was classified as "gleysolic leptic dystrophic regosol" (Tables 1 and 2; Figure 3(c)). The histic horizon (10YR 2/1) was less than 40 cm thick and had high TOC ($421\,g\,kg^{-1}$, Table 3). Below, two mineral horizons were observed, Big (10YR 5/2) and C (10YR 5/4), both with "mottles" in color (10YR 5/7). Due to the geomorphic position, the soils observed were "shallow" according to EMBRAPA (2006). The Big horizon signified the action of hydric fluxes. The "concave convergent" slope in "strong undulated" relief permitted the formation of mineral horizons through the combined action of morphogenetic (colluvium) and pedogenetic processes. These extremely acidic soils also presented with low base saturation, consistent with those found in other upper montane soils [7, 22–28].

As compared to the study by Scheer et al. [7], which was conducted in a mountain range 41 km south of our study site, the organic horizons of the upper montane soils in Caratuva Peak contained higher exchangeable base P concentrations, CEC, and potential acidity than the mineral horizons (Tables 1 and 2). CEC and acidity were due to the $H^+$ ions from the carboxylic and phenolic groups of the organic matter [9, 11]. The base of the desaturated Big horizon presented with high Al saturation (alitic character), suggesting that it was part of the forest cycle between the histic horizon and the litter. According to the literature, regosols have yet to be found and described in upper montane environments in southern Brazil.

*3.3. Common Characteristics of Upper Montane Soils.* The low fertility rates observed in vegetation of upper montane landscapes are related to the loss of nutrients caused by leaching from high rainfall amounts (horizontal precipitation, [15, 49]), in addition to generally low nutrient concentrations afforded by parental material [10]. According to these authors,

TABLE 1: Soil properties of the upper montane ecosystems in Caratuva, southern Brazil.

| Horizon | Depth. cm | Color | pH CaCl$_2$ | Al$^{+3}$ | H$^+$ + Al$^{+3}$ | Ca$^{+2}$ | Mg$^{+2}$ | K$^+$ | S | CEC | P mg dm$^{-3}$ | BS % | m % | Ca/Mg | *Factor cmol$_c$ dm$^{-3}$ to cmol$_c$ kg$^{-1}$ |
|---|---|---|---|---|---|---|---|---|---|---|---|---|---|---|---|
| | | | | | | cmol$_c$ dm$^{-3}$ | | | | | | | | | |
| Profile 1: grassland: lithic hemic/sapric folic histosol | | | | | | | | | | | | | | | |
| O1 | 0–20 | 10YR 3/3 | 3.3 | 1.4 | 22.5 | 1.5 | 1.3 | 0.6 | 3.4 | 25.9 | 13.0 | 13 | 29 | 1.2 | 2.6 |
| O2 | 20–30 | N2/ | 2.9 | 13.1 | 57.0 | 0.5 | 0.2 | 0.1 | 0.8 | 57.8 | 3.8 | 1 | 94 | 2.5 | 2.0 |
| Profile 2: grassland: typic hemic/sapric haplic histosol | | | | | | | | | | | | | | | |
| O1 | 0–20 | 10YR 3/3 | 3.1 | 1.8 | 29.3 | 0.7 | 0.8 | 0.5 | 2.0 | 31.3 | 13.9 | 6 | 48 | 0.9 | 2.6 |
| O2 | 22–49 | | 3.0 | 3.6 | 45.5 | 0.2 | 0.2 | 0.1 | 0.5 | 46.0 | 9.2 | 1 | 87 | 1.0 | 2.6 |
| H1 | 49–55 | | 2.9 | 15.2 | 53.0 | 0.2 | 0.1 | 0.02 | 0.3 | 53.3 | 3.8 | 1 | 98 | 2.0 | 2.0 |
| Cg | 55–70 | | 3.4 | 7.4 | 19.0 | 0.2 | 0.1 | 0.05 | 0.4 | 19.4 | 3.2 | 2 | 95 | 2.0 | 1.0 |
| Profile 3: forest: "gleysolic" leptic dystrophic regosol | | | | | | | | | | | | | | | |
| O | 0–15 | 10YR 2/1 | 2.6 | 3.1 | 41.6 | 0.4 | 0.6 | 0.3 | 1.3 | 42.9 | 16.6 | 3 | 70 | 0.7 | 2.7 |
| Big | 15–28 | 10YR 5/2 | 3.4 | 2.7 | 9.7 | 0.1 | 0.1 | 0.05 | 0.3 | 10.0 | 2.2 | 3 | 92 | 1.0 | 1.0 |
| C | 28–57 | 10YR 5/4 (mottles) 10YR 5/7 | 3.9 | 2.1 | 8.4 | 0.2 | 0.1 | 0.04 | 0.3 | 8.7 | 1.2 | 4 | 86 | 2.0 | 1.0 |

*Transforming factor from cmol$_c$ dm$^{-3}$ to cmol$_c$ kg$^{-1}$ calculated based on the inverse of the density of the prepared sample for the analysis of each horizon.

TABLE 2: Chemical properties of the soil horizons of the upper montane ecosystems in Caratuva Peak ($n$ = 2–6). Values in parentheses represent the standard errors.

| Horizon | Depth (average) cm | pH CaCl$_2$ | Al$^{+3}$ | H$^+$+Al$^{+3}$ | Ca$^{+2}$ | Mg$^{+2}$ | K$^+$ | S | CEC | P mg dm$^{-3}$ | BS % | m % | Ca/Mg | *Factor cmol$_c$ dm$^{-3}$ to cmol$_c$ kg$^{-1}$ |
|---|---|---|---|---|---|---|---|---|---|---|---|---|---|---|
| | | | | | cmol$_c$ dm$^{-3}$ | | | | | | | | | |
| Grassland: histosols | | | | | | | | | | | | | | |
| O1 $n$ = 6 | 0–20 | 3.2a (0.1) | 1.5b (0.1) | 27.7b (2.4) | 0.9a (0.2) | 0.9a (0.1) | 0.47a (0.0) | 2.3a (0.3) | 26.6b (1.7) | 15.6a (1.5) | 8.0a (1.2) | 40.2a (4.4) | 1.0a (0.2) | 2.6 |
| O2 $n$ = 6 | 20–40 | 3.1a (0.0) | 7.1a (1.7) | 46.8a (2.1) | 0.3b (0.1) | 0.2b (0.0) | 0.1b (0.0) | 0.6b (0.1) | 47.4a (2.2) | 7.7b (1.6) | 1.2b (0.2) | 90.2b (2.0) | 1.2a (0.3) | 2.6 |
| O3/H $n$ = 3 | 40–55 | 3.1a (0.1) | 9.9a (2.7) | 43.5ab (6.1) | 0.17b (0.03) | 0.17b (0.07) | 0.07b (0.04) | 0.4b (0.1) | 43.9ab (6.2) | 6.03b (1.24) | 1.0b (0.0) | 95.7b (1.4) | 1.2a (0.4) | 2.7 |
| Cg $n$ = 2 | 55–70 | 3.5 | 7.25 | 19.70 | 0.15 | 0.10 | 0.05 | 0.30 | 20.00 | 4.45 | 1.50 | 96.00 | 1.50 | 1.0 |
| Forest: regosols and gleysols | | | | | | | | | | | | | | |
| O1 $n$ = 5 | 0–20 | 2.9b (0.2) | 2.3a (0.3) | 35.9a (2.5) | 0.7a (0.3) | 1.0a (0.2) | 0.46a (0.0) | 2.1a (0.5) | 38.0a (2.5) | 26.0a (2.9) | 5.4a (1.2) | 53.6b (6.9) | 0.6a (0.2) | 2.7 |
| Cg1/Big $n$ = 5 | 20–40 | 3.4a (0.2) | 3.7a (0.5) | 13.6b (1.6) | 0.1b (0.0) | 0.1b (0.0) | 0.05b (0.0) | 0.3b (0.0) | 13.9b (1.6) | 2.4b (0.2) | 2.2b (0.4) | 92.8a (0.9) | 1.2a (0.2) | 1.0 |
| C $n$ = 2 | 40–60 | 3.8 | 3.2 | 11.3 | 0.2 | 0.1 | 0.05 | 0.4 | 11.6 | 1.5 | 3.0 | 89.0 | 2.0 | 1.0 |

*Transforming factor from cmol$_c$ dm$^{-3}$ to cmol$_c$ kg$^{-1}$ calculated based on the inverse of the density of the prepared sample for the analysis of each horizon. Averages followed by the same letter in vertical correspond the medians that do not differ statistically by the Kruskall-Wallis test ($P$ < 0.05).

acidity favors the dissolution of kaolinite and aluminosilicates, while the Al saturates the exchange complex. Therefore, nutrient cycling is essential because it leads to higher amounts of nutrients in superficial organic horizons [10] through their retention in high-density roots [9, 11], which increase the efficiency of hydrosoluble absorption [7]. Although the upper montane soils contained high Al saturation, considerable amounts of this element complexed with organic matter.

Therefore, its concentration was controlled in the soil solution, and toxicity to plants was avoided [9, 11, 13, 50, 51].

The genesis occurred in seasonal hydromorfisms of these soils was controlled by the elevated altitudes, facilitating cloud formation, lower temperatures, higher storage of organic matter by the low mineralization rate [24], and high rainfall amounts. The accumulation of organic matter surpassed the decomposition rate, because despite the existence

TABLE 3: Physicohydrical properties of soil in profiles in Caratuva Peak ($n = 2$).

| Horizon | Thickness | Clay/silt/sand** | Porosity (%) | | | Available water | Saturated permeability | TOC | Soil density | C stock | Potential water retention capacity |
|---|---|---|---|---|---|---|---|---|---|---|---|
| | cm | g kg$^{-1}$ | Total | Macro | Aeration | (%) | cm h$^{-1}$ | % | kg m$^{-2}$ | kg m$^{-2}$ | L m$^{-2}$ |
| Profile 1: grassland: lithic hemic/sapric folic histosol | | | | | | | | | | | |
| O1 | 20 | 175 | 98.4 | 35.1 | 37.5 | 5.8 | 319.5 | 41.9 | 0.117 | 9.80 | 121.8 |
| O2/H | 10 | 200 | 95.8 | 12.1 | 12.8 | 5.2 | 0.3 | 40.9 | 0.238 | 9.73 | 83.0 |
| Total | — | — | — | — | — | — | — | — | — | 19.54 | 204.8 |
| Average plus standard errors extrapolating to all sampling points ($n = 3$) | | | | | | | | | | (26.1 ± 3.3) | (338.2 ± 8.5) |
| Profile 2: grassland: typic hemic/sapric haplic histosol | | | | | | | | | | | |
| O1 | 22 | 100 | 99.9 | 34.4 | 35.3 | 7.4 | 158.2 | 42.3 | 0.108 | 10.05 | 142.1 |
| O2 | 27 | 125 | 98.3 | 12.6 | 13.5 | 5.7 | 36.2 | 37.9 | 0.133 | 13.61 | 229.0 |
| H | 6 | 125 | 87.3 | 9.6 | 10.2 | 7.3 | 3 | 20.8 | 0.302 | 3.77 | 46.3 |
| Cg | 15 | 125 | 62.8* | 9.2* | 9.3* | 5.3* | 7.2* | 1.37* | 1.089* | 2.24 | 80.3 |
| Total | — | — | — | — | — | — | — | — | — | 29.67 | 497.6 |
| Average plus standard errors extrapolating to all sampling points ($n = 3$) | | | | | | | | | | (46.3 ± 2.7) | (565.6 ± 7.7) |
| Profile 3: forest: "gleysolic" leptic dystrophic regosol | | | | | | | | | | | |
| O | 15 | 138 | 99.2 | 34.8 | 36.4 | 6.6 | 238.9 | 42.1 | 0.113 | 7.10 | 94.1 |
| Big | 13 | 125 | 62.8 | 9.2 | 9.3 | 5.3 | 7.2 | 1.37 | 1.089 | 1.94 | 69.6 |
| C | 29 | 125 | 62.8* | 9.2* | 9.3* | 5.3* | 7.2* | 1.37* | 1.089* | 4.33 | 155.2 |
| Total | — | — | — | — | — | — | — | — | — | 13.37 | 318.8 |
| Average plus standard errors extrapolating to all sampling points ($n = 5$) | | | | | | | | | | (15.9 ± 2.5) | (398.8 ± 8.9) |

*Values estimated based on horizons with similar characteristics.
**Limited estimates for granulometric analysis due to the high organic matter concentrations.

of concurrent biological transformation, humified organic carbon was highly preserved [23].

### 3.4. What Can Explain the Occurrence of Forest or Grasslands in Upper Montane Environments?

According to studies carried out in high-altitude grasslands of southern Brazil, the histosols and histic entisols are the most common soils [7]. Both soils contained histic horizons without mineral horizons or without their presence at 40 cm depth. Histic horizons between 20 to 40 cm deep that lacked mineral horizons (also known as a histosol, according to the Brazilian Classification, EMBRAPA [39]), seem to suppress the establishment of trees taller than 1 m [7]. Together, the depth of histic horizons (in some cases > 50 cm), their low global densities, the high density of roots from typic grassland species, and the exposure of mountaintops to strong winds constitute conditions that are refractory to the growth and establishment of large trees that could promote the advancement of upper montane forests over grasslands as suggested previously [7] and reinforced by our data. However, lithic folic histosols (histic horizons between 20 and 35 cm deep) and histic litholic entisols (histic horizons < 20 cm) were identified by Roderjan [22] and Rocha [24] in upper montane cloud forests. Despite being stunted (low stature, high density, and low basal area), these forests were classified as dwarf forests.

Future climate changes affecting rainfall levels may result in the disappearance of grasslands through considerable loss of area to forests or erosion of histic horizons caused by oxidation [7]. Other areas must be investigated to know whether the forests have already reached the maximum limit of expansion on the high-altitude grasslands of the Sea Mountain Range [7]. Studies of soils associated with high-altitude rocky complexes in southeastern Brazil [10] revealed that some peatlands (histosols) can dry sufficiently by a drainage system, leading to a forest growth but maintaining the large stocks of carbon. In addition to climate change, the fragility of upper montane ecosystems is challenged by the impact of human activity at nearby hiking trails and camping sites. The mountain range of Caratuva Peak (Paraná Peak State Park) attracts many visitors, and, in dryer periods, the environment has a high susceptibility to anthropic fires and a high risk of loss of these ecosystems [7].

The chemical composition of histic horizons below the depth of 20 cm depth at high-altitude grasslands was remarkably different from mineral horizons found at the same depth in soils of upper montane forests (Tables 1 and 2). Previously, other soils with relatively low densities and high basal areas were observed in upper montane rain/cloud forests in southern Brazil including cambisols [22, 24], argisols [24], and gleysols [7, 24] in valleys and peaks with higher colluvium. Upper montane forests in (typic humic and typic dystrophic) entisols were studied by Roderjan [22], Portes [25], and Vashchenko et al. [28]. The mineral horizons (A humic and C) of these soils as well as Big and C of regosols observed in Caratuva Peak seemed to promote higher mechanical

(a)

(b)

(c)

(d)

FIGURE 3: (a) Profile 1: histosol in high-altitude grassland near to the ridge of Caratuva Peak. (b) Profile 2: histosol in high-altitude grassland nearer to the grassland-forest ecotone. (c) Profile 3: regosol in upper montane rain/cloud forest. (d) Toposequence in Caratuva Peak (foreground). At the bottom, Paraná Peak, southern Brazil.

aggregation and sustainability of trees, common in high-altitude grasslands [7, 28], compared to histic horizons of histic litholic entisols observed by Portes [25]. Therefore, in these environments, geomorphic features and climate control the soils, which regulate the occurrence of forest or grassland vegetation.

3.5. Carbon Stocks in Upper Montane Soils of Caratuva Peak. The soil carbon stocks in the high-altitude grasslands ranged from 195.4 to 512.5 Mg C ha$^{-1}$ (mean ± standard error: 260.7 ± 32.7 Mg C ha$^{-1}$ for the lithic histosols near the ridge and 462.7 ± 26.8 Mg C ha$^{-1}$ for the typic histosols downslope). The soil carbon storage estimated for the upper montane cloud/rain forest on regosols and gleysols varied from 93.9 to 220.6 Mg C ha$^{-1}$ (158.6 ± 24.9 Mg C ha$^{-1}$). The storage was statistically higher in grassland than in forest ($U$ test; $P <$ 0.05; Table 3).

Based on soil data from Serra da Igreja [7], the potential C stocks estimated for the soils of high-altitude grasslands and the upper montane dense cloud/rain forest in the state of Paraná, southern Brazil, were 0.54 × 10$^6$ Mg C and 1.25 × 10$^6$ Mg C, respectively. These values, particularly grassland stocks, were in line with the observations presented in this study. Specifically, we found that potential C stocks from high-altitude grasslands and Upper montane dense rainforest in the state of Paraná based on Caratuva Peak soil data and PRO ATLÂNTICA vegetation cover data [45] contained 0.59 × 10$^6$ Mg C and 0.91 × 10$^6$ Mg C, respectively. This indicates that although these upper montane environments are separated by about 500 m a.s.l. (Caratuva Peak stretches between 1790 and 1840 m s.n.m and Igreja Mountain Range between 1325 and 1335 m s.n.m.), similar chemical characteristics, physical characteristics, and levels of carbon stocks were observed.

The small differences could be caused by many factors including geomorphic position (slope types, declivity, etc.), temperature differences, rainfall levels, variable decomposition rates, and primary productivity. Larger and more detailed samplings are needed to identify the main factors. Thus, the presence of histic horizons rather than altitude consideration could be a better parameter to help determine sites of preservation (in discussion for the current Brazilian Forest Code), as the same type of upper montane vegetation is commonly found below 1800 m a.s.l. in southern Brazil.

Because the Ibitiraquire Mountain Range was estimated to have 790 ha of high-altitude grasslands [44], the soil C stocks for these ecosystems range from 0.138 × 10$^6$ Mg C to

TABLE 4: Soil carbon stocks in different forest and grassland ecosystems.

| Location | Coord. (Aprox.) | Altitude (m a.s.l.) | Soil/lithology | Vegetation | Temp. annual average (°C) | Rainfall (mm) | Carbon stock (Mg ha$^{-1}$ ± SE) | Author |
|---|---|---|---|---|---|---|---|---|
| Sehestedt, Germany | 51°17'N; 10°39'E | 280 | Vertisols, limestone | Managed grassland | 8.0 | 550 | 86 ± 1.7 | Don et al. [13] |
| Kaltenborn, Germany | 50°47'N; 10°14'E | 335 | Quartzipsamments, sandstone | Managed grassland | 8.0 | 650 | 48 ± 1.3 | Don et al. [13] |
| Bale, Ethiopia Mountains | 6°45'N; 39°45'E | 2390–2800 | Andosols, volcanic rocks | Afromontane forest | 12.1 | 985 | 326 ± 2.2 | Yimer et al. [17] |
| | | 2800–3250 | Andosols, volcanic rocks | Afromontane forest | 9.2 a 12.1 | 985 | 364 ± 2.2 | Yimer et al. [17] |
| | | 3580+ | Andosols, volcanic rocks | Subalpine shrub physiognomy | 9.2 | 985 | 460 ± 2.2 | Yimer et al. [17] |
| Pyrenees Mountains, France and Spain | 42°50'N; 0°40'S | 1845–2900 | Granites, slates, and limestones | Alpine and subalpine grasslands | −0.7 a 5 | 1416 a 1904 | 153 ± 0.9 | Garcia-Pausas et al. [19] |
| Diverse | — | — | 1 m | Tropical forest | — | — | 186 ± 104** | Jobbágy and Jackson [2] |
| Diverse | — | — | 1 m | Tropical grassland and Savanna | — | — | 132 ± 87** | Jobbágy and Jackson [2] |
| Amazon region, Brazil | — | — | Mineral soils (lato. agri. cambi. gleysols, etc.) | Forests | — | — | 98 | Batjes and Dijkshoorn [16] |
| | — | — | Quartzipsamments | | — | — | 40 | Batjes and Dijkshoorn [16] |
| | — | — | Histosols | | — | — | 724 | Batjes and Dijkshoorn [16] |
| Juruema, Amazon, Brazil | 10°28'S; 58°28'W | 260 | Xingu complex, agri, and latosols | Dense rainforest | 24 | 2200 | 60 ± 10.3 | Novaes Filho et al. [18] |
| Amazon region, Brazil | — | — | | Dense rainforest | — | — | 103 | Britez et al. [42] |
| Many studies in South and Southeastern Brazil | — | — | | Grassland | — | — | 79 | Britez et al. [42] |
| | — | — | | Mixed rainforest | — | — | 71 | Britez et al. [42] |
| | — | — | | Dense rainforest | 20.9 | — | 61 | Britez et al. [42] |
| | — | — | | Seasonal semideciduous forest | — | — | 37 | Britez et al. [42] |
| Cachoeira reserve, Antonina, southern Brazil | 29°19''S; 48°42''W | 20 | Haplic cambisols | pasture Herbaceous | | | 98.5 ± 15.4** | Balbinot [43] |
| | | | | Early successional dense rainforest | 20.9 | 2400 | 118.5 ± 14.1** | Balbinot [43] |
| | | | | Medium successional dense rainforest | | | 113.7 ± 17.8** | Balbinot [43] |
| | | | | late successional submontane dense rainforest | | | 100.7 ± 18.1** | Balbinot [43] |
| | | | | | | | 102.0 ± 19.1 | Balbinot [43] |
| Igreja Mountain, Paraná range, southern Brazil | 25°36'S; 48°51'W | 1335 | Histosols, granite | High-altitude grassland | 13.4 | >2000 | 314 ± 17.2 | Scheer et al. [7] |
| | | 1325 | Gleysols, granite | Upper montane dense rain/cloud forest | 13.4 | >2000 | 217 ± 10.3 | Scheer et al. [7] |
| Ibitiraquire Mountain, Paraná range, southern Brazil | 25°14'S; 48°49'W | 1830 | Lythic histosols, granite | High-altitude grassland | 10.7 | >2000 | 260 ± 32.7 | This study* |
| | | 1810 | Typic histosols, granite | High-altitude grassland | 10.7 | >2000 | 462.3 ± 26.8 | This study* |
| | | 1790 | Regosols and gleysols, granite | Upper montane dense rain/cloud forest | 10.7 | >2000 | 158 ± 24.9 | This study* |

$0.361 \times 10^6$ Mg C. Additionally, the studies at the Sea Mountain Range of southern Brazil revealed that the soil carbon stocks by unit area in high-altitude grasslands are higher than the stocks of the upper montane forests. Moreover, both stocks are considerably higher when compared to soils of other ecosystems (Table 4). Therefore, to obtain precise estimates, more surveys are needed.

Differences in C stocks from various mountain aspects, topographic/geomorphic positions, and parent materials are very common in the same mountain [17, 19]. Although the C concentrations are 30-fold lower in mineral horizons than histic horizons of upper montane rain forests, they comprise almost half (46%) of the carbon soil stocks. This is explained by the 10-fold greater density of mineral horizons compared to histic horizons (Table 3). However, the mineral horizons of the observed histosols in grasslands contribute between 0 and 8% of total soil carbon due to their small thickness and to lower densities.

*3.6. Upper Montane Soil Carbon Stocks and the Carbon Storage in Above-Ground Forest Biomass.* The mean values of soil carbon stocks of high-altitude grasslands and upper montane forests of the Igreja Mountain Range were 314.7 and 217.2 Mg C ha$^{-1}$, respectively, according to Scheer et al. [7], and the corresponding estimates for Caratuva Peak were $361.4 \pm 48.9$ and $158.6 \pm 24.9$ Mg C ha$^{-1}$. These estimates were higher than the highest estimates for the above-ground biomass of submontane rainforests (about 200 Mg C ha$^{-1}$, mean near 120 Mg C ha$^{-1}$), which represents the highest above-ground productivity among the Atlantic Forests [52, 53].

Based on many forest structure surveys of the Atlantic Forest in the state of Paraná [42], carbon stocks in above-ground biomass are estimated at 26 Mg C ha$^{-1}$ (early secondary successional stage) and 153 Mg C ha$^{-1}$ (mature forest). As stated previously, such information highlights the great importance of upper montane soil carbon stocks. Namely, the high-altitude grassland soils have the capacity to stock more than twice the above-ground biomass of the lower montane Atlantic rainforest, where the highest primary productivity among the forests of southern and southeastern of Brazil is found.

*3.7. Upper Montane Soil Water Retention Potential in Caratuva Peak.* The high-saturated permeability of the histic horizons (reaching 319.5 cm h$^{-1}$, Table 3) was considerably reduced to 3 to 7.2 cm h$^{-1}$ with depth, favoring higher water retention in the soil volume. The hemic and fibric characteristics of superficial horizons, which include their low densities (0.108 to 0.117 kg dm$^{-3}$), high macroporosity and aeration, and considerable pore connectivity, contribute to very high permeability. Although the histic horizons had high water retention capacity, the available water concentrations were low (Table 3). The gleyic horizons, such as the Big horizon representing regosols, showed lower total porosities. They also presented with lower percentage of macropores leading to lower aeration capacity and saturated permeability, which is related to structural and textural changes (Table 3). Like

the estimated carbon stocks, the potential water retention capacities of the histosols (profiles 1 and 2; 2048 and 4976 m$^3$ ha$^{-1}$, resp.) were similar to those calculated for the Igreja Mountain Range [7]. The regosols of the forest showed a capacity of 3188 m$^3$ ha$^{-1}$.

## 4. Conclusions

Soils from the ridge of the Caratuva, as in other upper montane environments of southern Brazil, are dystrophic, extremely acidic, and highly saturated with Al and TOC. Upper montane forests of the Sea Mountain Range of southern Brazil were strongly established by physical impacts from valleys and peaks. Additionally, these forests were influenced by slopes susceptible to higher action of morphogenetic processes, which result in soils with mineral horizons containing higher amounts of clay (such as gleysols and regosols). The high density of these horizons also promoted better conditions to the development of tree species. Therefore, in the upper montane environments, the thickening of mineral horizons and the thinning of histic horizons contribute to the growth of forests. High-altitude grasslands occupy ridges of folic and haplic histosols found on divergent convex ramps, resulting in thicker histic horizons. The low density and lack of consistency of these horizons associated with seasonal hydromorphy and strong winds hinder the development of forests in these environments.

Considering the studies at Sea Mountain Range of southern Brazil, the soil carbon stocks in the high-altitude grasslands are higher by unit area than the stocks of the upper montane forests. However, both values are higher, when compared to soils from other ecosystems including the above-ground biomass of submontane dense rainforests, which exhibited high primary productivity. The high potential of soil water retention of upper montane environments is strongly related to the high total porosities of histic horizons associated with the gleyic horizons. These characteristics, in addition to the filter capacity of these soils, must also be recognized as environmentally functional and worthy of preservation.

## Acknowledgments

The authors would like to thank many colleagues for their substantial help in the field: Marcos Rachwal, Bruno Miranda, Eduardo Lozano, Letícia Sousa, Yury Vashchenko, Anette Bonnet, and Marcia de Lima. This research was supported by the Fundação Grupo Boticário de Proteção à Natureza (0800_2008_2). The authors are greatly indebted to the Sociedade Chauá and Programa de Pós-Graduação em Engenharia Florestal da UFPR.

## References

[1] N. H. Batjes, "Total carbon and nitrogen in the soils of the world," *European Journal of Soil Science*, vol. 47, no. 2, pp. 151–163, 1996.

[2] E. G. Jobbágy and R. B. Jackson, "The vertical distribution of soil organic carbon and its relation to climate and vegetation," *Ecological Applications*, vol. 10, no. 2, pp. 423–436, 2000.

[3] M. Bernoux, M. C. S. Carvalho, B. Volkoff, and C. C. Cerri, "Brazil's soil carbon stocks," *Soil Science Society of America Journal*, vol. 66, no. 3, pp. 888–896, 2002.

[4] K. Kitayama and S. I. Aiba, "Ecosystem structure and productivity of tropical rain forests along altitudinal gradients with contrasting soil phosphorus pools on Mount Kinabalu,Borneo," *Borneo. Journal of Ecology*, vol. 90, no. 1, pp. 37–51, 2002.

[5] M. Zimmermann, P. Meir, M. I. Bird, Y. Malhi, and A. J. Q. Ccahuana, "Climate dependence of heterotrophic soil respiration from a soil-translocation experiment along a 3000 m tropical forest altitudinal gradient," *European Journal of Soil Science*, vol. 60, no. 6, pp. 895–906, 2009.

[6] L. Roman, F. N. Scatena, and L. A. Bruijnzeel, "Global and local variations in tropical montane cloud forest soils," in *Tropical Montane Cloud Forests: Science For Conservation and Management*, L. A. Bruijnzeel, F. N. Scatena, and L. Hamilton, Eds., Cambridge University Press, Cambridge, UK, 2010.

[7] M. B. Scheer, G. R. Curcio, and C. V. Roderjan, "Environmental functionalities of upper montane soils in Serra da Igreja, Southern Brazil," *Revista Brasileira de Ciencia do Solo*, vol. 35, no. 4, pp. 1113–1126, 2011.

[8] L. A. Bruijnzeel, "Hydrology of tropical montane cloud forests: a reassessment," in *Proceedings of the Second International Colloquium*, J. S. Gladwell, Ed., pp. 353–383, UNESCO, Paris, France, 2000.

[9] F. N. B. Simas, C. E. G. R. Schaefer, E. I. Fernandes Filho, A. C. Chagas, and P. C. Brandão, "Chemistry, mineralogy and micropedology of highland soils on crystalline rocks of Serra da Mantiqueira, Southeastern Brazil," *Geoderma*, vol. 125, no. 3-4, pp. 187–201, 2005.

[10] V. M. Benites, C. E. G. R. Schaefer, F. N. B. Simas, and H. G. Santos, "Soils associated with rock outcrops in the Brazilian mountain ranges Mantiqueira and Espinhaço," *Revista Brasileira de Botanica*, vol. 30, no. 4, pp. 569–577, 2007.

[11] W. Zech, N. Senesi, G. Guggenberger et al., "Factors controlling humification and mineralization of soil organic matter in the tropics," *Geoderma*, vol. 79, no. 1-4, pp. 117–161, 1997.

[12] B. Volkoff, C. C. Cerri, and A. J. Melfi, "Húmus and mineralogy of superficial horizons of three soils of upland grass fields of Minas Gerais, Paraná and Santa Catarina states, Brazil," *Revista Brasileira de Ciencia do Solo*, vol. 8, pp. 277–283, 1984.

[13] A. Don, J. Schumacher, M. Scherer-Lorenzen, T. Scholten, and E. D. Schulze, "Spatial and vertical variation of soil carbon at two grassland sites—Implications for measuring soil carbon stocks," *Geoderma*, vol. 141, no. 3-4, pp. 272–282, 2007.

[14] M. L. Martínez, O. Pérez-Maqueo, G. Vázquez et al., "Effects of land use change on biodiversity and ecosystem services in tropical montane cloud forests of Mexico," *Forest Ecology and Management*, vol. 258, no. 9, pp. 1856–1863, 2009.

[15] L. A. Bruijnzeel and J. Proctor, "Hydrology and biogeochemistry of tropical montane cloud forests: what do we really know?" in *Tropical Montane Cloud Forests*, Hamilton, L. S et al., Eds., pp. 38–78, Springer, New York, NY, USA, 1995.

[16] N. H. Batjes and J. A. Dijkshoorn, "Carbon and nitrogen stocks in the soils of the amazon region," *Geoderma*, vol. 89, no. 3-4, pp. 273–286, 1999.

[17] F. Yimer, S. Ledin, and A. Abdelkadir, "Soil organic carbon and total nitrogen stocks as affected by topographic aspect and vegetation in the Bale Mountains, Ethiopia," *Geoderma*, vol. 135, pp. 335–344, 2006.

[18] J. P. N. Novaes Filho, E. C. Selva, E. G. Couto, J. Lehmann, M. S. Johnson, and S. J. Riha, "Spatial distribution of soil carbon under primary forest cover in southern Amazônia," *Sociedade de Investigações Florestais*, vol. 31, no. 1, pp. 83–92, 2007.

[19] J. Garcia-Pausas, P. Casals, L. Camarero et al., "Soil organic carbon storage in mountain grasslands of the Pyrenees: Effects of climate and topography," *Biogeochemistry*, vol. 82, no. 3, pp. 279–289, 2007.

[20] A. Becker, C. Körner, J. J. Brun, A. Guisan, and U. Tappeiner, "Ecological and land use studies along elevational gradients," *Mountain Research and Development*, vol. 27, no. 1, pp. 58–65, 2007.

[21] G. A. Arteaga, N. E. G. Calderón, P. V. Krasilnikov, S. N. Sedov, V. O. Targulian, and N. V. Rosas, "Soil altitudinal sequence on base-poor parent material in a montane cloud forest in Sierra Juárez, Southern Mexico," *Geoderma*, vol. 144, no. 3-4, pp. 593–612, 2008.

[22] C. V. Roderjan, *A floresta ombrafila densa altomontana no morro Anhangava. Quatro Barras. PR: aspectos climlticos, pedologicos e fitossociolagicos [Ph.D. thesis]*, Universidade Federal do Paraná, 1994.

[23] N. L. B. Ghani, *Caracterização morfológica. física. química. mineralógica. gênese e classificação de solos altimontanos derivados de riólito e migmatito da Serra do Mar. PR [M.S. thesis]*, Universidade Federal do Paraná, 1996.

[24] M. R. L. Rocha, *Caracterização fitossociológica e pedológica de uma floresta ombrófila densa no parque estadual do pico do Marumbi-Morretes. PR [M.S. thesis]*, Universidade Federal do Paraná, 1999.

[25] M. C. G. de O. Portes, *Deposição de serapilheira e decomposição foliar em floresta ombrófila densa altomontana. Morro Anhangava. Serra da Baitaca. Quatro Barras. PR [M.S. thesis]*, Universidade Federal do Paraná, 2000.

[26] D. B. Falkenberg, *Matinhas nebulares e vegetação rupícola dos Aparados da Serra Geral (SC/RS). sul do Brazil [Ph.D. thesis]*, Universidade de Campinas, 2003.

[27] C. Wisniewski, P. B. Tempesta, and V. M. R. Rodrigues, "Solos e vegetação em uma topossequência do morro Mãe Catira. alto da Serra da Graciosa. Morretes PR," in *Proceedings of the 30th Congresso Brasileiro de Ciência do Solo*, Sociedade Brasileira de Ciências do Solo, Viçosa, Brazil, 2005.

[28] Y. Vashchenko, R. P. Piovesan, M. R. Lima, and N. Favaretto, "Solos e Vegetação dos Picos Camacuã. Camapuã e Tucum—Campina Grande Do Sul-PR," *Scientia Agraria*, vol. 8, no. 4, pp. 411–419, 2007.

[29] A. C. Silva, I. Horák, A. M. Cortizas et al., "Peat bogs of the serra do espinhaço meridional—Minas Gerais, Brazil. I—Characterization and classification," *Revista Brasileira de Ciencia do Solo*, vol. 33, no. 5, pp. 1385–1398, 2009.

[30] Mineropar Minerais do Paraná S. A., *Atlas Geológico do Estado do Paraná*, Curitiba, Brazil, 2001.

[31] Melo, P. C. F. Giannini, and L. C. R. Pessenda, "Gênese e Evolução da Lagoa Dourada. Ponta Grossa. PR," *Revista do Instituto Geológico*, vol. 21, no. 2, pp. 17–31, 2000.

[32] A. Y. Mocochinski, *Campos de altitude na serra do mar paranaense: aspectos florísticos e estruturais [M.S. thesis]*, Universidade Federal do Paraná, 2006.

[33] A. Y. Mocochinski and M. B. Scheer, "Campos de altitude na serra do mar paranaense: aspectos florísticos," *Revista Floresta*, vol. 38, no. 4, pp. 625–640, 2008.

[34] M. B. Scheer and A. Y. Mocochinski, "Floristic composition of four tropical upper montane rain forests in Southern Brazil," *Biota Neotropica*, vol. 9, no. 2, pp. 51–69, 2009.

[35] M. B. Scheer and A. Y. Mocochinski, "Tree component structure of tropical upper montane rain forests in Southern Brazil," *Acta Botanica Brasilica*, vol. 25, no. 4, pp. 735–750, 2011.

[36] C. V. Roderjan and L. Grodski, "Acompanhamento meteorológico em um ambiente de Floresta Ombrófila Densa Altomontana no morro Anhangava. Mun. De Quatro Barras. PR," *Cadernos da Biodiversidade*, vol. 2, no. 1, pp. 27–34, 1999.

[37] R. Maack, *Geografia Física do Estado do Paraná*, Imprensa Oficial, Curitiba, Brazil, 2002.

[38] R. D. Santos, R. C. Lemos, H. G. Santos, J. C. Ker, and L. H. C. Anjos, *Manual de Descrição e Coleta de Solo no Campo*, Sociedade Brasileira de Ciências do Solo, Viçosa, Brazil, 2005.

[39] Empresa Brasileira de Pesquisa Agropecuária, *Sistema Brasileiro de Classificação de Solos*, EMBRAPA, Rio de Janeiro, Brazil, 2nd edition, 2006.

[40] Empresa Brasileira de Pesquisa Agropecuária, *Manual de Métodos de Análises de Solos*, EMBRAPA, Rio de Janeiro, Brazil, 1997.

[41] C. L. Prevedello, *Física do Solo: Com Prolemas Resolvidos*, Curitiba, Brazil, 1996.

[42] R. M. Britez, M. Borgo, G. Tiepolo, A. Ferreti, M. Calmon, and R. Higa, *Estoque e Incremento de Carbono em Florestas e Povoamentos de Espécies Arbóreas Com ênfase na Floresta Atlântica do Sul do Brazil*, EMBRAPA, Colombo, Sri Lanka, 2006.

[43] R. Balbinot, *Carbono nitrogênio e razões isotópicas $^{13}C$ e $^{15}N$ no solo e vegetação de estágios sucessionais de Floresta Ombrófila Densa Submontana [Ph.D. thesis]*, Universidade Federal do Paraná, 2009.

[44] C. V. Roderjan, *Caracterização da vegetação dos Refúgios Vegetacionais Altomontanos nas serras dos órgãos e do Capivari no Estado do Paraná*, Relatório Técnico do CNPq., Curitiba, Brazil, 1999.

[45] Programa de Proteção da Floresta Atlântica—PRÓ-ATLÂNTICA, *Projeto Carta Geológica, Cartas: MI-2844-2 Ariri. MI-2843-1 Represa do Capivari. MI-2843-3 Morretes. MI-2858-1 Mundo Novo e MI-2858-3 Pedra Branca do Araraquara, Escala 1:50000*, SEMA, Curitiba, Brazil, 2002.

[46] V. M. Benites, A. N. Caiafa, E. S. Mendonça, C. E. G. R. Schaeffer, and J. C. Ker, "Soil and vegetation on the high altitude rocky complexes of the Mantiqueira and Espinhaço mountain," *Revista Floresta e Ambiente*, vol. 10, pp. 76–85, 2003.

[47] V. M. Benites, E. S. Mendonça, C. E. G. R. Schaefer, E. H. Novotny, E. L. Reis, and J. C. Ker, "Properties of black soil humic acids from high altitude rocky complexes in Brazil," *Geoderma*, vol. 127, no. 1-2, pp. 104–113, 2005.

[48] B. M. Serrat, K. I. Krieger, and A. C. V. Motta, "Considerações sobre a interpretação de análise de solos," in *Recomendações de Manejo do Solo: Aspectos Tecnológicos e Metodológicos*, M. R. Lima et al., Ed., pp. 125–143, UFPR, Curitiba, Brazil, 2006.

[49] L. S. Hamilton, J. O. Juvik, and F. N. Scatena, "The Puerto Rico tropical cloud forest symposium: introduction and workshop synthesis," in *Tropical Montane Cloud Forests*, L. S. Hamilton et al., Ed., pp. 1–23, Springer, New York, NY, USA, 1995.

[50] F. J. Stevenson, *Humus Chemistry: Genesis, Composition, Reactions*, John Wiley & Sons, New York, NY, USA, 1982.

[51] M. A. Pavan, "Alumínio em solos ácidos do Paraná: relação entre o alumínio não-trocável, trocável e solúvel, com o pH, CTC, porcentagem de saturação de Al e matéria orgânica," *Revista Brasileira de Ciencia do Solo*, vol. 7, pp. 39–46, 1983.

[52] S. A. Vieira, L. F. Alves, M. Aidar et al., "Estimation of biomass and carbon stocks: the case of the Atlantic Forest," *Biota Neotropica*, vol. 8, no. 2, pp. 21–29, 2008.

[53] G. Tiepolo, M. Calmon, and A. R. Feretti, "Measuring and monitoring carbon stocks at the guaraqueçaba climate action project, Paraná, Brazil," in *International Symposium on Forest Carbon Sequestration and Monitoring*, vol. 153, pp. 98–115, Taiwan Forestry Research Institute, 2002.

**4**

# Rapid Screening of Berseem Clover (*Trifolium alexandrinum*) Endophytic Bacteria for Rice Plant Seedlings Growth-Promoting Agents

**H. Etesami, H. Mirsyedhosseini, and H. A. Alikhani**

*Department of Soil Science, College of Agriculture & Natural Resources, Tehran University, Karaj 31587-77871, Iran*

Correspondence should be addressed to H. A. Alikhani; halikhan@ut.ac.ir

Academic Editors: G. Benckiser, L. A. Dawson, J. A. Entry, D. Jacques, and D. Lin

A simple screening method to detect berseem clover (*Trifolium alexandrinum*) endophytic bacteria for rice plant growth-promoting agents on the basis of a root colonization bioassay and a plant growth promoting trait is characterized. Firstly, 200 isolates (80 endophytes and 120 rhizospheric isolates) isolated from berseem clover were inoculated as 10 mixtures of 20 strains each on two rice varieties under gnotobiotic conditions. Then, the reisolated endophytic strains from two rice varieties were characterized for plant growth promoting (PGP) traits. Secondly, the colonization and growth promoting effects of endophytic strains were compared in inoculated rice plantlets as single-strain inoculants. A significant relationship among indole-3-acetic acid (IAA) producing isolates, the size of root colonization, and plant growth was observed. Our results suggest that the ability of IAA production by the endophytic bacteria which may have a stimulatory effect on plant development may be the first plant growth promoting trait for screening bacteria isolated from clover plant for rice plant growth promoting agents. In addition, this study indicates that the selected bacterial isolates based on their IAA producing trait have the potential for PGP and more colonization of rice plant.

## 1. Introduction

Research into how plant growth can be promoted has mainly concentrated on rhizobacteria. More recently, however, attention has been focused on the plant growth-promoting capacity of endophytes [1]. The study of plant bacterial endophytes is important for understanding ecological interactions and to develop biotechnological applications. Endophytic bacteria can be defined as those bacteria that can colonize the internal tissue of the plant without showing negative symptoms on their host [2]. Endophytes promote the growth of plants in various ways, for example, through secretion of plant growth regulators, such as indole-3-acetic acid (IAA) [3], via phosphate solubilizing activity [4], by enhancing hyphal growth and mycorrhizal colonization [5], production of siderophores [6], and supplying biologically fixed nitrogen [7]. In addition, endophytic bacteria supply essential vitamins to plants [7]. The production of auxin-like compounds increases seed production and germination [8] along with

increased shoot growth and tillering [9]. Other effects of endophyte infection on the host plant include osmotic adjustment, stomatal regulation, modification of root morphology, enhanced uptake of minerals, and alteration of nitrogen accumulation and metabolism [10]. During the last few years, there has been an increased interest in exploring the possibility of extending the beneficial interactions between rice and some of plant growth-promoting bacteria including $N_2$-fixing bacteria. Many reports found in literature strongly suggest that these endophytes have an excellent potential to be used as plant growth promoters with legumes and nonlegumes [11]. Yanni et al. [12] showed that rhizobia can naturally colonize the interior of rice roots rotated with clover berseem and assessed the potential impact of this novel plant-microbe association on rice production. In Iran, the second cropping in rotation with rice is berseem clover. Our interest has been to assess the possible existence and agronomic importance of naturally occurring endophytes within rice roots in Iran where rice production is significantly benefited by rotation with berseem clover that could sustain the populations of the corresponding bacteria at a high inoculum potential for the next rice growing season. PGB traits can be assessed under laboratory conditions and allow the selection of strains that could lead to increased plant growth [12]. Nature selects endophytes that are competitively fit to occupy compatible niches within this nutritionally enriched and protected habitat of the root interior without causing pathological stress on the host plant. But when we intended to screen these bacteria for other plant growth-promoting agents, it is better to screen these bacteria for achieving the most promising isolates having suitable colonization and plant growth-promoting traits. In most researches, it has been seen that, following incubation, bacterial flora are taken at random from Petri plates or morphological representatives are selected for further study. But this type of selection may remove some superior bacteria regarding plant growth-promoting traits and with high colonization ability. The Gram reaction test and other phenotypic characteristics could not definitively determine the classification for the isolates. So, it is essential to study all the bacteria isolated in an economic way. On the other hand, if we test all strains for all PGP traits, this process will take a long time and will be costly. Since the ability to colonize roots is a necessary condition for a rhizobacteria to be considered a true plant growth-promoting rhizobacteria (PGPR) [13], we would not need to identify the PGP traits of all of the isolates, just those that are able to establish themselves in rice roots. In addition, the first screening on the basis of the most promising PGP traits is very important and economic too. So, we were interested in finding an important auxiliary tool for helping in selection of the isolates with colonization and PGPR potentiality. Yanni et al. [12] showed that in spite of developing higher populations within roots of Indica rice, the endophytic bacteria elicited higher short-term PGP responses on the Japonica rice. They indicated that identification of superior combinations of rhizobia and rice genotypes for optimal growth responses will likely require PGP bioassays rather than just an assessment of the bacterial endophyte's ability to colonize the root interior, and manipulations to increase the endophyte population above the natural level achievable within rice per se may not necessarily improve the resultant growth promotion response. Several methods have been used to demonstrate that root colonization is taking place, including use of fluorescence techniques, antibiotic-resistant mutants, and marker genes, such as LUX and GUS. However, these methods are relatively expensive and time consuming [13, 14].

Hence, the present study was undertaken to (i) isolate indigenous endophytic and rhizoshpere isolates from berseem clover plants grown in rotation with rice plants, (ii) evaluate which bacteria can colonize the endophytic sites of two rice varieties by batch inoculations onto rice in the presence of low or moderate levels of N under gnotobiotic conditions, (iii) evaluate the PGP traits of the isolates, (iv) reapply the endophytic isolates in single-strain inoculations onto rice in the presence of N and measure rice growth to evaluate if they promote the growth of the rice, and (v) evaluate efficiency of the relationship between root colonization and the most promising PGP trait on rice seedlings inoculated with endophytic isolates.

## 2. Materials and Methods

*2.1. Isolation of Endophytic and Rhizosphere Bacteria.* Rhizosphere soil and roots of the berseem clover plants (*Trifolium alexandrinum*) were collected from the Dashte naz Research Farm, Māzandarān Province, Iran. Bacterial endophytes were isolated based on the method described by the Sturz et al. [15]. Plants (at flowering) were collected randomly from a field plot where rice has been rotated with clover and conveyed to the laboratory in coolers. Roots were washed thoroughly with tap water, rinsed with deionized water, and drained on absorbent towels. Roots were cut into 2-3 cm long slices with a sterile scalpel, and ten grams of roots was shaken for 30 minutes in 500 mL Erlenmeyer flask containing 250 mL sterile deionized water and 25 grams of glass beads. Clover roots were washed in running tap water and commercial detergent to remove soil and then rinsed for 30 s in a 95% ethanol solution. Surface sterilization was with 10% sodium hypochlorite solution for 4 min, followed by three rinses in sterile distilled water. Aseptic technique was used in every part. Root slices were placed in a sterilized commercial blender in quarter-strength Ringer's solution and macerated for 3 min. The macerate was decanted into conical flasks, shaken on a wrist-action shaker for 45 min, a dilution series made and the diluent plated onto tryptic soy agar (TSA). For isolating rhizosphere bacteria, soil adhered to roots was collected. After proper mixing, 10 g of soil was transferred to a 250 mL Erlenmeyer flask containing 90 mL sterile distilled water and shaken (120 rpm) for 30 min. Serial dilutions (up to $10^{-7}$) were made, and 0.1 mL aliquots were spread on to TSA plates. All the plates were incubated at 28°C for 3–5 days at which time the number of colony-forming units (cfu) was counted. Numbers of endophytic bacteria cells recovered were expressed as cfu $g^{-1}$ fresh tissue weight. Three replicates per dilution were made. To confirm that the surface-sterilization process was successful, tissue pieces were pressed onto or rolled over TSA, and aliquots of water from the final rinse solutions were plated onto TSA and tested for contaminants. No contaminants were found, and the surface-sterilization procedure was considered effective. Following incubation of bacterial flora, bacterial isolates identified as individual cfu were selected and subcultured onto TSA. Similar bacterial isolates were grouped on the basis of phenotypic characteristics (shape, motility, color, rate of growth, and culture morphology) and Gram-staining reaction and stored in a refrigerator at 4°C for further studies.

*2.2. Colonization Assay.* We used the method described by Yanni et al. [12] with some modification for colonization assay of rice roots with all the strains isolated from clover plant in presence and absence of nitrogen. Two rice varieties, Khazar a bred variety and Hashemi a nonbred variety obtained from the rice research institute in Iran, both currently cultivated in Māzandarān in rotation with berseem clover, were used for this study. Separate tests were established that the seeds of these two rice cultivars harbored no endophytic bacteria

Rapid Screening of Berseem Clover (Trifolium alexandrinum) Endophytic Bacteria for Rice Plant Seedlings Growth-Promoting Agents

35

that would survive surface sterilization. 8 mM $(NH_4)_2SO_4$ was used in Hoagland's plant growth medium. 100 $\mu$L of the bacterial inocula grown separately on NB medium for 2 days at 30°C was suspended in sterile Hoagland's medium as 10 mixtures of 20 strains each. The mixtures were selected at random. Each seedling root was inoculated with $10^8$ cells mL$^{-1}$ (12 replicates were used for each treatment) and incubated in a growth chamber. Rice plants in culture tubes were gently uprooted 20 days after inoculation and then excised at the stem base. Roots were rinsed free of agar and sand, blotted, weighed, surface-sterilized with 70% ethanol followed by 10% sodium hypochlorite solution, rolled over plates of TSA to check for surface sterility, and then macerated in 5 mM Na-phosphate buffer as described above. Viable plate counts of the rice endophyte populations were made after 3 days of incubation of diluted root macerates plated on TSA as an indicator of bacterial invasion capacity in the presence or absence of nitrogen source. Numbers of CFU/mL in the external rooting medium were also analyzed to determine bacterial viability in the presence or absence of a nitrogen source. Then colonies were picked, restreaked on TSA, and stocked in pure culture. For counting rhizoplane bacteria, roots were washed with sterile distilled water several times and subsequently immersed in 0.8% saline in a 125 mL Erlenmeyer, added with 3.0 g of glass beads and shaken at 200 rpm for 30 min, at 28°C. The resulting supernatants were diluted and cultured on TSA, and bacterial growth observed after three-day incubation at 28°C was regarded as rhizoplane population.

*2.3. In Vitro Screening of Reisolated Bacterial Isolates for Their PGP Activities.* All strains, reisolated from inside roots of rice, were characterized for PGP traits. Production of siderophore, IAA, salicylic acid (SA), and HCN was determined according to the methodology described by Schwyn and Neilands [16], Patten and Glick [17], Meyer et al. [18], and Lorck [19], respectively, whereas all the isolates were screened for their phosphate-solubilizing ability on medium proposed by Sperber [20]. The ability of the isolates to produce ACC deaminase was also screened on minimal media containing ACC as their sole nitrogen source as described by Glick et al. [21].

*2.4. In Vitro Antifungal Activity.* Bacterial isolates were evaluated for their antifungal activity against fungal rice pathogens by dual culture assay as described by Jinantana and Sariah [22]. Fungal rice pathogens *Fusarium proliferum*, *Fusarium verticillioides*, *Fusarium fujikuroi*, *Magnaporthe salvinii*, and *Magnaporthe grisea* were kindly provided by laboratory of phytopathology, Department of plant protection, University of Tehran, Iran.

*2.5. Extracellular Hydrolytic Enzymes Activity Assay.* Chitinase activity was measured according to Jung and Kim [23]. Cellulase and pectinase activities were assayed on indicator plates as described by Mateos et al. [24].

*2.6. Efficient Establishment of Endophytic Colonization of Rice Plants by 9 Endophytes.* Endophytic colonization of rice

roots and assessment of their potential to promote plant growth were performed using the same procedures as described above, but, in this assay, the bacterial inocula were suspended in sterile Hoagland's medium containing 8 mM $(NH_4)_2SO_4$ as single-strain inoculants. Shoot biomass (stem plus leaves) and root mass were measured as dry and fresh weights, respectively. In addition, root height was measured. Reisolation and viable plate counting of endophytic isolates populations were also carried out as described above.

*2.7. Confirmation Test.* Ten isolates isolated from berseem clover plant grown at field, which had been screened only based on the production of IAA, were retested, using the same procedures as described previously in order to confirm the efficiency of the bioassay. As additional negative controls, two IAA nonproducing isolates were also included.

*2.8. Statistical Analysis.* All the experiments were arranged in randomized complete design with four replications in each treatment and repeated twice. Analysis of variance (ANOVA) was performed, and means were compared by the Tukey test at 5% probability level using the SAS (V. 8) software package (SAS Institute, Cary, NC, USA).

## 3. Results

*3.1. Isolation of Bacteria from Clover Root.* A total of 200 bacterial strains were isolated from the rhizosphere (120 isolates) and surface-sterilized roots (80 isolates) of berseem clover grown in rotation with rice in region of Dashte Naz, Mazandaran, Iran. Colony and cell morphology and Gram-staining tests were performed on the isolates. 54% of the isolates were Gram positive, and the rest were Gram negative. No further attempts for identification were made. All isolates were used for detailed investigation.

*3.2. Colonization of Rice Plants.* To determine which of the endophytic bacteria had the ability to colonize and persist at high levels in two rice varieties, we carried out studies with a gnotobiotic system, using axenic rice plantlets. In all colonization studies, controls were included to verify that the inoculated bacteria were recovered. No bacteria could be isolated from noninoculated plants. All 200 isolates were inoculated as 10 mixtures of 20 strains each into the rice hosts and grown under gnotobiotic conditions. Of these, 34 strains (9 endophytes and 25 rhizoplane isolates) colonized the plants at levels ranging from 4.3 to 7.9 log 10 CFU/g (fresh weight) at 20 days postinoculation (data not shown). The 34 isolates were all good root colonizers. The highest proportions of endophytes that were able to colonize the hosts were obtained from variety Hashemi grown in Hoagland + N growth medium (data not shown). The results indicated that none of the mixtures produced any disease symptoms or abnormalities in two rice varieties. Interestingly, 20 days after inoculation of the isolates in gnotobiotically cultured plantlets, the number of viable bacteria in the external rooting medium range in $10^{10}$ CFU, regardless the absence or presence of nitrogen.

TABLE 1: Plant growth promoting activities of nine clover endophytic strains reisolated from the root interior of rice and twenty-five strains isolated from the rhizoplane of rice grown in gnotobiotic tube culture for 20 days.

| Isolate | Siderophore production | Phosphorous solubilization | IAA production ($\mu$g mL$^{-1}$) | HCN production | ACC deaminase activity | Chitinase activity | Pectinase activity | Cellulase activity | Salicylic acid production | Antagonistic to all of the pathogens |
|---|---|---|---|---|---|---|---|---|---|---|
| E1 | − | − | + | − | + | + | + | + | − | − |
| E2 | − | + | + | − | + | − | + | + | − | − |
| E3 | + | − | + | − | − | − | + | − | − | + |
| E4 | + | + | + | + | − | − | − | + | − | − |
| E5 | − | + | + | − | + | − | + | − | + | − |
| E6 | + | + | + | − | − | − | + | + | − | + |
| E7 | + | − | + | − | + | − | − | + | − | − |
| E8 | + | − | + | + | − | − | + | + | − | − |
| E9 | + | − | + | − | − | − | + | + | − | − |
| R1 | − | + | + | − | + | − | + | − | − | − |
| R2 | + | + | + | − | − | − | + | + | − | − |
| R3 | + | + | + | − | − | − | + | + | − | − |
| R4 | + | + | + | + | − | − | − | + | − | − |
| R5 | + | + | + | − | + | − | − | + | − | − |
| R6 | − | − | + | − | − | − | − | + | − | − |
| R7 | + | − | + | − | − | − | − | + | − | − |
| R8 | + | − | + | − | − | − | − | + | − | − |
| R9 | + | − | + | + | + | − | − | + | + | + |
| R10 | + | + | + | − | + | − | + | − | − | − |
| R11 | + | − | + | − | − | − | + | − | − | − |
| R12 | + | + | + | − | + | + | + | − | − | − |
| R13 | + | − | + | − | − | − | − | − | − | − |
| R14 | + | − | + | + | + | − | − | + | − | − |
| R15 | + | − | + | − | + | − | + | + | + | − |
| R16 | − | − | + | − | − | − | − | + | − | − |
| R17 | − | − | + | + | + | + | − | + | − | − |
| R18 | + | + | + | + | + | − | + | − | − | − |
| R19 | + | − | + | − | − | − | + | + | − | + |
| R20 | + | − | + | − | − | − | + | + | − | − |
| R21 | + | + | + | − | + | − | + | + | − | − |
| R22 | + | − | + | − | + | + | + | + | − | − |
| R23 | + | + | + | − | − | − | + | − | − | − |
| R24 | + | + | + | − | + | − | + | − | − | − |
| R25 | + | − | + | − | − | − | + | − | − | − |

−: represents no activity, +: represents activity.
E: represents isolates reisolated from the root interior of rice seedlings.
R: represents isolates reisolated from the rhizoplane of rice seedlings.

*3.3. Plant Growth-Promoting Traits of Test Isolates.* A total of 34 bacterial isolates were reisolated from the roots of two rice varieties under gnotobiotic conditions. When the bacterial isolates were evaluated for their plant growth-promoting traits, all isolates produced IAA, 27, 7, 16, 4, 22, 24, and three isolates produced siderophore, HCN, ACC deaminase, chitinase, pectinase, cellulase, and SA, respectively, whereas 15 isolates solubilized phosphorous. 4 isolates inhibited the growth of fungal rice pathogens. Of the 9 isolates isolated from within rice plants, all isolates produced IAA, 6, 2, 4, 1, 6, 7, and one isolate produced siderophore, HCN, ACC deaminase, chitinase, pectinase, cellulase, and SA, respectively, whereas 4 isolates solubilized phosphorous. Two isolates inhibited the growth of fungal rice pathogens in the dual culture assay (Table 1). The isolates produced IAA between 3 and 17 $\mu$g mL$^{-1}$.

Rapid Screening of Berseem Clover (Trifolium alexandrinum) Endophytic Bacteria for Rice Plant Seedlings
Growth-Promoting Agents

37

*3.4. Reapply the Endophytic Isolates in Single-Strain Inoculations.* Efficient establishment of endophytic colonization of rice plants by 9 endophytes was demonstrated with a gnotobiotic system using axenic rice plantlets. Bacterial strains were successfully reisolated from roots of two rice varieties, and a considerably high recovery was recorded from rice plants of 20 days after inoculation. The average number of strains endophytically colonizing roots of Khazar and Hashemi varieties was, respectively, $7.5 \times 10^6$ and $2.6 \times 10^7$ CFU/g of fresh weight plantlet tissue. The influence of nitrogen fertilization in the experiments conducted with bacterial strains was performed with axenic plantlets cultured on Hoagland's solution supplemented with 8 Mm $(NH_4)_2SO_4$. Colonization extent was considerably higher in roots in presence of N, compared to roots in absence of N (Table 2). Plantlets cultivated for 20 days in the presence of $(NH_4)_2SO_4$ showed a significant increase in bacterial invasion levels. Plating experiments of macerates from surface-sterilized roots of the inoculated plants indicated substantial populations of the endophytic bacteria, the magnitude of which varied the rice cultivar and the plant growth medium (Table 2). The inoculated roots appeared healthy without development of nodule-like hypertrophies or obvious symptoms of disease. Under these experimental conditions, internal root colonization by the isolates was not suppressed in Hoagland's plant growth medium, which contains $NH_4^+$ as a source of N. Colonization by the strains induced, in all inoculated plants, a large increase in root hair number and length. The colonization and growth-promoting the effects of nine IAA producing endophytic strains were compared in inoculated rice plantlets as single-strain inoculants. The presence of a higher number of lateral roots and more abundant root hairs was observed in response to colonization by the strains when compared with noninoculated plants. All the IAA producing isolates increased shoot dry mass by 22–44%, root fresh mass by 8–48%, and root length by 3–52%, over the control. The isolate E9 demonstrated to be the best plant growth-promoting bacteria, with an increase of 48% in root fresh weight and 44% in shoot dry weight as compared with control (Table 3). A significant relationship among IAA producing isolates, rate of root colonization, and plant growth was observed (Figures 1 and 2).

*3.5. Confirmation Test.* Screening to detect isolates with good potential as rice growth-promoting agents indicated that 7 of 10 endophytes, inoculated on rice varieties, were able to colonize within roots and promote plant growth (data not shown). According to the results obtained, 7 of 10 isolates tested (70%) behaved as potentially good plant growth-promoting and colonizing agents. Seedlings inoculated with IAA producing isolates yielded more shoot biomass and colonization than the control plants inoculated with IAA nonproducing strains. Plants inoculated with both IAA producing isolates and other PGP traits producing isolates did not yield more biomass and colonization than plants inoculated with IAA producing isolates alone. Rate of colonization of rice plants with IAA Producing isolates showed that inoculation with the strains yielded a 10–20% increase in root weight as compared with

TABLE 2: CFU counts of endophytic bacteria colonizing internal part of roots of the Iranian rice variety grown in gnotobiotic tube culture for 20 days in Hoagland liquid with or without nitrogen (8 mM as $(NH_4)_2SO_4$).

| Endophytic isolate | Plant growth medium (Hoagland) | Rice endophyte population (log 10 CFU/g root fresh wt) | |
|---|---|---|---|
| | | Khazar | Hashemi |
| E1 | +N | 6.30 | 7.03 |
| E2 | +N | 6.45 | 6.87 |
| E3 | +N | 7.68 | 8.54 |
| E4 | +N | 7.98 | 8.15 |
| E5 | +N | 6.92 | 7.32 |
| E6 | +N | 8.34 | 8.56 |
| E7 | +N | 8.65 | 8.96 |
| E8 | +N | 8.54 | 9.13 |
| E9 | +N | 9.21 | 9.76 |
| E1 | −N | 5.92 | 6.43 |
| E2 | −N | 5.87 | 6.32 |
| E3 | −N | 6.21 | 7.12 |
| E4 | −N | 6.01 | 7.04 |
| E5 | −N | 5.66 | 6.66 |
| E6 | −N | 7.87 | 7.98 |
| E7 | −N | 7.08 | 8.45 |
| E8 | −N | 7.87 | 8.32 |
| E9 | −N | 8.90 | 8.96 |

the noninoculated control and IAA non producing isolates. Interestingly, IAA producing isolates developed higher culturable endophytic populations in the roots. In this study, a significant relationship among IAA producing isolates, rate of root colonization, and plant growth was also confirmed. The screening procedure appears to be very effective and less time consuming.

## 4. Discussion

To determine whether bacteria isolated from within clover plant tissue can have rice plant growth-promotion potential, colonize rice roots, and survive under flooded conditions, since rice is grown at such condition, seeds of two rice varieties were inoculated with individual bacterial isolates or mixtures of bacteria. We found IAA producing isolates that not only significantly improved rice plant growth but also, when used for colonization assay, significantly increased the rate of colonization than IAA nonproducing isolates (Figures 1 and 2). These results indicated that the endophytes had potential for promoting plant growth. Since the final aim after selecting the best isolate will be to introduce these isolates as a biofertilizer (suitable for pudding) for farmers, to find how much of chemical fertilizers (maximum yield) those bacteria can replace, we should select bacteria that have been isolated in the presence of N. We inoculated rice plant with both endophytic isolates and rhizosphere isolates isolated from the clover roots because microfloral populations already resident

TABLE 3: Evaluation of various morphological responses of rice (variety Hashemi) in gnotobiotic tube culture for 20 days in Hoagland liquid with nitrogen (8 Mm as $(NH_4)_2SO_4$) after inoculation with nine endophytes.

| Isolate code | Root length (cm) | Root fresh weight (mg) | Shoot dry weight (mg) |
|---|---|---|---|
| Control | $2.6 \pm 0.03^g$ | $2.50 \pm 0.06^e$ | $27.16 \pm 0.2^c$ |
| E1 | $2.7 \pm 0.07^f$ | $2.55 \pm 0.08^e$ | $27.82 \pm 0.1^c$ |
| E2 | $2.6 \pm 0.08^g$ | $2.58 \pm 0.04^e$ | $27.93 \pm 0.3^c$ |
| E3 | $3.6 \pm 0.08^d$ | $2.74 \pm 0.09^c$ | $34.62 \pm 0.2^b$ |
| E4 | $3.2 \pm 0.05^e$ | $2.78 \pm 0.04^c$ | $34.29 \pm 0.4^b$ |
| E5 | $2.7 \pm 0.04^f$ | $2.63 \pm 0.09^d$ | $27.80 \pm 0.4^c$ |
| E6 | $3.6 \pm 0.06^d$ | $2.82 \pm 0.07^c$ | $36.18 \pm 0.7^b$ |
| E7 | $3.8 \pm 0.07^c$ | $2.91 \pm 0.06^b$ | $40.30 \pm 0.5^a$ |
| E8 | $3.9 \pm 0.05^b$ | $3.74 \pm 0.04^a$ | $40.85 \pm 0.5^a$ |
| E9 | $4.0 \pm 0.04^a$ | $3.76 \pm 0.06^a$ | $40.38 \pm .06^a$ |

In each column, values followed by the same letter are not significantly different as determined by Tukey's mean comparison test ($P \leq 0.05$; $n = 4$).

within the host plant may well influence and be influenced by rhizosphere bacteria [25, 26]. In addition, several studies have reported that endophytic microbial communities originate from the soil and rhizosphere [27–29]. Since aseptic technique was used throughout, and the surface-sterilization procedure was considered effective, we may verify that the same bacteria inoculated to sterile rice seedlings could be reisolated from these seedlings (to fulfill Koch's postulate) and examine their endophytic competence (infection and persistence characteristics). 34 strains had colonized the plants at levels ranging from 4.3 to 7.9 log 10 CFU/g (fresh weight) at 20 days after inoculation. The number of CFU of these bacteria in roots fell in the range of endophytic bacteria [30, 31]. Interestingly, the endophytic isolates developed higher culturable endophytic populations in roots of varieties Hashemi than in those of varieties Khazar (Table 2). Thus, the degree to which endophytic isolates establish endophytic populations within rice varies among different varieties [12]. The establishment of plant growth promotion by rhizobia strains has been observed in crops such as wheat [32], rice [12], canola, lettuce [33], and sunflower [34, 35]. Colonization extent was considerably higher in roots in the presence of N, compared to roots in the absence of N (Table 2). Plantlets cultivated for 20 days in the presence of $(NH_4)_2SO_4$ showed a significant increase in bacterial invasion levels. Muthukumarasamy et al. [36] showed that the N-fertilization did not affect the diazotrophic bacterial population in all stages of growth of rice. Nitrogen alters the physiological state of the plant, and this subsequently affects its association with the bacterial population [37]. Sensitivity of many diazotrophic bacteria to nitrogen fertilizers could affect their endophytic ability [38, 39]. But, we have demonstrated that nitrogen fertilization was not a limiting factor for the recovery of the strains from internal plant tissues. Mechanisms of plant growth promotion by plant-associated bacteria vary greatly and can be broadly categorized into direct and indirect effects. The biological role of endophytes in supplying nitrogen to their host plant has not yet been confirmed. Yanni et al. [12] showed that rice plants grown under N-free conditions in gnotobiotic tube culture were not consistently increased in N-content nor did they have detectable acetylene reduction

activity when examined 32 days after inoculation with these rhizobia; hence, we did not study the ability of producing N fixation by this isolates. Phosphorus is the second most limiting mineral nutrient affecting terrestrial plant growth. Bacteria can solubilize inorganic and organic phosphates by different ways. [1, 40]. In the present study since phosphorus existed in a soluble form (Hoagland's growth medium), the isolates do not have any role in solubilizing phosphate and subsequently rice plant growth. Many bacteria produce organic compounds, so-called siderophores. Siderophores are produced by PGPR under iron-limited conditions [1, 41]. Iron also existed in a soluble form in the used growth medium. So, the isolates do not have any role in making Fe available for rice seedlings and subsequently rice plant growth. Of 34 isolates evaluated for siderophore production in this study, 27 isolates produced siderophore, indicating probably the ability of producing siderphore as the second PGP traits for screening the isolates isolated from clover for rice PGP agents. Under flooding conditions, iron availability is higher due to the reduction of ferric oxide hydrate complexes, releasing the more soluble Fe (II) that can even be toxic for the rice roots. The role of siderphore producing endophytes is may be to capture Fe (III), generated by oxidation of Fe (II) in oxic microniches into the plant or in the rihzosphere, increasing the iron availability locally or to reduce Fe (II) toxicity towards the plant by accumulation of the sequestered metal into the bacterial cells [42]. Ethylene is a phytohormone that increases in plants when they are exposed to both abiotic and biotic environmental stress conditions. The most commonly observed mechanism that reduces levels of ethylene production is via the activity of bacterial ACC deaminase [1]. But in our study due to the presence of $NH_4^+$ as $(NH_4)_2SO_4$, none of the isolates was able to utilize ACC as sole nitrogen. Hence, the isolate also could not have had any role in producing rice seedlings [43]. Because bacterial endophytes colonize an ecological niche that is similar to that of plant pathogens, they could potentially benefit plant growth indirectly by competing with pathogens for space and nutrients, thus suppressing the growth or activity of these pathogens [1]. We also examined the potential of the isolates in inhibiting fungal rice

Rapid Screening of Berseem Clover (Trifolium alexandrinum) Endophytic Bacteria for Rice Plant Seedlings
Growth-Promoting Agents

39

pathogens. Since the rice seedlings have not been inoculated with fungal pathogens, the isolates could not have any role in rice seedlings growth. Many microorganisms produce and release lytic enzymes that can hydrolyze a wide variety of polymeric compounds, including chitin, proteins, cellulose, hemicellulose, and DNA [44]. When endophytes colonize on the plant surface, they produce enzymes to hydrolyze plant cell walls. As a result, these enzymes also have the function to suppress plant pathogen activities directly and have the capability of degrading the cell walls of fungi and oomycetes [45–47]. Several studies showed that production of cell wall degrading enzymes such as cellulase, phosphatase, or pectinase in PGPR was important in facilitating entry of the bacteria into the intercellular spaces of plant root hairs [48]. The present investigation also showed the presence of different levels of cellulase and pectinase activities in different isolates suggesting their potential for inter- and intracellular colonization. But the cellulase and pectinase activities of the inoculated strain did not affect the growth or health of the seedlings [49]. Systemic acquired resistance (SAR) induced by pathogen infection is mediated by salicylic acid and associated with the accumulation of pathogenesis-related (PR) proteins [50]. In the study, one of 9 endophytes produced SA, and there is no relationship between rice growth and SA producing isolates [51]. Indole-3-acetic acid (IAA), the most studied phytohormone produced by plant-associated bacteria, contributes to plant growth and development by increasing root growth and root length and has also been associated with proliferation and elongation of root hairs. Our results showed that there is a significant relationship among IAA producing isolates, the rate of root colonization, and plant growth (Figures 1 and 2). An alternative working hypothesis is that endophytic colonization by these native isolates modulates growth physiology of rice (possibly by hormone action) enabling the plant root system to utilize the existing resources of available nutrients and water more efficiently in ways that may be independent of biological $N_2$ fixation [12]. IAA producing endophytes reisolated from the surface sterilized roots of rice appear to be very competent plant growth-promoting endophytes. Our findings are consistent with earlier reports where similar trend was reported in different bacteria [52–57]. Since The initial step of bacteria invasion in plant root consists of the attachment of bacteria onto epidermal cells of the root surface, where root hair zone represents one of the major sites of primary colonization, mainly on the basal region of emerging hairs, it is possible that IAA producing strains by increased root system can colonize plant roots better than other strains. The role of bacterial IAA in different microorganism-plant interactions highlights the fact that bacteria use this phytohormone to interact with plants as a part of their colonization strategy, including phytostimulation and circumvention of basal plant defense mechanisms [58]. Results obtained by in vitro screenings for the assessment of bacterial PGP cannot fully reflect the reality, for example, at the field or in the greenhouse. However, in vitro screenings for bacterial PGP can provide a tool to select strains out of the vast amount of bacteria living in plant-associated habitats that fulfill in situ what they promise in vitro [53]. The findings described here represent

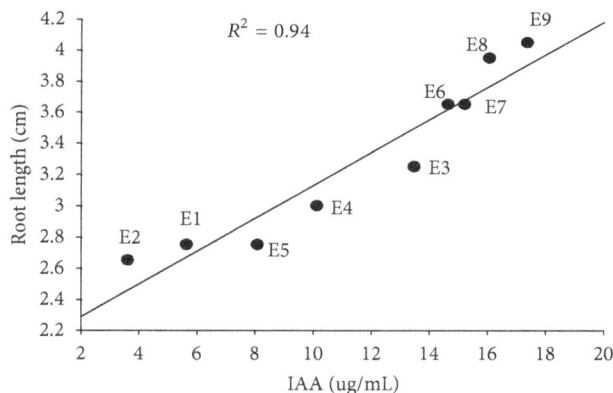

FIGURE 1: Correlation between IAA producing isolates with root length by nine endophytic isolates reisolated from rice 20 days after inoculation under gnotobiotic conditions.

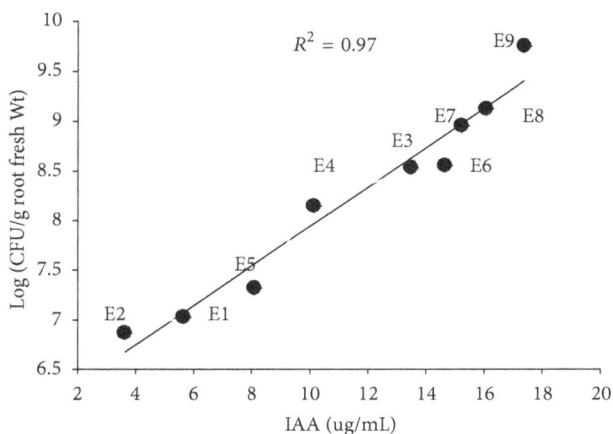

FIGURE 2: Correlation between IAA producing isolates with the size of population by nine endophytic isolates reisolated from rice 20 days after inoculation under gnotobiotic conditions.

a step forward in achieving the technically challenging goal of increasing rice productivity and screening superior bacteria by reducing its dependence of the need for fertilizer-N through enhancement of its natural association with rhizobia without requiring as a highly developed system as the root nodule rhizobium-legume symbiosis.

## 5. Conclusion

IAA production confers to bacteria competitive advantages to colonize plant tissues. Bacterial IAA loosens plant cell walls and as a result facilitates an increasing amount of root exudation that provides additional nutrients to support the growth of rhizosphere bacteria. The screening procedure appears to be very effective and less time consuming. Therefore, proper screening of rice growth-promoting bacteria can be useful for future agricultural applications, providing higher production yields, reduced input costs, and negative environmental impact due to the use of nitrogen fertilizers.

## Acknowledgment

The authors wish to thank the chairman of the Department of Soil Science of Tehran University, Iran, for providing the necessary facilities for this study.

## References

[1] N. Weyens, D. Van der Lelie, S. Taghavi, L. Newman, and J. Vangronsveld, "Exploiting plant-microbe partnerships to improve biomass production and remediation," *Trends in Biotechnology*, vol. 27, no. 10, pp. 591–598, 2009.

[2] R. P. Ryan, K. Germaine, A. Franks, D. J. Ryan, and D. N. Dowling, "Bacterial endophytes: recent developments and applications," *FEMS Microbiology Letters*, vol. 278, no. 1, pp. 1–9, 2008.

[3] S. Lee, M. Flores-Encarnación, M. Contreras-Zentella, L. Garcia-Flores, J. E. Escamilla, and C. Kennedy, "Indole-3-acetic acid biosynthesis is deficient in *Gluconacetobacter diazotrophicus* strains with mutations in cytochrome c biogenesis genes," *Journal of Bacteriology*, vol. 186, no. 16, pp. 5384–5391, 2004.

[4] S. A. Wakelin, R. A. Warren, P. R. Harvey, and M. H. Ryder, "Phosphate solubilization by *Penicillium* spp. closely associated with wheat roots," *Biology and Fertility of Soils*, vol. 40, no. 1, pp. 36–43, 2004.

[5] M. E. Will and D. M. Sylvia, "Interaction of rhizosphere bacteria, fertilizer, and vesicular-arbuscular mycorrhizal fungi with sea oats," *Applied and Environmental Microbiology*, vol. 56, no. 7, pp. 2073–2079, 1990.

[6] J. M. Costa and J. E. Loper, "Characterization of siderophore production by the biological control agent *Enterobacter cloacae*," *Molecular Plant-Microbe Interactions*, vol. 7, no. 4, pp. 440–448, 1994.

[7] B. Rodelas, V. Salmerón, M. V. Martinez-Toledo, and J. González-López, "Production of vitamins by *Azospirillum brasilense* in chemically-defined media," *Plant and Soil*, vol. 153, no. 1, pp. 97–101, 1993.

[8] K. Clay, "Effects of fungal endophytes on the seed and seedling biology of *Lolium perenne* and *Festuca arundinacea*," *Oecologia*, vol. 73, no. 3, pp. 358–362, 1987.

[9] V. J. Kevin, "Plant growth promoting rhizobacteria as biofertilizers," *Plant and Soil*, vol. 255, no. 2, pp. 571–586, 2003.

[10] D. P. Belesky and D. P. Malinowski, "Abiotic stresses and morphological plasticity and chemical adaptations of Neotyphodium-infected tall fescue plants; in Microbial endophytes," C. W. Bacon and J. F. White Jr., Eds., pp. 455–484, Marcel Dekker, New York, NY, USA, 2000.

[11] H. Antoun, C. J. Beauchamp, N. Goussard, R. Chabot, and R. Lalande, "Potential of Rhizobium and *Bradyrhizobium species* as plant growth promoting rhizobacteria on non-legumes: effect on radishes (*Raphanus sativus* L.)," *Plant and Soil*, vol. 204, no. 1, pp. 57–67, 1998.

[12] Y. G. Yanni, R. Y. Rizk, V. Corich et al., "Natural endophytic association between *Rhizobium leguminosarum* bv. *trifolii* and rice roots and assessment of its potential to promote rice growth," *Plant and Soil*, vol. 194, no. 1-2, pp. 99–114, 1997.

[13] H. S. Alves Silva, R. Da Silva Romeiro, and A. Mounteer, "Development of a root colonization bioassay for rapid screening of rhizobacteria for potential biocontrol agents," *Journal of Phytopathology*, vol. 151, no. 1, pp. 42–46, 2003.

[14] J. W. Kloepper, "Plant growth-promoting rhizobacteria (other systems)," in *Azospirillum/Plant Associations*, Y. Okon, Ed., pp. 137–166, CRC Press, Boca Raton, Fla, USA, 1997.

[15] A. V. Sturz, B. R. Christie, B. G. Matheson, and J. Nowak, "Biodiversity of endophytic bacteria which colonize red clover nodules, roots, stems and foliage and their influence on host growth," *Biology and Fertility of Soils*, vol. 25, no. 1, pp. 13–19, 1997.

[16] B. Schwyn and J. B. Neilands, "Universal chemical assay for the detection and determination of siderophores," *Analytical Biochemistry*, vol. 160, no. 1, pp. 47–56, 1987.

[17] C. L. Patten and B. R. Glick, "Bacterial biosynthesis of indole-3-acetic acid," *Canadian Journal of Microbiology*, vol. 42, no. 3, pp. 207–220, 1996.

[18] J.-M. Meyer, P. Azelvandre, and C. Georges, "Iron metabolism in Pseudomonas: salicylic acid, a siderophore of *Pseudomonas fluorescens* CHA0," *BioFactors*, vol. 4, no. 1, pp. 23–27, 1992.

[19] H. Lorck, "Production of hydrocyanic acid by bacteria," *Physiol Plant*, vol. 1, pp. 142–146, 1948.

[20] J. I. Sperber, "The incidence of apatite-solubilizing organisms in the rhizosphere and soil," *Australian Journal of Agricultural Research*, vol. 9, no. 6, pp. 778–781, 1995.

[21] B. R. Glick, D. M. Karaturovic, and P. C. Newell, "A novel procedure for rapid isolation of plant growth promoting pseudomonads," *Canadian Journal of Microbiology*, vol. 41, no. 6, pp. 533–536, 1995.

[22] J. Jinantana and M. Sariah, "Antagonistic effect of Malaysian isolates of *Trichoderma harzianum* and *Gliocladium virens* on *Sclerotium rolfsii*," *Pertanika Journal of Tropical Agricultural Science*, vol. 20, pp. 35–41, 1997.

[23] K.-H. Jung and H. J. Kim, "Development of an agar diffusion method to measure elastase inhibition activity using Elastin-Congo red," *Journal of Microbiology and Biotechnology*, vol. 16, no. 8, pp. 1320–1324, 2006.

[24] P. F. Mateos, J. I. Jimenez-Zurdo, J. Chen et al., "Cell-associated pectinolytic and cellulolytic enzymes in *Rhizobium leguminosarum* biovar trifolii," *Applied and Environmental Microbiology*, vol. 58, no. 6, pp. 1816–1822, 1992.

[25] J. H. Li, E. T. Wang, W. F. Chen, and W. X. Chen, "Genetic diversity and potential for promotion of plant growth detected in nodule endophytic bacteria of soybean grown in Heilongjiang province of China," *Soil Biology and Biochemistry*, vol. 40, no. 1, pp. 238–246, 2008.

[26] P. Nejad and P. A. Johnson, "Endophytic bacteria induce growth promotion and wilt disease suppression in oilseed rape and tomato," *Biological Control*, vol. 18, no. 3, pp. 208–215, 2000.

[27] M. Elvira-Recuenco and J. W. L. Van Vuurde, "Natural incidence of endophytic bacteria in pea cultivars under field conditions," *Canadian Journal of Microbiology*, vol. 46, no. 11, pp. 1036–1041, 2000.

[28] J. Hallmann, A. Quadt-Hallmann, W. F. Mahaffee, and J. W. Kloepper, "Bacterial endophytes in agricultural crops," *Canadian Journal of Microbiology*, vol. 43, no. 10, pp. 895–914, 1997.

[29] A. V. Sturz, B. R. Christie, and J. Nowak, "Bacterial endophytes: potential role in developing sustainable systems of crop production," *Critical Reviews in Plant Sciences*, vol. 19, no. 1, pp. 1–30, 2000.

[30] L. L. Wang, E. T. Wang, J. Liu, Y. Li, and W. X. Chen, "Endophytic occupation of root nodules and roots of *Melilotus dentatus* by *Agrobacterium tumefaciens*," *Microbial Ecology*, vol. 52, no. 3, pp. 436–443, 2006.

Rapid Screening of Berseem Clover (Trifolium alexandrinum) Endophytic Bacteria for Rice Plant Seedlings
Growth-Promoting Agents

41

[31] A. M. Abdel Wahab, H. H. Zahran, and M. H. Abd-Alla, "Root-hair infection and modulation of four grain legumes as affected by the form and the application time of nitrogen fertilizer," *Folia Microbiologica*, vol. 41, no. 4, pp. 303–308, 1996.

[32] K. V. B. R. Tilak, N. Ranganayaki, K. K. Pal et al., "Diversity of plant growth and soil health supporting bacteria," *Current Science*, vol. 89, no. 7, pp. 869–885, 2005.

[33] T. C. Noel, C. Sheng, C. K. Yost, R. P. Pharis, and M. F. Hynes, "Rhizobium leguminosarum as a plant growth-promoting rhizobacterium: direct growth promotion of canola and lettuce," *Canadian Journal of Microbiology*, vol. 42, no. 3, pp. 279–283, 1996.

[34] Y. Alami, W. Achouak, C. Marol, and T. Heulin, "Rhizosphere soil aggregation and plant growth promotion of sunflowers by an exopolysaccharide-producing *Rhizobium* sp. strain isolated from sunflower roots," *Applied and Environmental Microbiology*, vol. 66, no. 8, pp. 3393–3398, 2000.

[35] K. A. Mattos, V. L. M. Pádua, A. Romeiro et al., "Endophytic colonization of rice (*Oryza sativa* L.) by the diazotrophic bacterium *Burkholderia kururiensis* and its ability to enhance plant growth," *Anais da Academia Brasileira de Ciencias*, vol. 80, no. 3, pp. 477–493, 2008.

[36] R. Muthukumarasamy, G. Revathi, and C. Lakshminarasimhan, "Influence of N fertilisation on the isolation of *Acetobacter diazotrophicus* and *Herbaspirillum* spp. from Indian sugarcane varieties," *Biology and Fertility of Soils*, vol. 29, no. 2, pp. 157–164, 1999.

[37] J. Prakamhang, K. Minamisawa, K. Teamtaisong, N. Boonkerd, and N. Teaumroong, "The communities of endophytic diazotrophic bacteria in cultivated rice (*Oryza sativa* L.)," *Applied Soil Ecology*, vol. 42, no. 2, pp. 141–149, 2009.

[38] V. M. Reis, J. I. Baldani, V. L. D. Baldani, and J. Dobereiner, "Biological dinitrogen fixation in Gramineae and palm trees," *Critical Reviews in Plant Sciences*, vol. 19, no. 3, pp. 227–247, 2000.

[39] L. E. Fuentes-Ramirez, T. Jimenez-Salgado, I. R. Abarca-Ocampo, and J. Caballero-Mellado, "Acetobacter diazotrophicus, an indoleacetic acid producing bacterium isolated from sugarcane cultivars of México," *Plant and Soil*, vol. 154, no. 2, pp. 145–150, 1993.

[40] B. E. Ramey, M. Koutoudis, S. B. Vonbodman, and C. Fuqa, "Biofilm formation in plant microbe associations," *Current Opinion in Microbiology*, vol. 7, pp. 602–609, 2004.

[41] J. Kuklinsky-Sobral, W. L. Araújo, R. Mendes, I. O. Geraldi, A. A. Pizzirani-Kleiner, and J. L. Azevedo, "Isolation and characterization of soybean-associated bacteria and their potential for plant growth promotion," *Environmental Microbiology*, vol. 6, no. 12, pp. 1244–1251, 2004.

[42] I. Loaces, L. Ferrando, and A. F. Scavino, "Dynamics, diversity and function of endophytic siderophore-producing bacteria in rice," *Microbial Ecology*, vol. 61, no. 3, pp. 606–618, 2011.

[43] E. Dell'Amico, L. Cavalca, and V. Andreoni, "Analysis of rhizobacterial communities in perennial Graminaceae from polluted water meadow soil, and screening of metal-resistant, potentially plant growth-promoting bacteria," *FEMS Microbiology Ecology*, vol. 52, no. 2, pp. 153–162, 2005.

[44] A. A. Belimov, V. I. Safronova, T. A. Sergeyeva et al., "Characterization of plant growth promoting rhizobacteria isolated from polluted soils and containing 1-aminocyclopropane-1-carboxylate deaminase," *Canadian Journal of Microbiology*, vol. 47, no. 7, pp. 642–652, 2001.

[45] S. Tripathi, S. Kamal, I. Sheramati, R. Oelmuller, and A. Varma, "Mycorrhizal fungi and other root endophytes as biocontrol agents against root pathogens," *Mycorrhiza*, vol. 3, pp. 281–306, 2008.

[46] T. Andro, J. P. Chambost, and A. Kotoujansky, "Mutants of *Erwinia chrysanthemi* defective in secretion of pectinase and cellulase," *Journal of Bacteriology*, vol. 160, no. 3, pp. 1199–1203, 1984.

[47] K. E. Germaine, G. Keogh, B. Garcia-Cabellos et al., "Colonisation of poplar trees by gfp expressing bacterial endophytes," *FEMS Microbiology Ecology*, vol. 48, no. 1, pp. 109–118, 2004.

[48] R. M. Teather and P. J. Wood, "Use of Congo red-polysaccharide interactions in enumeration and characterization of cellulolytic bacteria from the bovine rumen," *Applied and Environmental Microbiology*, vol. 43, no. 4, pp. 777–780, 1982.

[49] P. F. Mateos, D. L. Baker, M. Petersen et al., "Erosion of root epidermal cell walls by Rhizobium polysaccharide-degrading enzymes as related to primary host infection in the Rhizobium-legume symbiosis," *Canadian Journal of Microbiology*, vol. 47, no. 6, pp. 475–487, 2001.

[50] P. Mylona, K. Pawlowski, and T. Bisseling, "Symbiotic nitrogen fixation," *Plant Cell*, vol. 7, no. 7, pp. 869–885, 1995.

[51] V. Ramamoorthy, R. Viswanathan, T. Raguchander, V. Prakasam, and R. Samiyappan, "Induction of systemic resistance by plant growth promoting rhizobacteria in crop plants against pests and diseases," *Crop Protection*, vol. 20, no. 1, pp. 1–11, 2001.

[52] R. F. White, "Acetylsalicylic acid (aspirin) induces resistance to tobacco mosaic virus in tobacco," *Virology*, vol. 99, no. 2, pp. 410–412, 1979.

[53] F. Chi, S.-H. Shen, H.-P. Cheng, Y.-X. Jing, Y. G. Yanni, and F. B. Dazzo, "Ascending migration of endophytic rhizobia, from roots to leaves, inside rice plants and assessment of benefits to rice growth physiology," *Applied and Environmental Microbiology*, vol. 71, no. 11, pp. 7271–7278, 2005.

[54] F. Ahmad, I. Ahmad, and M. S. Khan, "Screening of free-living rhizospheric bacteria for their multiple plant growth promoting activities," *Microbiological Research*, vol. 163, no. 2, pp. 173–181, 2008.

[55] R. Mendes, A. A. Pizzirani-Kleiner, W. L. Araujo, and J. M. Raaijmakers, "Diversity of cultivated endophytic bacteria from sugarcane: genetic and biochemical characterization of *Burkholderia cepacia* complex isolates," *Applied and Environmental Microbiology*, vol. 73, no. 22, pp. 7259–7267, 2007.

[56] R. H. Chabot, H. Antoun, J. W. Kloepper, and C. J. Beauchamp, "Root colonization of maize and lettuce by bioluminescent *Rhizobium leguminosarum* biovar phaseoli," *Applied and Environmental Microbiology*, vol. 62, no. 8, pp. 2767–2772, 1996.

[57] M. Fürnkranz, H. Müller, and G. Berg, "Characterization of plant growth promoting bacteria from crops in Bolivia," *Journal of Plant Diseases and Protection*, vol. 116, no. 4, pp. 149–155, 2009.

[58] S. Spaepen, J. Vanderleyden, and R. Remans, "Indole-3-acetic acid in microbial and microorganism-plant signaling," *FEMS Microbiology Reviews*, vol. 31, no. 4, pp. 425–448, 2007.

# 5

# Remote Sensing of Soil Moisture

**Venkat Lakshmi**

*Department of Earth and Ocean Sciences, University of South Carolina, Columbia, SC 29208, USA*

Correspondence should be addressed to Venkat Lakshmi; vlakshmi@geol.sc.edu

Academic Editors: G. Benckiser, J. A. Entry, and Z. L. He

Soil moisture is an important variable in land surface hydrology as it controls the amount of water that infiltrates into the soil and replenishes the water table versus the amount that contributes to surface runoff and to channel flow. However observations of soil moisture at a point scale are very sparse and observing networks are expensive to maintain. Satellite sensors can observe large areas but the spatial resolution of these is dependent on microwave frequency, antenna dimensions, and height above the earth's surface. The higher the sensor, the lower the spatial resolution and at low elevations the spacecraft would use more fuel. Higher spatial resolution requires larger diameter antennas that in turn require more fuel to maintain in space. Given these competing issues most passive radiometers have spatial resolutions in 10s of kilometers that are too coarse for catchment hydrology applications. Most local applications require higher-spatial-resolution soil moisture data. Downscaling of the data requires ancillary data and model products, all of which are used here to develop high-spatial-resolution soil moisture for catchment applications in hydrology. In this paper the author will outline and explain the methodology for downscaling passive microwave estimation of soil moisture.

## 1. Introduction

Soil moisture is an important variable in land surface hydrology. Soil moisture has very important implications for agriculture, ecology, wildlife, and public health and is probably (after precipitation) the most important connection between the hydrological cycle and life—animal, plant, and human.

Land surface hydrology is a well-studied portion of the terrestrial water cycle. The main variables in land-surface hydrology are soil moisture, surface temperature, vegetation, precipitation, and streamflow. Of these, surface temperature, vegetation, and precipitation are currently observed using satellites, and streamflow is routinely observed at in situ watershed locations. Soil moisture remains the only variable not observed (or observed very sparsely) either in situ or via remote sensing. Due to this very reason, in the past decade, satellite soil moisture has been increasingly used in hydrological, agricultural, and ecological studies due to its spatial coverage, temporal continuity, and (now) easiness of use.

Numerous studies have shown the influence of soil moisture on the feedbacks between land-surface and climate that has a profound influence on the dynamics of the atmospheric boundary layer and a direct relationship to weather and global climate [1–4]. Chang and Wetzel [5] have shown the influence of spatial variations of soil moisture and vegetation on the development and intensity of severe storms, whereas Engman, 1997 [6], demonstrated the ability of soil moisture to influence surface moisture gradients and to partition incoming radiative energy into sensible and latent heat. In large-scale modeling, the soil moisture and surface temperature are key variables in deciding the depth of the planetary boundary layer and circulation and wind patterns [7–9]. It has been demonstrated that the assimilation of soil moisture observations in hydrologic models can improve the accuracy of estimated hydrological variables such as evaporation, surface temperature, and root-zone soil moisture [10–12]. The various atmospheric processes that affect the land surface and in turn the influence of the land surface on the atmosphere need to be clearly quantified. In order to accomplish these tasks, it is necessary to intimately understand the relationship of soil moisture to these phenomena on small and large spatial scales. Unfortunately we are limited in our ability to completely observe large-scale hydrologic land-surface interactions. Satellite and aircraft remote sensing enable us to estimate large-scale soil moisture for the purpose of modeling the interactions between land and atmosphere, helping us to model weather and climate with higher accuracy.

Recognizing the need for soil moisture observations on large spatial scales for continental scale hydrological modeling investigators in the 1980s used the special sensor microwave Imager (SSM/I) [13] and scanning multichannel microwave radiometer [14] data sets. The SMMR data has been used to study soil moisture retrievals, sensitivity, and scaling on continental scales [15–17]. The SSM/I [15, 18] has been used in catchment scale studies [19] and in continental scale studies in conjunction with a hydrological model [20–22]. More recently, successful retrieval has been carried out using several missions including WindSat [23], tropical rainfall measuring mission (TRMM) microwave imager (TMI) [24]. Recently a promising soil moisture data set has been developed jointly by researchers of NASA Goddard Space Flight Center and the Vrije Universiteit Amsterdam [25]. It utilizes C-band AMSR-E microwave brightness temperatures in a Land Parameter Retrieval Model (LPRM) to obtain soil moisture. This product has been tested in a series of validation studies (e.g., [26–28]) and shows high correlations with field observations [29]. The Global Change Observation Mission-Water (GCOM-W) launched by the Japanese Space Agency is now providing us with microwave data sets of the land surface [30].

Active and passive microwave remote sensing provides a unique capability to obtain observations of soil moisture at global and regional scales that help satisfy the science and application needs for hydrology [31–33]. The emissive and scattering characteristics of soil surface depend on soil moisture among other variables, that is, surface temperature, surface roughness, and vegetation. The electromagnetic response of the land surface is modified by soil moisture and modulated by surface roughness, vegetation canopy effects, and interaction with the atmosphere before being received by a sensor. These ancillary (non-soil-moisture) effects increase at higher frequencies making low-frequency observations desirable for observation of soil moisture [34, 35]. Longer wavelengths also sense deeper soil layers (2–5 cm) at the L-band, the penetration depth being of the order of one tenth of the wavelength [36]. Retrieval of soil moisture using ground-based or aircraft-mounted radiometer operating at the L-band has been demonstrated in several prior studies [37–42].

An important experiment for investigating the capability of the PALS (passive and active L-and S-bands radiometer/radar) was conducted in the Southern Great Plains region of United States in July 1999. The SGP99 experiments included bare, pasture, and agricultural crop surface cover with field averaged vegetation water contents mainly in the $0–2.5\,\mathrm{kg\,m^{-2}}$ range. Studies based on the SGP99 experiments exhibited varying soil moisture retrieval potential, with a 2-3% accuracy using passive channels and 2–5% accuracy using the active channels of the PALS instrument [40, 42]. There was a need to conduct similar studies under higher vegetation water contents in order to evaluate the performance of soil moisture retrieval algorithms under these conditions.

The Soil Moisture Experiments in 2002, SMEX02 were conducted in Iowa over a one-month period between mid-June and mid-July, 2002. A major focus of SMEX02 was extension of instrument observations and algorithms to more challenging vegetation conditions and understanding the implications on soil moisture retrieval. In situ measurements of gravimetric soil moisture, soil temperature, soil bulk density, and vegetation water content were carried out coincidently with PALS observations in active and passive channels. This study evaluated the performance of existing algorithms and models for soil moisture retrieval using active and passive measurements under the moderately high to very high vegetation water content conditions.

Field scale validation networks exist in Oklahoma [43–45], Illinois [46], and the USDA operational SCAN (Soil Climate Analysis Network) [47]. Soil observing networks have their challenges especially when used to validate satellite data sets [48].

The NASA soil moisture active passive (SMAP) mission [53], is set for launch in 2014. SMAP will utilize a very large antenna and combined radiometer/radar measurements to provide soil moisture at higher resolutions than radiometers alone can currently achieve. SMAP [53] consists of both passive and active microwave sensors. The passive radiometer will have a nominal spatial resolution of 36 km and the active radar will have a resolution of 1 km. The active microwave remote sensing data can provide a higher spatial resolution observation of backscatter than those obtained from a radiometer (order of magnitude: radiometer ∼40 km and radar ∼1 km or better). Radar data are more strongly affected by local roughness, microscale topography, and vegetation than a radiometer, meaning that it is difficult to invert backscatter to soil moisture accurately, thus limiting the development of such algorithms. Therefore, it can be difficult to use radar data alone. SMAP will use high-resolution radar observations to disaggregate coarse resolution radiometer observations to produce a soil moisture product at 3 km resolution. The soil moisture has been retrieved from radiometer data successfully using various sensors and platforms and these retrieval algorithms have an established heritage [31, 54].

There have been methods integrating the use of active sensors that have a higher spatial resolution to downscale passive microwave soil moisture retrievals [55–57]. Recent studies have addressed the soil moisture downscaling problem using MODIS sensor derived temperature, vegetation, and other surface ground variables. The major publications in this area of study include the following. (i) A method based on a "universal triangle" concept was used to retrieve soil moisture from Normalized Difference Vegetation Index (NDVI) and land surface temperature (LST) data [50]. (ii) A relationship between fractional vegetation cover and soil evaporative efficiency was explored for catchment studies in Southeastern Australia by Merlin et al., 2010 [49] while Merlin et al., 2008 [51], developed a simple method to downscale soil moisture by using two soil moisture indexes: evaporative fraction (EF) and the actual EF (AEF) [58]. (iii) A sequential model which used MODIS as well as ASTER (Advanced Scanning Thermal Emission and Reflection Radiometer) data

TABLE 1: Studies on downscaling soil moisture using various remote sensing and modeling techniques [41].

| Author | Methodology | Time and region | Result |
| --- | --- | --- | --- |
| Merlin et al., [49] | Based on the relationship between soil evaporative efficiency and soil moisture | NAFE 2006 (Oct-Nov), Yanco, Southeastern Australia | Mean correlation slope between simulated and measured data is 0.94, the most accuracy with an error of 0.012 |
| Piles et al., [50] | Build model between LST, NDVI, and soil moisture | Jan-Feb 2010, Murrumbidgee catchment, Yanco, Southeastern Australia | $R^2$ is between 0.14~0.21 and RMSE 0.9~0.17 |
| Merlin et al., [51] | Downscaling algorithm is derived from MODIS and physical-based soil evaporative efficiency model | NAFE 2006 (Oct-Nov), Murrumbidgee catchment, Yanco, Southeastern Australia | Overall RMSE is between 1.4%~1.8% v/v |
| Merlin et al., [51] | Based on two soil moisture indices EF and AEF | June and August 1990 (Monsoon' 90 experiment), USDA-ARS WGEW in southeastern Arizona | Total accuracy is 3% vol. for EF and 2% vol. for AEF, and correlation coefficient is 0.66~0.79 for EF and 0.71~0.81 for AEF |
| Merlin et al., [52] | Sequential model | NAFE 2006 (Oct-Nov), Yanco, Southeastern Australia | RMSE is −0.062 vol./vol. and the bias is 0.045 vol./vol. |

was proposed for downscaling soil moisture [52, 59, 60]. Table 1 lists these studies, the methods, and significant results of the soil moisture downscaling.

This paper is organized as follows. Section 2 outlines the theory of passive microwave radiative transfer. Section 3 explains results from the Soil Moisture Experiment 2002 (SMEX02) using aircraft-based passive and active sensors. Section 4 describes two methods used for disaggregation of soil moisture: (1) use of active sensors to detect change in soil moisture at a finer scale and use that information for construction of higher spatial resolution estimates of soil moisture and (2) use of visible and near infrared satellite observations to disaggregate passive microwave satellite soil moistures. Section 5 discusses the future of remote sensing of soil moisture.

## 2. The Radiative Transfer Model

The earth's surface as seen by a spaceborne or airborne radiometer may include bare or vegetated soil, varying amounts of roughness, similar or varying soil types, and variation in the soil moisture content.

Any model of radiative transfer from the earth's surface must incorporate the effects of these factors observed brightness temperatures. The radiative transfer model described here begins with the bare soil emissivity and then is modified for roughness and vegetation. Most of the studies using the model have been done in the 1.4–6.6 GHz range. Atmospheric effects on the brightness temperatures and the effects of volume scattering are not considered, as they are negligible at these frequencies (1.4–6.6 GHz).

2.1. Emissivity of a Smooth Surface. The relationship between the brightness temperature $T_B$ of a radiating body and its thermodynamic temperature $T_s$ is given by the expression

$$T_B = eT_s,\tag{1}$$

where $e$ is the emissivity of the body $T_B$ expressed in Kelvins. The emissivity is related to the reflectivity $r$ of the surface by

$$e = 1 - r.\tag{2}$$

For a smooth surface and a medium of uniform dielectric constant, the expressions for reflectivity at horizontal and vertical polarizations may be derived from electromagnetic theory [1] as

$$r_v = \left| \frac{\varepsilon_r \cos\theta - \sqrt{\varepsilon_r - \sin^2\theta}}{\varepsilon_r \cos\theta + \sqrt{\varepsilon_r - \sin^2\theta}} \right|^2,$$

$$r_h = \left| \frac{\cos\theta - \sqrt{\varepsilon_r - \sin^2\theta}}{\cos\theta + \sqrt{\varepsilon_r - \sin^2\theta}} \right|^2,\tag{3}$$

where $\theta$ is the incidence angle and $\varepsilon_r$ is the complex dielectric constant of the medium.

2.2. Emissivity of a Bare Smooth Soil. Water has a much higher dielectric constant compared to soil. An increase in the soil water content increases both the real and imaginary parts of the dielectric constant of the soil-water mix. The dielectric properties of wet soils have been studied by several investigators [61, 62] (e.g., [63, 64]). The texture of the soil also plays an important role in determining its dielectric constant. The dielectric constant for wet soil is evaluated using an empirical mixing model from Fang et al. [61]. For the purposes of this study the moisture content of the soil is assumed to be uniform to the penetration depth of the sensor and the effects of nonuniformity of moisture with depth are not considered.

### 2.3. Effect of Surface Roughness.

Surface roughness causes the emissivity of natural surfaces to be somewhat higher [1, 35, 65, 66]. This is attributed mainly to the increased surface area of the emitting surface. A semiempirical expression for rough surface reflectivity from Ni-Meister et al. [12] is used to account for surface roughness:

$$r_p = \left[ Q r_{0q} + (1 - Q) r_{0p} \right] \exp(-h), \qquad (4)$$

where $r_{0q}$ and $r_{0p}$ are the reflectivities the medium would have if the surface were smooth. The expression utilizes two parameters, which are dependent on the surface conditions. $Q$ is the polarization mixing parameter and $h$ is the height parameter. These two parameters depend on the frequency, look angle of the sensor, and the roughness of the surface and have to be determined experimentally.

The values of $Q$ and $h$ are useful for fitting and modeling experimental data, but recent theoretical calculations have indicated that $Q$ and $h$ are not directly related to the parameters, rms height, and horizontal correlation length, measured in the field and used to characterize rough surface reflectivity [67].

### 2.4. Effect of Vegetation.

The presence of vegetation canopy in natural areas and crop canopy in agricultural fields has a significant effect on the remotely sensed microwave emission from soils [68, 69]. The sensitivity of measured microwave emission in vegetation-covered fields will be different from bare fields. Vegetation is modeled as a single homogenous layer above the soil. The brightness temperature $T_B^p$ corresponding to polarization $p$, vertical ($v$), or horizontal ($h$) over such a dual vegetation-soil layer is given by Das et al. [57]:

$$T_B^p = e_p T_s \exp(-\tau) + T_C \left[ 1 - \exp(-\tau) \right] \left[ 1 + r_p \exp(-\tau) \right], \qquad (5)$$

where $T_C$ is the vegetation temperature, $T_s$ is the soil temperature, $\tau$ is the vegetation opacity, and $e_p$ and $r_p$ are the soil emissivity and reflectivity, respectively. The value of $\tau$ is dependent on frequency, vegetation type, and vegetation water content. The relationship between $\tau$ and the vegetation water content $W_e$ can be described by the following equation from [16, 35]:

$$r = \frac{bW_e}{\cos \theta}, \qquad (6)$$

where $b$ is a function of canopy type, polarization, and wavelength and $\cos \theta$ accounts for the nonvertical/slant path through the vegetation.

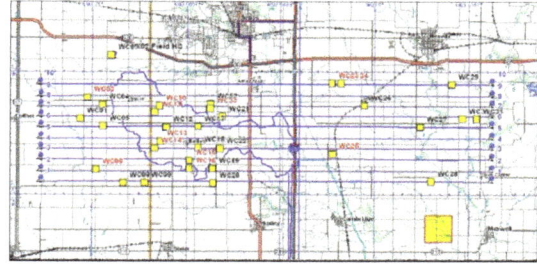

FIGURE 1: The Walnut Creek watershed region and PALS flight lines are shown by the blue lines [41].

The land surface as seen by the satellite sensor is a heterogeneous combination of vegetated and bare areas. The vegetated and bare areas have to be disaggregated to find the proportion of the radiation that reaches the sensor from the different types of land cover. The land surface can be disaggregated into a completely shadowed fraction $M$ and a bare soil fraction $M$ by utilizing the leaf area index (LAI). Leaf area index (LAI) defines an important structural property of a plant canopy, the number of equivalent layers of leaves the vegetation possesses relative to a unit ground area:

$$M = 1 - e^{-\mu \text{LAI}}, \qquad (7)$$

where $\mu$ is the extinction coefficient. The proportion of microwave brightness temperatures contributed by bare soils and vegetated regions within a certain area can be calculated using

$$T_B = M T_B^{\text{canopy}} + (1 - M) T_B^{\text{bare}}, \qquad (8)$$

where $T_B^{\text{canopy}}$ is from (5) and $T_B^{\text{bare}}$ corresponds to brightness temperature of bare soil ((1) with emissivity corresponding to bare soil). This gives us ability to model microwave brightness temperatures that would be detected by a sensor.

## 3. Results from the SMEX02 Field Experiment

### 3.1. SMEX02 Experiment.

The Soil Moisture Experiments in 2002 (SMEX02) were conducted in Iowa between June 25 and July 12th, 2002. The study site chosen was Walnut Creek, a small watershed in Iowa. This watershed has been studied extensively by the USDA and hence was well instrumented for in situ sampling of hydrologic parameters. The terrain is undulating and the land lover type for the watershed region is primarily agricultural with corn and soybeans being the major crops. Figure 1 shows a map of the study region indicating the location of the data collection sites and the layout of the aircraft flight lines. The experiments were conducted from 25 June to 12th July, 2002, during which the soybean fields grew from essentially bare soils to vegetation water content of 1–1.5 kg/m$^2$ while the corn fields grew from 2-3 kg/m$^2$ to 4-5 kg/m$^2$ (Figures 2 and 3). SMEX02 provided unique conditions of high initial biomass content and significant change in biomass over the course of the experiment. The fields selected for in situ sampling

TABLE 2: Sensitivity of radiometer channel to 0–6 cm layer gravimetric soil moisture evaluated over corn and soy fields, respectively. The LH channel has the maximum sensitivity with the sensitivity being much greater over soy fields than corn fields [41].

| Period | LH | LV | SH | SV |
|--------|------|------|------|------|
| | | Corn | | |
| 176–178 | −0.386 | −0.242 | −0.131 | −0.076 |
| 186-187 | −0.831 | −0.284 | −0.644 | −0.070 |
| 187-188 | −0.316 | −0.165 | −0.164 | −0.124 |
| 188-189 | −0.357 | −0.132 | −0.132 | −0.084 |
| | | Soy | | |
| 176–178 | −1.182 | −0.333 | 0.183 | −0.054 |
| 186-187 | −1.013 | −0.426 | 0.787 | −0.425 |
| 187-188 | −1.117 | −0.409 | −0.379 | −0.133 |
| 188-189 | −1.050 | −0.848 | −1.047 | −0.665 |

FIGURE 2: Field WC05 on July 8 (VWC ~ 4.8 kg/m$^2$) [41].

FIGURE 3: Field WC03 on July 8 (VWC ~ 0.85 kg/m$^2$) [41].

were approximately 800 m × 800 m and the sampling points were distributed throughout the field to take into account the variations in soil types and therefore soil moisture within each field.

The PALS instrument was flown over the SMEX02 region on June 25, 27 and July 1st, 2nd, 5, 6, 7, and 8, 2002 (corresponding to DOY 176, 178, 182, 183, 186, 187, 188, and 189, resp.). Initial soil conditions were dry but scattered thunderstorms occurred between July 4 and 6 enabling a wetting and subsequent dry down to be observed. The PALS radiometer and radar provided simultaneous observations of horizontally and vertically polarized L- and S-bands brightness temperatures, radar backscatter measured in VV, HH and VH configurations, and nadir-looking thermal infrared surface temperature. The instrument was flown on a C-130 (velocity ~ 70 m/s) aircraft at a nominal altitude of ~3500 feet with the angle of incidence on the surface being ~45 degrees. In this configuration the instantaneous 3 dB footprint on the surface was 330 m × 470 m. The instrument thus sampled a single line footprint track along the flight path. Aircraft location and navigation data and downward looking thermal infrared (IR) temperatures were also recorded in addition to the radiometer and radar measurements.

*3.2. Sensitivity.* Sensitivity for the passive sensor is evaluated as the ratio of the change in emissivity to the change in gravimetric soil moisture $\Delta e/\Delta m_g$ and as the change in radar backscatter to the change in percentage gravimetric soil moisture $\Delta\sigma/\Delta(m_g\%)$ for the active sensor. The estimation of emissivity was carried out by dividing the brightness temperature in a channel by the surface temperature. Sensitivity of the radiometer is expected to be negative as an increase in soil moisture results in a decrease in the emissivity while the sensitivity of the radar is supposed to be positive as the value of backscatter increases with increase in soil moisture. Overlying vegetation tends to attenuate the radiometric response of soil thereby reducing sensitivity of the microwave sensor to soil moisture, the effect of which is seen in the reduced sensitivity over corn fields as compared to soy fields. Some of the sensitivity values presented in the study have positive values for radiometer channels and negative values for radar channels indicating that overlying vegetation completely mask sout the effect of soil moisture on observed brightness temperatures and backscatter coefficients. These measures of sensitivity allow an intercomparison between different channels of the radar or the radiometer.

There was major precipitation event in the watershed region on July 7 which provided around 24 mm of rain on the SMEX02 study region. Sensitivity computation was done by averaging PALS observations in a particular channel for corn and soy fields separately and comparing the change in averaged PALS observations to the change in the 0–6 cm layer soil moisture before and after the precipitation event. The results have been presented in Tables 2 and 3. As expected, the L band horizontal polarization channel had the maximum sensitivity among passive channels with a maximum sensitivity of 0.831 over corn fields and 1.182 over soy fields. For the radar channels the L band vertically copolarized backscatter was most sensitive to soil moisture with sensitivities of 0.421 for corn and 0.639 for soy fields. The S-band was less sensitive to soil moisture than the L band with the maximum sensitivities for passive and active PALS observations being 0.644 (SH) and 0.12 (SVV) over corn fields and 1.047 (SH) and 0.380 (SVV) over soy fields. These findings illustrate the lower capability of the S-band to penetrate denser vegetation canopies as in case of corn fields as opposed to soybean fields which had significantly lower vegetation water content. In general, the PALS sensor exhibited greater sensitivity over the less vegetated soy fields

TABLE 3: Sensitivity of radar channel to 0–6 cm layer gravimetric soil moisture evaluated over corn and soy fields. L band vertically copolarized channel is seen to be most sensitive to soil moisture and the sensitivity is seen to be higher for soy fields as compared to corn fields. A number of negative sensitivity values during the dry period DOY 176–178 indicate that the contribution of soil moisture to the radar backscatter was masked out by vegetation and surface roughness effects [41].

| Period | lhh | lvv | lvh | shh | svv | svh |
|---|---|---|---|---|---|---|
| | | | Corn | | | |
| 176–178 | 0.237 | 0.115 | −0.214 | 0.138 | −0.173 | 0.051 |
| 187-188 | 0.222 | 0.302 | 0.199 | 0.114 | 0.120 | 0.122 |
| 188-189 | 0.309 | 0.421 | 0.374 | 0.078 | 0.103 | 0.113 |
| | | | Soy | | | |
| 176–178 | 0.019 | −0.114 | 0.329 | −0.001 | −0.521 | −0.086 |
| 187-188 | 0.454 | 0.639 | 0.360 | 0.387 | 0.380 | 0.324 |
| 188-189 | 0.625 | 0.504 | 0.666 | 0.335 | 0.105 | 0.219 |

Active channels (sensitivity = $(\Delta\sigma/\Delta m_g) * 0.01$).

TABLE 4: Correlations between PALS horizontal polarization brightness temperatures and soil moisture in the 0-1 cm and 0–6 cm layers. Correlations generally reduce with increasing vegetation water content. The $2.5\,kg/m^2$ < VWC < $3.5\,kg/m^2$ and $1.0\,kg/m^2$ < VWC < $2.5\,kg/m^2$ classes consist mostly of measurements made in the corn fields for the dry days of June 25th and 26th when the contribution of radiation from soil is masked out by the overlying vegetation and the coefficient of correlation is seen to be positive [41].

| VWC (kg/m²) | No. of points | GSM (0-1 cm) | | GSM (0–6 cm) | |
|---|---|---|---|---|---|
| | | R (LH) | R (SH) | R (LH) | R (SH) |
| VWC < 1.0 | 52 | −0.881 | −0.885 | −0.808 | −0.846 |
| 1.0 < VWC < 2.5 | 6 | 0.142 | 0.278 | 0.420 | 0.562 |
| 2.5 < VWC < 3.5 | 21 | −0.094 | −0.197 | 0.258 | 0.161 |
| 3.5 < VWC < 4.5 | 13 | −0.950 | −0.863 | −0.847 | −0.833 |
| VWC > 4.5 | 48 | −0.690 | −0.641 | −0.676 | −0.626 |

(VWC ~ $1.0\,Kg/m^2$) in comparison to the cornfields (VWC ~ $3.5\,Kg/m^2$) across all channels. These findings support previous studies showing the LH channel to be more sensitive to near-surface soil moisture than LV, SH, or SV and the LVV channel to be more sensitive to soil moisture than other active channels [36].

### 3.3. Statistical Analysis.

Numerous prior studies have focused on either regression between radiometer or radar observations and in situ observations of surface soil moisture [71, 72] under conditions of low vegetation water content. The relationship between soil moisture and microwave emission is nearly linear. The present study evaluates the performance of linear regression technique for soil moisture estimation under the conditions of high vegetation water content encountered during the SMEX02 campaign.

Soil moisture retrieval was carried out by a simple multilinear regression procedure. As the best correlation between brightness temperatures and soil moisture was given by a band combination of LH, LV, and SH bands a multilinear regression was done using two combinations of PALS channels (LH, LV, and SH) and (LH, SH). Soil moisture retrieval was also carried out using the brightness temperatures from a single LH channel. Individual observations were classified into five subclasses depending on the vegetation water content. Table 4 presents the correlation coefficients ($R$) obtained from a linear regression applied to

the colocated data for all available days of PALS observations. Significant correlation was seen between the L band horizontal polarization brightness temperature and in situ soil moisture for three of the subclasses with a negative value indicating that the brightness temperature decreases as soil moisture increases. Regression equations were developed for each class using 2-3 days of observations (calibration period) and the soil moisture was retrieved for all other days (validation period) using these (calibration) equations. Average root mean square error (RMSE) was computed for the soil moisture retrieval in each case. Tables 5 and 10 present the errors associated with retrieval of soil moisture from the LH band and the band combination LH, LV, and SH. The retrieval errors reduce significantly when multichannel regression retrieval is done, especially in the case of the highest vegetation water content class (VWC > 4.5) where the lowest retrieval error of 0.045 g/g (gravimetric soil moisture) is obtained for the band combination LH, LV, and SH as compared to 0.062 g/g for retrieval from the LH band only. Similarly, retrieval error for the class with VWC < $1 kg/m^2$ is also the least in case of LH, LV, and SV band combination, the error being 0.034 g/g. As expected the retrieval accuracies decrease with increase in vegetation water content indicating that the sensor becomes less sensitive to soil moisture as the vegetation water content increases.

Retrieval of soil moisture was also done by employing linear or multiple regressions between the colocated radar

TABLE 5: Root mean square errors associated with soil moisture retrieval from the PALS LH channel by the statistical regression technique. Note that the retrieval errors are greatest for the VWC > 4.5 kg/m² class [41].

| Calibration days | Note | VWC < 1.0 | 1 < VWC < 2.5 | 2.5 < VWC < 3.5 | 3.5 < VWC < 4.5 | VWC > 4.5 |
|---|---|---|---|---|---|---|
| 178, 189 | 1 dry, 1 wet | 0.0371 | 0.0326 | 0.0579 | 0.0417 | 0.0623 |
| 176, 186, 188 | 1 dry, 2 wet | 0.0371 | 0.0302 | 0.0578 | 0.0409 | 0.0738 |
| 178, 188, 189 | 1 dry, 2 wet | 0.0374 | 0.0322 | 0.0492 | 0.0410 | 0.0643 |
| 176, 178, 189 | 2 dry, 1 wet | 0.0375 | 0.0271 | 0.0609 | 0.0417 | 0.0623 |
| 176, 178, 187 | 2 dry, 1 wet | 0.0383 | 0.0328 | 0.0531 | 0.0450 | 0.0635 |
| 188, 189 | 2 wet | 0.0403 | 0.0337 | — | 0.0402 | 0.0643 |
| 176, 178 | 2 dry | 0.1459 | 0.1228 | 0.0774 | — | — |

TABLE 6: Correlations between PALS vertically copolarized radar backscatter and soil moisture in the 0-1 cm layer. Correlations generally reduce with increasing vegetation water content. The $1.0 \, \text{kg/m}^2 < \text{VWC} < 2.5 \, \text{kg/m}^2$ and $2.5 \, \text{kg/m}^2 < \text{VWC} < 3.5 \, \text{kg/m}^2$ classes consist mostly of measurements made in the corn fields for the dry days of June 25th and 26th when the contribution of radiation from soil is masked out by the overlying vegetation and the coefficient of correlation is seen to be negative [41].

| VWC (kg/m²) | No. of points | GSM (0-1 cm) | | GSM (0-6 cm) | |
|---|---|---|---|---|---|
| | | R (LVV) | R (SVV) | R (LVV) | R (SVV) |
| VWC < 1.0 | 57 | 0.858 | 0.814 | 0.79 | 0.73 |
| 1.0 < VWC < 2.5 | 8 | 0.443 | 0.222 | 0.17 | 0.389 |
| 2.5 < VWC < 3.5 | 28 | −0.15 | −0.146 | −0.346 | −0.434 |
| 3.5 < VWC < 4.5 | 14 | 0.742 | 0.794 | 0.457 | 0.482 |
| VWC > 4.5 | 47 | 0.811 | 0.519 | 0.793 | 0.503 |

backscatter and in situ soil moisture dataset classified on the basis of vegetation water content. Table 6 presents the coefficient of correlation (R) values obtained by single channel linear regression for five different subclasses of vegetation water content. Active channels also give acceptable retrieval errors, with the lowest prediction error for the VWC > 4.55 kg/m² class being 0.0481 g/g for the LVV, SVV, and LHH band combination. The retrieval errors associated with soil moisture retrieval from PALS backscatter coefficients are tabulated in Tables 7 and 8.

It is important that the calibrating dataset fairly represents the conditions encountered during the course of the SMEX02 experiments. It is seen that the errors increase significantly if only dry or only wet days are used for calibration. Table 8 shows that the RMS error increases to 0.08 g/g from 0.05 g/g in case of fields with VWC between 2.5 and 3.5 kg/m², when 2 dry days were used for developing the regression equation rather than 1 dry 2 wet or 2 dry and 1 wet days. The discrepancy seen in the 2.5 kg/m² < VWC < 3.5 kg/m² and 1.0 kg/m² < VWC < 2.5 kg/m² classes in the form of a positive value of coefficient of correlation for the passive case and negative value for the active case (Tables 4 and 6) is due to the fact that these classes consist mostly of measurements made in the corn fields for the dry days of June 25 and 26 when the contribution of radiation from soil was masked out by the overlying vegetation. This effect is also reflected by

the higher soil moisture retrieval errors for this class. Soil moisture retrieval by a multiple channel regression produces more prediction errors than single channel regression as the variance in both vegetation and soil moisture is taken into account by a multiple regression.

*3.4. Forward Model for Passive Sensor.* The forward model for simulation of brightness temperatures considers a uniform layer of vegetation overlying the soil surface. The dielectric behavior of the soil water mixture is modeled using a semiempirical, four-component, dielectric-mixing model [64]. The upwelling radiation from the land surface observed from above the canopy was expressed in terms of radiative brightness temperature and is described by the radiative transfer model [69, 73, 74].

A physical model for passive radiative transfer was used for simulation of PALS-observed brightness temperatures and subsequent soil moisture retrieval. In situ measurements of soil moisture, land surface temperature, bulk density, and vegetation water content were made during the SMEX02 experiments. The in situ measurements of vegetation water content were used to calibrate a model that used a Landsat TM derived parameter, NDWI, which was used to derive the vegetation water contents for the SMEX02 region for the entire study period. A summary of the parameters that were used for calibration and soil moisture retrieval from PALS-observed L band brightness temperatures has been presented in Table 9.

Soil surface roughness, $h$, polarization mixing parameter, $q$, and single scattering albedo were not available as measured quantities. Polarization mixing was taken as zero and single scattering albedo was taken as 0.1 for both vertical and horizontal polarizations for corn as well as soy canopies. Surface roughness was used as a free parameter for calibrating the model. Vegetation opacity was taken to be 0.086 for soy canopy and 0.12 for corn canopy as reported in previous studies [17]. For deriving the optimum values, $h$ was varied between 0 and 0.6 separately for corn and soy fields and the estimation was done on the criteria of minimum root mean square error between PALS-observed average brightness temperature $(T_{BH} + T_{BV})/2$, in the L band and the modeled average brightness temperature for the L band. The average brightness temperatures were normalized by $T_{LST}$ to account for surface temperature contribution. Calibration

TABLE 7: Root mean square errors associated with soil moisture retrieval from the PALS LVV, SVV, and LHH band combination. The errors generally increase with vegetation water content [41].

| Calibration days | Note | VWC < 1.0 | 1 < VWC < 2.5 | 2.5 < VWC < 3.5 | 3.5 < VWC < 4.5 | VWC > 4.5 |
|---|---|---|---|---|---|---|
| 178, 189 | 1 dry, 1 wet | 0.0401 | 0.0340 | 0.0477 | 0.0447 | 0.0490 |
| 176, 186, 189 | 1 dry, 2 wet | 0.0373 | 0.0286 | 0.0576 | 0.0551 | 0.0501 |
| 178, 188, 189 | 1 dry, 2 wet | 0.0395 | 0.0293 | 0.0502 | 0.0439 | 0.0481 |
| 176, 178, 189 | 2 dry, 1 wet | 0.0398 | 0.0319 | 0.0493 | 0.0443 | 0.0490 |
| 176, 178, 187 | 2 dry, 1 wet | 0.0382 | 0.0340 | 0.0465 | 0.0450 | 0.0987 |
| 188, 189 | 2 wet | 0.0571 | 0.1747 | — | 0.0443 | 0.0481 |
| 176, 178 | 2 dry | 0.0421 | 0.0598 | 0.0497 | — | — |

TABLE 8: Root mean square errors associated with soil moisture retrieval from the PALS LVV band by the statistical regression technique [41].

| Calibration days | Note | VWC < 1.0 | 1 < VWC < 2.5 | 2.5 < VWC < 3.5 | 3.5 < VWC < 4.5 | VWC > 4.5 |
|---|---|---|---|---|---|---|
| 178, 189 | 1 dry, 1 wet | 0.0430 | 0.0360 | 0.0474 | 0.0517 | 0.0505 |
| 176, 186, 189 | 1 dry, 2 wet | 0.0424 | 0.0348 | 0.0511 | 0.0454 | 0.0505 |
| 178, 188, 189 | 1 dry, 2 wet | 0.0430 | 0.0360 | 0.0498 | 0.0511 | 0.0507 |
| 176, 178, 189 | 2 dry, 1 wet | 0.0447 | 0.0375 | 0.0478 | 0.0517 | 0.0505 |
| 176, 178, 187 | 2 dry, 1 wet | 0.0441 | 0.0419 | 0.0465 | 0.0485 | 0.0517 |
| 188, 189 | 2 wet | 0.0485 | 0.0665 | — | 0.0761 | 0.0507 |
| 176, 178 | 2 dry | 0.0637 | 0.0731 | 0.0474 | — | — |

TABLE 9: Summary of parameters input to the radiative transfer model for simulation of PALS brightness temperatures [41].

| (a) Media and sensor parameters | |
|---|---|
| Vegetation: | |
| Single scattering albedo, $\omega$ | 0.1 |
| Opacity coefficient, $b$ | 0.086 (soy), 0.12 (corn) |
| Soil: | |
| Roughness coefficients, $h$ (cm) and $Q$ | $h$-calibrated, $Q = 0.$ |
| Bulk density (g cm$^{-3}$) | in-situ |
| Sand and clay mass fractions, $s$ and $c$ | CONUS-SOIL dataset |
| Sensor: | |
| Viewing angle, $\theta$ (deg) | 45 |
| Frequency, $f$ (GHz) | 1.41, 2.69 |
| Polarization | H, V |
| (b) Media variables | |
| Land surface: | |
| Surface soil moisture, $m_v$ (g cm$^{-3}$) | In-situ |
| Vegetation water content, $w_c$ (kg m$^{-2}$) | Landsat TM |
| Surface temperature, $t$ (K) | In-situ |

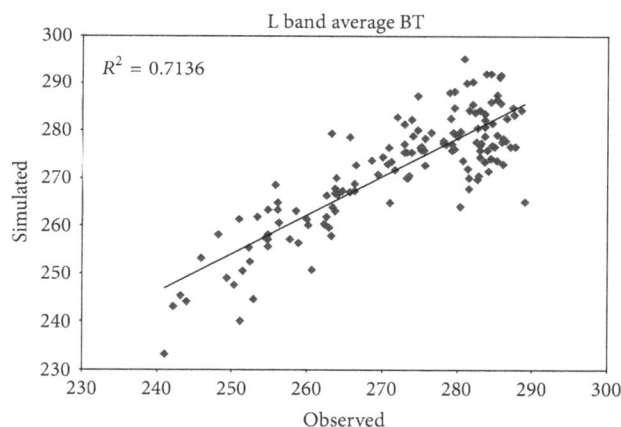

FIGURE 4: Model-predicted versus PALS-observed L band brightness average temperatures. The prediction is better at high soil moisture conditions (lower average brightness temperature) when contribution of vegetation and roughness is not as significant as that of soil moisture. Model calibrated using PALS and in situ data for July 7 and July 8 [41].

was performed for different combinations of 2 or 3 days of PALS observations out of the 7 days available.

The calibrated values of $h$ for each field and in situ measurements of model soil moisture, bulk density, surface temperature, and vegetation water content were used to simulate horizontal and vertical polarization brightness temperatures for the L- and S-bands. Gravimetric soil moisture in the 0–6 cm layer was used. Simulations were run for various combinations of calibration days. The root mean square error (RMSE) for simulation of average brightness

temperature in the L band was 7.1 K and 8.0 K for the S-band when the calibration was done for DOY's 176, 188, and 189. Simulation results were better for soy fields with an RMSE of 6.8 K as opposed to corn fields with an RMSE of 7.2 K for the L band. Figures 4 and 5 present the plots for PALS observed versus model-predicted average brightness temperature for the L- and S-bands. Retrieval of soil moisture was performed by using a simple retrieval algorithm that arrives at a soil moisture estimate by minimizing the error between simulated and observed L band average brightness temperature normalized with the land surface temperature

TABLE 10: Root mean square errors (g/g) associated with retrieval of soil moisture (gravimetric) by physical modeling of PALS L band observations considering the retrieved soil moisture is representative of the in situ 0–6 cm layer soil moisture [41].

| Calibration days | Note | vwc < 1 | 1 < vwc < 2.5 | 2.5 < vwc < 3.5 | 3.5 < vwc < 4.5 | 4.5 < vwc |
|---|---|---|---|---|---|---|
| 178, 189 | 1 dry, 1 wet | 0.0375 | 0.0343 | 0.0287 | 0.0327 | 0.0439 |
| 176, 186, 189 | 1 dry, 2 wet | 0.0404 | 0.0310 | 0.0365 | 0.0504 | 0.0436 |
| 178, 188, 189 | 1 dry, 2 wet | 0.0475 | 0.0338 | 0.0309 | 0.0271 | 0.0398 |
| 176, 178, 189 | 2 dry, 1 wet | 0.0398 | 0.0297 | 0.0230 | 0.0519 | 0.0518 |
| 176, 178, 187 | 2 dry, 1 wet | 0.0442 | 0.0042 | 0.0252 | 0.0661 | 0.0465 |
| 188, 189 | 2 wet | 0.0326 | 0.0341 | 0.0403 | 0.0334 | 0.0283 |
| 176, 178 | 2 dry | 0.0455 | 0.0033 | 0.0221 | 0.0737 | 0.0472 |
| 176, 188, 189 | 1 dry, 2 wet | 0.0333 | 0.0286 | 0.0299 | 0.0391 | 0.0454 |

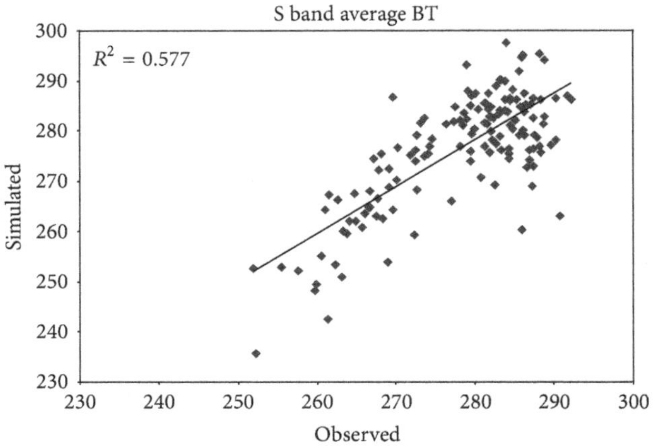

FIGURE 5: Model-predicted versus PALS-observed S-band average brightness temperatures. Model calibrated using PALS and in situ data for July 7 and July 8 [41].

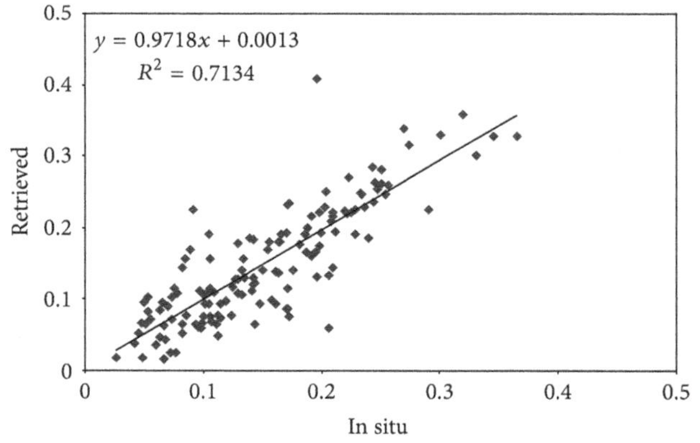

FIGURE 6: Model-retrieved gravimetric soil moisture (g/g) versus soil moisture measured in situ in the 0–6 cm soil layer. Model calibrated using PALS and in situ data for July 7 and July 8 [41].

$(T_B(\text{modeled}) - T_B(\text{observed})/T_{\text{LST}})$, where $T_{\text{LST}}$ is the land surface temperature in Kelvins. Soil moisture retrieval was done for various combinations of calibration days and the RMSE for the retrieval was evaluated in each case. Table 10 presents the errors between in situ soil moisture in the 0–6 cm soil layer and the model retrieved soil moisture values for five classes based on vegetation water content. The retrieval errors increase as the vegetation water content increases, being around 0.03 g/g for the VWC < 1 kg/m$^2$ fields and greater than 0.04 g/g for the fields with VWC > 4.5 kg/m$^2$. Figure 6 is a plot of soil moisture retrieved from PALS L band horizontal polarization brightness temperatures versus the in situ soil moisture in the 0–6 cm soil layer with the calibration being done using DOY's 176, 188, and 189.

Physical modeling of brightness temperatures proved to be more accurate than statistical regression technique with an overall soil moisture retrieval accuracy of around 0.036 g/g as compared to 0.05 g/g for the statistical regression technique. Error may have been introduced in the soil moisture retrieval process due to improper in-situ sampling, measurement of gravimetric soil moisture and bulk density and the assumption that single scattering albedo and vegetation opacity are independent of look angle and polarization. Assignment of a single value of $h$ and $q$ for all fields of a particular crop type is also a source of error and field measurements of surface roughness are desirable. In the present study both east-to-west and west-to-east flight lines were considered to maximize the number of fields observed by PALS on each day. If a field was observed during both the east-to-west and west-to-east passes, the value from the east-to-west flight line was chosen. This also introduces a source of error in both the statistical regression and physical modeling techniques of soil moisture retrieval as the PALS overpass times may not coincide with the in situ sampling time for soil moisture in a particular field, and data from two flight lines may be different due to factors such as sun glint.

## 4. Disaggregation of Soil Moisture

### 4.1. Radar-Radiometer Method

*4.1.1. The Importance of Radar for Soil Moisture.* Radars have a higher spatial resolution than radiometers. Retrieval of soil moisture using radar backscattering coefficients is difficult due to more complex signal target interaction associated with measured radar backscatter data, which is highly influenced by surface roughness and vegetation canopy structure and water content. Several empirical and semiempirical algorithms for retrieval of soil moisture from radar backscattering coefficients have been developed but they are valid mostly in the low vegetation water content conditions [75–77]. On the other hand, the retrieval of soil moisture from radiometers is well established and has a better accuracy with limited requirements for ancillary data [78, 79]. Radiometer measurements are less sensitive to uncertainty in measurement and parameterization of surface roughness and vegetation canopy interaction. However, the spatial resolution of radiometer is much lower compared to radar operating in the same band. An optimal soil moisture retrieval algorithm that combines the higher spatial resolution of radar with higher sensitivity of a radiometer might result in improved soil moisture products.

Temporal evolution of soil moisture can be potentially monitored through change detection. Change detection methods [70] have been implemented as a convenient way to determine relative soil moisture or the change in soil moisture [42, 80]. Both brightness temperature and radar backscatter change depend approximately linearly on soil moisture and hence sensitivity can be assumed to be independent of soil moisture. However, quantification of sensitivity requires soil moisture measurements, which is difficult in the case of radar in the presence of moderate to high vegetation cover. It may be possible to estimate the radar sensitivity to soil moisture by using radiometer-estimated soil moisture measurements if the impact of vegetation on sensitivity and spatial heterogeneity issues can be accounted for.

This section proposes a simple algorithm that uses higher resolution radar observations along with coarser resolution radiometer observations to determine the change in soil moisture at the spatial resolution of radar operation, without using any in situ soil moisture measurements. The present study simplifies the problem of spatial disaggregation of soil moisture by considering that the spatial variability of bare soil properties (texture, roughness) that influence radar sensitivity to soil moisture is not significant and hence the variability of radar signal within the radiometer footprint is due to soil moisture and canopy vegetation water content variability only.

The next section explains the theoretical basis and assumptions behind the algorithm for spatial disaggregation used in this study. Section 4.1.3 presents the data and the methods that are applied to evaluate the performance of the algorithm presented in the study. Section 4.1.4 presents the results in terms of comparison of in situ measurements of soil moisture with the disaggregated estimates obtained from the algorithm.

*4.1.2. Theory for Change Detection.* Brightness temperature and radar backscatter have a nearly linear relationship to surface soil moisture, for uniform vegetation and land surface characteristics. The radiative transfer model for estimation of soil moisture from brightness temperature is well established and needs few ancillary parameters for soil moisture estimation. The C-band radiometer AMSR-E has a global soil moisture product and future L-band radiometers such as SMOS and HYDROS will have radiometer-only soil moisture products. However, the radiometer-only soil moisture product is limited in application by the low spatial resolution of the radiometer instrument. Higher spatial resolution is possible with radar soil moisture estimation; however, estimation of absolute soil moisture from radar backscattering coefficients requires modeling a complex signal target interaction. Even in empirical and semiempirical studies, vegetation canopy and soil parameters may be needed to classify a heterogeneous target area into subclasses that are fairly uniform in terms of those parameters. Several studies based on the approach of classification and linear parameterization of L-band radar backscatter measurements with respect to soil moisture within each class have been performed in the past [65, 76, 81, 82].

The approach taken by the present study is change estimation, which takes advantage of the approximately linear dependence of radar backscatter change on soil moisture change [42]. Njoku et al. demonstrated the feasibility of a change detection approach using the PALS radar and radiometer data obtained during the SGP99 campaign. The PALS and in situ soil moisture data were classified into 3 different classes based on the vegetation water content. For each class linear least square fits of PALS brightness temperature and radar backscatter to soil moisture were developed. The linear relationships were modeled as

$$T_{Bp} = A + Bm_v, \tag{9}$$

$$\sigma_{pp}^0 = C + Dm_v. \tag{10}$$

The PALS data used for the SGP99 study had the same footprint size for both radar and radiometer. Hence, $A$, $B$, $C$, and $D$ are parameters for each pixel in the coincident radar and radiometer images and were assumed to primarily be functions of surface vegetation and roughness (and temperature for the passive case). Difference images were obtained by subtracting the sensor data on the first day from the sensor data on the consecutive days. They were able to calibrate $C$ and $D$ parameters in (10) using 2 days of radiometer estimates of soil moisture under wet and dry soil conditions. Further using $C$, $D$, and $\sigma_{vv}^0$ they derived radar-estimated soil moisture with satisfactory results. Our study is aimed at estimation of soil moisture change at the spatial resolution of radar by combining radar and radiometer data. The approach and assumptions are similar to the Njoku et al. study; however, in our case the radar is at higher spatial resolution of 100 m as compared to the 400 m spatial resolution of the radiometer. We assume that the changes in vegetation canopy parameters are insignificant as compared to the change in soil moisture when considering the resulting change in copolarized radar backscatter, given a sufficiently high revisit rate of the sensor over the target. Using this assumption, the difference image obtained by subtracting consecutive radar backscatter images acquired over an area would be given by

$$\Delta\sigma_{pp}^0 = D\Delta m_v. \tag{11}$$

$\Delta\sigma_{pp}^0$ is the change in copolarized radar backscatter (dB) and $\Delta m_v$ is the change in soil moisture. The parameter $D$ is expected to depend on the attenuation characteristics of the vegetation canopy and the surface roughness characteristics of the soil surface. Results from Friedl et al. [62] indicate that relative sensitivity of the L-band copolarized channels of radar should depend primarily on the vegetation canopy opacity. Relative sensitivity is defined as the ratio of the radar sensitivity in the presence of a vegetation canopy ($D$) to the sensitivity if there was only bare soil ($D_0$):

$$\frac{D}{D_0} = f(\tau). \tag{12}$$

Figure 12 in Du et al. shows the variation of the relative sensitivity for a medium rough soil surface having a soybean canopy. The plot suggests that relative sensitivity can be estimated from optical thickness once the canopy type and surface type are known. The vegetation opacity $\tau$ can be estimated using the ($\tau = b\text{VWC}/\cos\theta$) relationship for vegetation canopies that follow the electrically thin scatterer approximation. $b$ is a parameter that depends on vegetation structure and type, $\theta$ is the incidence angle, and VWC is the vegetation water content which can be estimated operationally using proxies such as NDWI [83]. Now, combining (11) and (12) we can write

$$\Delta\sigma^0_{pp} = f(\tau)D_0\Delta m_v. \tag{13}$$

The bare soil sensitivity $D_0$ depends only weakly on soil roughness variability for a given sensor configuration (frequency, polarization, and viewing angle). To substantiate this further, the author conducted a simulation in which the Integral Equation Model [65] was used to generate plots of vertically copolarized radar backscatter versus soil moisture for various root mean square soil roughness values (Figure 7). It is seen in the figure that the line plots for $s = 0.4$ cm to $s = 2.4$ cm can be approximated as a series of parallel lines with the same slope and different intercepts. The result indicates that for the L-band surface roughness variability has only a small effect on the soil moisture sensitivity of radar. Now, from (13) we obtain

$$\Delta m_v = \frac{\Delta\sigma^0}{S_0}, \tag{14}$$

where $S_0 = f(\tau) * D_0$. Let us assume that the radar backscatter is available at a finer resolution of "$x$" whereas radiometer data is available at a coarser resolution of "$\mathbf{X}$". The problem, then, is to estimate $\Delta m_{v,x}$, that is, the change in soil moisture from time step $t_0$ to $t_0 + \Delta t$, given $m_{v,\mathbf{X}}$, $\sigma^0_x$, and $\tau_x$ at times $t_0$ and $t_0 + \Delta t$ using (12). The change in the soil moisture at the coarser spatial resolution ($\mathbf{X}$) can be evaluated as the sum of all the changes at the finer resolution ($x$), that is,

$$\Delta m_{v,\mathbf{X}} = \frac{1}{N}\sum m_{v,x}. \tag{15}$$

The summation is over all the "$N$" smaller radar footprints within the larger radiometer footprint, and $\Delta m_{v,\mathbf{X}}$ is the change in soil moisture as measured by the radiometer at the lower spatial resolution, $\Delta m_{v,x}$ is the change in soil moisture

as measured by the radar at the higher spatial resolution given by (12). Combining (12) and (13) leads to

$$\Delta m_{v,\mathbf{X}} = \frac{1}{N}\sum\frac{\Delta\sigma^0_x}{S_0}. \tag{16}$$

The unknown $S_0$ will be the same for all the radar pixels within a radiometer pixel given uniform vegetation characteristics within the radiometer footprint, that is, $f(\tau)$ is the same for all radar pixels within the radiometer pixels. The author recognize the fact that the spatial variability of $f(\tau)$ will not be low in a real world setting. However, in the case of the SMEX02 experiments, each radiometer footprint was contained completely within an agricultural field with fairly uniform vegetation characteristics. In the present study we do not attempt to model for the vegetation canopy variability within the radiometer footprint. $S_0$ is evaluated for each radiometer pixel by dividing the summation of change in radar backscattering coefficients with the change in radiometer scale soil moisture. The summation is for all radar pixels that lie within the radiometer pixel:

$$S_0 = \frac{1}{N}\frac{\sum\Delta\sigma^0_x}{\Delta m_{v,\mathbf{X}}}. \tag{17}$$

$S_0$ for a particular radiometer pixel can be resampled to the radar spatial resolution and for each radar pixel within the radiometer pixel we write the change in soil moisture at resolution "$x$" as

$$\Delta m_{v,x} = \frac{\Delta\sigma^0_x}{S_0}. \tag{18}$$

Another important issue in the low spatial variability of $S_0$ assumption is the implicit assumption that multiple days of radar data were obtained over the region at the same angle of incidence. At different incidence angles, corresponding AIRSAR pixels will exhibit different sensitivity to soil moisture. During SMEX02, the near and far look angles were 22.8° and 71.3° and July 5, 22.0° and 71.2° for July 7, and 24.1° and 71.3° for July 8. This indicates that AIRSAR acquired data over each of the fields at approximately similar incidence angles for the three days. The variation of incidence angles between two fields is not important as the relative change in radar backscatter is considered with the relative change in the soil moisture on a field-wise basis. The low-resolution sensitivity parameter $S_0$ is derived separately for each field. As a result only the variation of incidence angle within each field will be important and this variation is small. Within the dimensions of the PALS footprint, this variation in AIRSAR incident angles will be small and its effect on sensitivity negligible.

Accurate estimation of the change in soil moisture at the lower spatial resolution ($\Delta m_{v,\mathbf{X}}$) is crucial to the accuracy of computed radar sensitivity to soil moisture (15) and hence the accuracy of the high-resolution soil moisture change estimated by the approach presented in this paper. In an operational setting $\Delta m_{v,\mathbf{X}}$ can be estimated from single-or multichannel passive remote sensing observations. Estimation of soil moisture from passive remote sensing

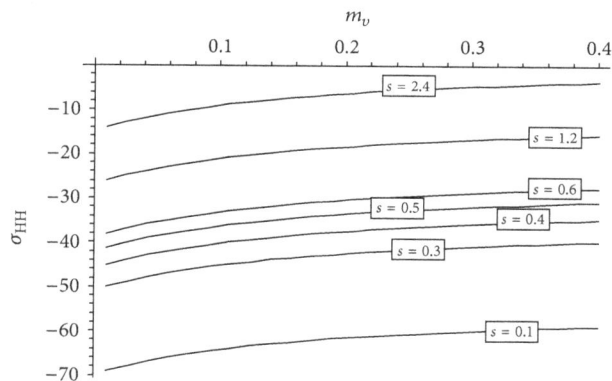

FIGURE 7: Simulation of L band horizontally copolarized radar backscattering coefficients using the integral equation model [70] for various values of volumetric soil moisture $m_v$ and rms surface roughness $s$ (cm). Surface correlation length has been taken as 8 cm, % sand = 30, and % clay = 30. For various roughness values, moisture versus backscatter curves can be approximated as straight lines with the same slope. L band vertically copolarized radar backscattering coefficients show a similar behaviour too [55].

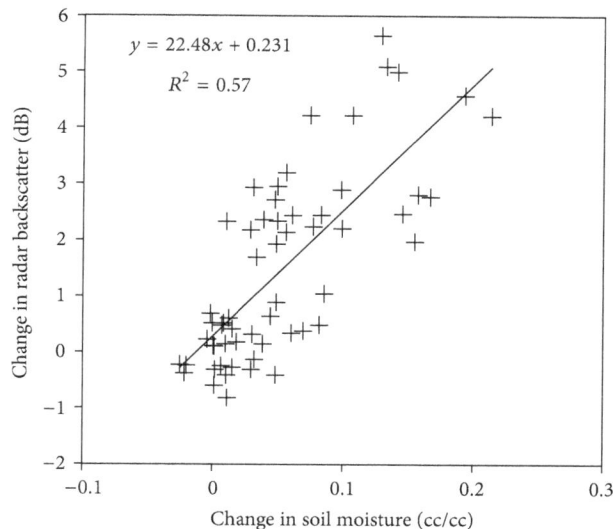

FIGURE 8: Plot of change in AIRSAR LVV backscatter at 800 m resolution versus change in in situ volumetric soil moisture at a 800 m resolution for the periods July 5 to July 7 and July 5 to July 8 [55].

has been studied in several prior works and the author do not attempt to retrieve soil moisture from PALS radiometer data used in this study. A previous study [41] demonstrated PALS radiometer to be sensitive to soil moisture during the SMEX02 experiments and retrieval of soil moisture from PALS L-band brightness temperature data was done with estimation errors of approximately 4%. In the current study, in situ soil moisture measurements have been upscaled to 400 m resolution by averaging and randomly varying noise is added to the upscaled soil moisture values. This provides a simple way to simulate soil moisture retrievals using passive remote sensing. PALS data are used to demonstrate the sensitivity of AIRSAR L-band copolarized channels to soil moisture only.

*4.1.3. Data from SMEX02.* The algorithm discussed above was tested on data obtained from the Soil Moisture Experiments in 2002 (SMEX02). The SMEX02 campaign was conducted in Walnut Creek, a small watershed in Iowa, over a one-month period between mid-June and mid-July 2002. An extensive dataset of in situ measurements of soil moisture (0–6 cm soil layer), soil temperature (surface, 5 cm depth) soil bulk density, and vegetation water content was collected. Aircraft-mounted instruments—the passive and active, L- and S-band sensor (PALS) and the NASA/JPL airborne synthetic aperture radar (AIRSAR)—were flown with supporting ground-sampling data. The PALS instrument was flown over the SMEX02 region on June 25, 27 and July 1st, 2nd, 5, 6, 7, and 8, 2002 [84, 85]. PALS radiometer and radar provided simultaneous observations of horizontally and vertically polarized L- and S-bands brightness temperatures, radar backscatter measured in VV, HH and VH configurations, and thermal infrared surface temperature at a resolution of ~400 m. The AIRSAR instrument has P-, L-, and C-bands with H/V dual microstrip polarizations and

spatial resolutions of 5 m in slant range and 1 m in azimuth [86]. AIRSAR instrument was flown on July 1st, 5, 7, 8, and 9. The algorithm proposed in this study needs simultaneous observations of the radar and radiometer. As the PALS spatial coverage of the watershed on July 1st was partial, the data sets for PALS and AIRSAR for July 5, July 7, and July 8 were used. AIRSAR data used in this study was L band HH and VV polarization and provided at a spatial resolution of 30 m after processing. The AIRSAR images were geolocated by registration to a Landsat TM7 image. The ground-based soil moisture data used in this study was the volumetric soil moisture measured using a theta probe at 14 locations in each field site for all the 31 fields. There were 10 soy fields and 21 corn fields. The representative area for each field site was $800 \times 800$ m$^2$. Seven measurements of volumetric soil moisture made along each of two parallel transects, 600 m in length and placed 400 m apart. Higher-resolution estimates of in situ soil moisture for each field were estimated by an inverse-distance-weighted spatial interpolation using all the 14 measurements for each field and a cell size of 100 m.

*4.1.4. Results from Radar-Radiometer.* As an initial evaluation of radar sensitivity to soil moisture, the AIRSAR L band vertically copolarized backscattering coefficients were aggregated to 800 m and then collocated with the field sites that were sampled for 0–6 cm volumetric soil moisture. Figure 8 presents a plot of the change in 800 m resolution field averages of AIRSAR LVV backscattering coefficient and soil moisture for the periods July 5 to July 7 and July 5 to July 8. The change in soil moisture between July 7 and July 8 was not very high. In order to see a greater change in soil moisture the difference in July 5–July 8 was selected. The $R^2$ value of 0.57 indicates that $\Delta\sigma^0_{LVV}$ is sensitive to the change in soil moisture; even under the dense vegetation conditions encountered in the SMEX02

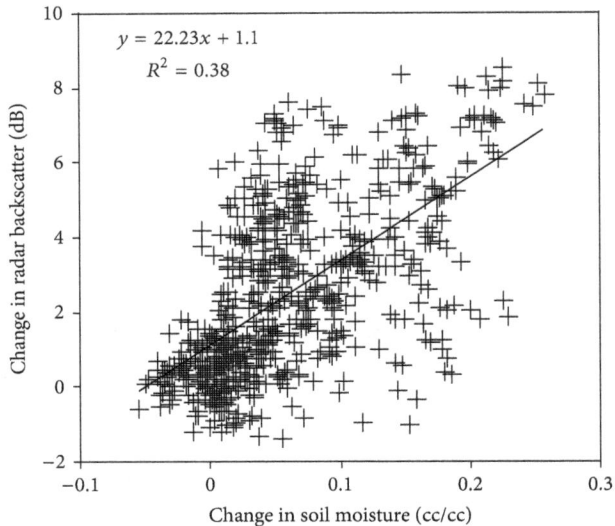

FIGURE 9: Plot of change in AIRSAR LHH backscatter at 100 m resolution versus change in in situ volumetric soil moisture at 100 m resolution for the periods July 5 to July 7 and July 5 to July 8 [55].

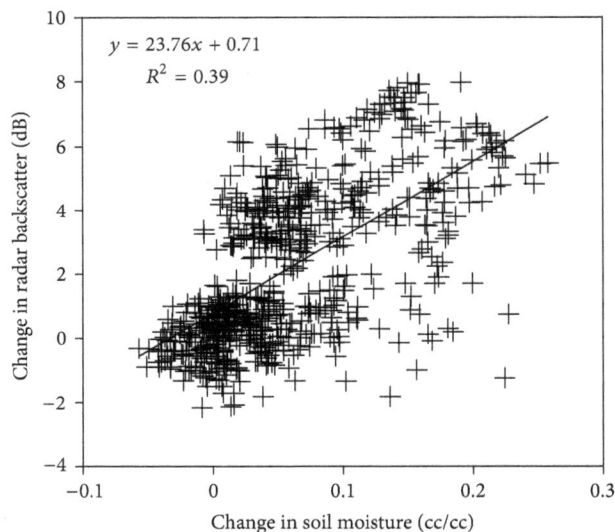

FIGURE 10: Plot of change in AIRSAR LVV backscatter at 100 m resolution versus change in in situ volumetric soil moisture at 100 m resolution for the periods July 5 to July 7 and July 5 to July 8 [55].

experiments with the vegetation water content of corn fields being around 4-5 kg/m$^2$.

The sensitivity of the AIRSAR LVV channel to soil moisture was also analyzed at a higher spatial resolution of 100 m. In Figures 9 and 10 the change in radar backscatter (LHH and LVV, resp.) is compared to the corresponding change in gravimetric soil moisture at a resolution of 100 m for the time periods July 5 to July 7 and July 5 to July 8. $R^2$ values of 0.38 and 0.39 are obtained for LHH and LVV, respectively, indicating that radar sensitivity to soil moisture is significant at the higher spatial resolution of 100 m also. The sensitivities are approximately the same for both LHH

and LVV channels with values of 22.2 and 23.8 dB/(cc/cc), respectively. It should be noted that there are several data points in Figures 8, 9, and 10 with negative backscatter change corresponding to positive change in soil moisture. At the 800 m spatial resolution (Figure 8) we see data points with a negative change in the range 0 to −1 dB for change in moisture in the range 0 to 0.05 cc/cc. At the 100 m spatial resolution (Figures 9 and 10) several data points undergo a negative change in the range 0 to −2 dB for change in moisture in the range 0 to 0.1 cc/cc. The authors believe that this effect is primarily due to change in vegetation water content. For example, in Figure 8, pixels increased in moisture from July 5 and July 7 by a small amount (<0.05) but still underwent a negative change in backscatter as the vegetation water content increased from July 5 to July 7 causing a greater attenuation of radar backscatter on July 7 as compared to July 5 to resulting in a negative net change in radar backscatter even though the moisture increased. Soil roughness may also change after a rainfall event causing the soil surface to reduce in roughness and result in lower radar backscatter values after the rainfall event. The assumption made by the author that vegetation and soil roughness do not change between consecutive observations is weak but it also simplifies the problem significantly with satisfactory results in terms of estimated soil moisture change.

It was shown in a previous study that for the SMEX02 field experiment PALS LV channel brightness temperatures were well correlated to soil moisture [41]. A further demonstration of the AIRSAR LVV channel sensitivity to change in soil moisture is done by comparison of the change in PALS LV channel brightness temperatures to the change in AIRSAR LVV channel backscattering coefficients. Figure 11 presents the difference images produced by the change in AIRSAR LVV backscattering coefficients from July 5 to July 7 compared to the change in PALS LV channel brightness temperature for the same period. The spatial patterns corresponding to wet and dry regions in the watershed are very similar for both the PALS and AIRSAR difference images. Regions that became wetter from July 5 to July 7 underwent a reduction in brightness temperature and an increase in backscattering coefficient. (The scales for the two images in Figure 11 are hence inverted to represent the positive change in radar backscatter and negative change in brightness temperature with an increase in soil moisture.) The correlation between PALS LV channel brightness temperature change and AIRSAR LVV channel radar backscatter change is further brought out in Figure 12. Change in AIRSAR backscattering coefficients (LVV) have been aggregated to 400 m resolution and compared with the change in PALS brightness temperatures (LV) at its resolution of 400 m. An excellent agreement is seen between the two with an $R^2$ value of 0.81 indicating that AIRSAR LVV channel has a significant sensitivity to soil moisture.

The algorithm discussed in the previous section was applied to the 3 days of AIRSAR LVV, PALS LV, and ground-based soil moisture data. 400 m resolution estimates of soil moisture ($m_{v,X}$) were calculated using the in situ measurements of soil moisture within each field. As mentioned earlier, 14 theta probe measurements of soil moisture were made at each watershed-sampling site that had dimensions of 800 m by 800 m. The in situ measurements were gridded to a 100 m dimension grid using inverse distance interpolation and then they were upscaled to 400 m corresponding to the dimensions of the PALS radiometer footprint. A uniformly distributed random noise of 0–0.016 g/cc was added to the upscaled in situ soil moisture values in order to simulate 4% maximum error that would be obtained if the soil moisture estimates were obtained by inverting soil moisture estimates from L-band brightness temperatures using a radiative transfer model. AIRSAR LVV data was also aggregated to a resolution of 100 m and the difference images were used to compute $\Delta\sigma_x^0$ where "$x$" denotes the lower cell size of 100 m. Using the algorithm discussed in the previous section, 100 m resolution estimates of the change in soil moisture were obtained for two periods of July 5 to July 7 and July 7 to July 8 for each field site. Figures 13(a) and 13(b) compare estimated change in soil moisture with measured change in soil moisture at a 100 m resolution for the period July 5 to July 7 and Figures 14(a) and 14(b) present the same comparison for the period July 5 to July 8. The root mean square error for the prediction (RMSE) in both cases is 0.046 cc/cc volumetric, a major portion of which seems to be contributed by a few outliers. It was noted that on removing 7 outliers from the plot in Figure 13(a) and 32 outliers from the plot in Figure 14(a), the RMSE improves to 0.032 cc/cc and 0.024 cc/cc, respectively. The outliers seen in Figures 13 and 14 are caused by the invalidity of the assumption that vegetation and surface roughness do not change significantly during consecutive time steps for few data points. For example, the encircled data point in Figure 13(a) is observed on field WC11 where the observed change in radar backscatter (LVV) was −1.871 dB and with no change in the corresponding change in in situ soil moisture. Table 11 presents the observed change in soil moisture and corresponding change in LVV radar backscatter from July 5 to July 7 for the field WC11. It is seen that although change in soil moisture was low for all data points, the change in corresponding radar backscatter was not low for all pixels. This may be a result of several factors such as significant localized change in vegetation or surface roughness, heterogeneity within the 100 m pixel, such as present of ponded water within the pixel but not at the soil moisture sampling location, human errors in sampling of soil moisture, and errors in geolocation of the AIRSAR data. Such pixels are not numerous and hence were not analyzed in complete detail in the present study. Computation of radar sensitivity when the soil moisture change is low is numerically unstable as $\Delta m_{v,X}$ appears in the denominator equation (15). It may be possible to develop an alternate approach for computing sensitivity where a time series of radar backscatter and soil moisture observations is used to compute sensitivity using the lowest and highest soil moisture values observed during a period of say 7–10 days during which the change in vegetation and surface roughness can be assumed to be insignificant. Having a greater range of soil moisture values will lead to a more accurate computation of radar sensitivity to soil moisture. The suggested approach has not been attempted in the current study; only 3 days of data were available.

## 4.2. Downscaling Using Visible and Near Infrared Satellite

### 4.2.1. Background.
In this approach we used the relationship between soil moisture and surface temperature modulated by vegetation. This approach has a background with past studies of Mallick et al. [87] who used the triangular relationship [68, 88–90] between surface temperature and the vegetation index (TvX) derived from the MODIS Aqua sensor data. From this they derived the soil wetness index that was converted to soil moisture at a 1 km scale. Minacapilli et al. [91] used thermal infrared observations from an airborne platform to estimate soil moisture using the thermal inertia principle for a bare soil field. They found that the estimated soil moisture correlated very well with in situ observations. Gillies and Carlson [67] devised a method that derived the fractional vegetation and spatial patterns of soil moisture using the AVHRR data set and demonstrated this method in a region of England. In our method, the diurnal temperature range (DTR) was used [92], which was affected by vegetation [93], soil moisture, and clouds [94].

This section is organized as follows: Section 4.2.2 provides the list of datasets and the locations of our study; Section 4.2.3 outlines the downscaling algorithm methodology. In Section 4.2.4 we discuss our findings and validation of the results. The results and discussion are presented in Section 4.2.4.

### 4.2.2. Data.
Oklahoma was selected as the study area due to the long history of soil moisture research focused on this region. The Oklahoma Mesonet and Little Washita River Experimental Watershed are two long-term in situ soil moisture networks which provide a solid foundation for soil moisture remote sensing research (shown in Figure 15). The Little Washita has been the location for various soil moisture field experiments including SGP97, SGP99, and SMEX03 [95, 96] and has been a key element in satellite validation studies [97, 98]. In addition to the ground resources, a variety of spaceborne sensors also contributed to this study. Descriptions and maps of the datasets used in this article are shown in Table 12 and Figure 16. Table 12 lists the spatial resolution and temporal repeat of these sensors and their data products.

(1) *NLDAS Data.* The NLDAS (North American Land Data Assimilation System, http://ldas.gsfc.nasa.gov/nldas/) phase 2 hourly mosaic data is used in this study. NLDAS is run hourly on a geographical grid with a spatial resolution of 1/8° (12.5 km). The NLDAS-2 data output includes various surface variables, such as radiation flux, surface runoff, surface temperature, vegetation indices, and soil moisture [99]. Soil, vegetation, and elevation are parameterized using high-resolution datasets (1 km satellite data in the case of vegetation). The forcing data [100, 101] and outputs have been extensively validated [102–104]. The soil moisture downscaling model in this study utilized two variables: surface skin temperature and soil moisture content at 0–10 cm depths. The data used in this study correspond to the closest local

overpass times of Aqua satellite for the Oklahoma region, which are approximately 8:00 and 20:00 PM (in UTC time).

(2) *AMSR-E Data.* The Advanced Scanning Microwave Radiometer on board the EOS Aqua platform (AMSR-E) collected microwave observations at 6, 10, 19, 37, and 85 GHz from 2002 to 2011 [105]. The AMSR-E instrument provided global passive microwave measurements of terrestrial, oceanic, and atmospheric parameters for hydrological studies from 2002–2011 [105, 106]. The soil moisture product derived from the AMSR-E sensor on board the Aqua satellite has 1/4° (25 km) spatial resolution. The estimation of AMSR-E soil moisture accuracy is approximately 10% and cannot be estimated in the area of vegetation biomass which is greater than 1.5 kg/m$^2$ [73].

In this study, the AMSR-E soil moisture was estimated by using the single-channel algorithm (SCA) [97, 107]. The single-channel algorithm uses the X-band observations at h-polarization (the most sensitive channel). The C-band observations cannot be used for land surface applications because they are significantly affected by manmade radio frequency interference (RFI). The land surface temperature was estimated using the 37 GHz v-pol observations. AVHRR-derived climatological dataset was used to correct for vegetation effects. For matching with other georeferenced datasets, a drop-in-the-bucket method was applied to the AMSR-E data and it was gridded to a 25 × 25 km EASE grid cell size. This method averaged all the AMSR-E points by determining if their center coordinates were within the borders of a particular EASE grid cell.

(3) *MODIS Data.* The surface temperature data which corresponded to the local time of 1:30 and 13:30, as well as the NDVI from MODIS/Aqua, were used in this study. MODIS has 36 spectral bands including visible, near infrared, and thermal infrared spectrum and provides 44 global data products [108]. The algorithms to derive the MODIS products are well established and have been extensively evaluated, including NDVI [109, 110], LAI [111], land cover classification [62, 112], and surface temperature [113]. In the current study surface temperature and NDVI products at two different spatial resolutions were used for downscaling soil moisture. The datasets included 1 km daily surface temperature (MYD11A1), 1 km biweekly NDVI (MYD13A2), and 5600 m biweekly Climate Modeling Grid (CMG) NDVI (MYD13C1). In addition, the dry down curves of soil moisture during May 2004, July 2005, and August 2005 in Oklahoma were examined. During these three months, clear days (due to the requirement of surface temperature in our algorithm) of 1 km surface temperature along with low cloud cover were selected for the downscaling algorithm application.

(4) *AVHRR Data.* For the years 1981–2001, prior to the launch of the Aqua satellite and the availability of MODIS data, the 5 km CMG daily NDVI data from AVHRR sensor (AVH13C1) was used. The AVHRR sensor is on board the NOAA satellites, including N07, N09, N11, and N14, and provides global and long-term surface ground measurements. Daily AVHRR NDVI data is available between 1981–1999 (http://ltdr.nascom.nasa.gov/ltdr/ltdr.html). After 2000, the

TABLE 11: Change in in-situ soil moisture (cc/cc) and corresponding change in radar backscatter (LVV) from July 5th to July 7th for the field WC11 at a 100 m spatial resolution [55].

| Field | $\Delta\theta$ | $\Delta\sigma$ |
|---|---|---|
| WC11 | −0.009 | 1.431 |
| WC11 | −0.007 | 0.356 |
| WC11 | −0.039 | −0.049 |
| WC11 | 0.000 | −1.871 |
| WC11 | −0.004 | −1.081 |
| WC11 | −0.005 | −0.546 |
| WC11 | −0.034 | −0.842 |
| WC11 | −0.011 | −0.816 |
| WC11 | −0.013 | 0.028 |
| WC11 | −0.001 | −1.419 |

N14 orbit drifted greatly, which can degrade the data quality. Therefore, the years between 2000–2002 were not used in this soil moisture downscaling exercise.

(5) *Oklahoma Mesonet Data.* The Oklahoma Mesonet is a network of 120 automated environmental monitoring stations with at least one site in each of the 77 counties of Oklahoma [114]. The environmental variables are obtained at intervals spanning every 5 to 30 minutes depending on the variable. The data quality is verified by a series of automated and manual checks via the Oklahoma Climatological Survey [45]. In this investigation, 5 cm soil water content measurement from 116 stations was extracted and geolocated for comparison with the 1 km downscaled, AMSR-E, and NLDAS soil moisture values. The locations of the Oklahoma Mesonet stations are denoted by open yellow circles in Figure 15.

(6) *Little Washita Watershed Data.* The Little Washita Watershed is located in the southwestern portion of Oklahoma and 20 stations are located within a 25 km by 25 km region referred to as the Little Washita Micronet. The watershed soil moisture estimates from the most reliable stations with the closest time to the Aqua overpass time were extracted and then averaged for validation [84, 97]. The locations of these stations are denoted by red dots in Figure 15.

*4.2.3. Methodology*

(1) *Daily NDVI Interpolation.* Because the MODIS sensor is influenced by cloud cover, only biweekly MODIS radiances were used to calculate the NDVI products at different spatial resolutions. The daily NDVI varies in a near-sinusoidal fashion through all the days every year. To provide NDVI estimates on a daily basis, 13 daily NDVI values of each year (except 2002) and the NDVI obtaining day of the years between 2003–2011 were fitted by using the sinusoidal method as

$$\text{NDVI}_d = a_0 \sin\left(a_1 * D + a_2\right) + a_3, \tag{19}$$

where $a_0, a_1, a_2,$ and $a_3$ are the regression coefficients, $\text{NDVI}_d$ is the daily NDVI value, and $D$ is the day of the year.

TABLE 12: Sources of land surface data used in the downscaling of soil moisture and their spatial resolution and temporal repeat [61].

| Source | Data | Spatial resolution | Temporal repeat |
|---|---|---|---|
| NLDAS | Soil moisture content (0–10 cm layer, kg/m$^2$) | 1/8 degree (12.5 km) | Hourly |
| | Surface skin temperature (K) | 1/8 degree (12.5 km) | Hourly |
| AVHRR | Normalized difference vegetation index (NDVI) | 5 km | Daily |
| MODIS | Normalized difference vegetation index (NDVI) | 5 km | Biweekly |
| | Land surface temperature (K) | 1 km | Daily |
| AMSR-E | Soil moisture content (m$^3$/m$^3$) | 1/4 degree (25 km) | Daily |
| Mesonet | Surface soil water content (0–5 cm layer) | 116~117 stations | 5 minutes |
| Little Washita Watershed | Soil moisture measurement (0–5 cm layer) | 9 stations | Hourly |

This equation was applied to the 5 km NDVI data for all years to obtain daily 5 km NDVI values. Then, the interpolated results were resampled to 12.5 km to match up with NLDAS pixels. Similarly, the daily 1 km NDVI maps for the three months studied in this paper were generated using this method.

*(2) Thermal Inertia Theory.* The concept of thermal inertia is, namely, the resistance of a material to temperature change, which is indicated by the time-dependent variations in temperature during a full heating/cooling cycle. It is defined as the square root of the product of the material's bulk thermal conductivity and volumetric heat capacity, where the latter is the product of density and specific heat capacity:

$$I = \sqrt{k\rho c}, \qquad (20)$$

where $k$ is the coefficient of thermal conductivity, $\rho$ is density, and $c$ is the specific heat capacity.

An approximation to thermal inertia can be obtained from the amplitude of the diurnal temperature curve. The temperature of a material with low thermal inertia will change significantly during the day, while the temperature of a material with high thermal inertia will not change as much. The volumetric heat capacity depends on soil moisture. There have been many such attempts in the past since HCMM (Heat Capacity Mapping Mission), the first of a series of Applications Explorer Missions (AEM) [115]. The objective of the HCMM was to provide comprehensive, accurate, high-spatial-resolution thermal surveys of the surface of the earth to determine thermal inertia.

Therefore, it is our assertion that lower values of daily average soil moisture $\theta^{av}$ will correspond to higher value of daily temperature difference $\Delta T_s$ and vice versa. $\Delta T_s$ can be described as

$$\Delta T_s = T_{\max} - T_{\min}, \qquad (21)$$

where $T_{\max}$, $T_{\min}$ are the daily highest and lowest temperatures, respectively. The two local overpass times of MODIS approximately correspond to the highest and lowest temperatures.

*(3) Construction of the Downscaling Model.* The MODIS sensor provides two very important products: NDVI and surface temperature ($T_s$). In this study, these two variables were extracted for each 1 km MODIS pixel $(i, j)$ in the 1/4° gridded AMSR-E radiometer data. We denote these variables by NDVI $(i, j)$ and $T_s(i, j)$. The radiometer-derived soil moisture $\theta$ corresponds to the daytime 13:30 overpass $\theta^a$ and nighttime 1:30 overpass $\theta^p$ for the entire 1/4° pixel. The daytime and the nighttime overpass soil moisture values for each of the MODIS pixels are referred as $\theta^a(i, j)$ and $\theta^p(i, j)$. The average value of the pixel soil moisture is denoted by $\theta^{av}(i, j)$ which refers to the arithmetic mean of the soil moisture for the 1 km pixel for the morning and night overpass. These are depicted in Figure 17. The MODIS sensor on Aqua was used because it matched the time of AMSR-E soil moisture estimates.

There are three theories that motivate the pixel-based downscaling algorithm. First, we must consider that the soil moisture history of each pixel is unique with regard to precipitation, surface overflow, and runoff and can be summarized by the average soil moisture $\theta^{av}(i, j)$. Also, based on the thermal inertia theory, the thermal inertia and soil moisture depend on soil thermal conductivity, which for a wet pixel will show a smaller change while a dry pixel will show larger change in surface temperature [91]. Finally, vegetation biomass of each pixel will vary and can modulate the change of surface temperature, which is represented by the daily temperature difference $\Delta T_s$ [49, 116]. The comparison of the pixel sizes between the three datasets and the look-up curve building method is shown in Figure 17.

The key to the proposed disaggregation procedure is establishing the relationship between the change in surface temperature and the average soil moisture for the 1 km pixel.

Since the NLDAS-2 data are at 1/8° resolution while the AMSR-E data are at 1/4° resolution, 4 NLDAS-2 pixels are within each AMSR pixel. To construct the look-up curves, we used the NLDAS-2 output plotted separately for each month (12 plots for each 1/4° pixel). For each plot, we used data from all the months of the study period (i.e., the July plot will have data for the surface temperature change and the average soil moisture for all years from 1979 to present). The data for equal NDVI lines at increments of 0.3 in NDVI was subsequently organized. For example, during some months (e.g., January), the vegetation growth was limited and few NDVI curves in the Upper Midwest were constructed (maybe one corresponding to NDVI of 0 and another corresponding to NDVI of 0.3). On the other hand, during other periods,

TABLE 13: Comparison statistics between the 1 km downscaled, NLDAS, and AMSR-E soil moisture compared to the Oklahoma Mesonet for 6 relatively clear days. RMSE, bias, and standard deviations are in $(m^3/m^3)$ [61].

| Day | Dataset | $R^2$ | RMSE | Standard deviation | Bias |
|---|---|---|---|---|---|
| May 9, 2004 | 1 km downscaled | 0.223 | 0.119 | 0.043 | −0.112 |
| | AMSR-E | 0.050 | 0.129 | 0.053 | −0.114 |
| | NLDAS | 0.171 | 0.105 | 0.074 | −0.077 |
| May 22, 2004 | 1 km downscaled | 0.360 | 0.128 | 0.044 | −0.123 |
| | AMSR-E | 0.161 | 0.114 | 0.059 | −0.100 |
| | NLDAS | 0.264 | 0.108 | 0.077 | −0.084 |
| July 17, 2005 | 1 km downscaled | 0.020 | 0.168 | 0.055 | −0.155 |
| | AMSR-E | * | 0.160 | 0.063 | −0.136 |
| | NLDAS | 0.005 | 0.099 | 0.059 | −0.068 |
| July 21, 2005 | 1 km downscaled | 0.031 | 0.130 | 0.047 | −0.115 |
| | AMSR-E | 0.001 | 0.143 | 0.053 | −0.126 |
| | NLDAS | 0.002 | 0.106 | 0.056 | −0.081 |
| August 9, 2005 | 1 km downscaled | 0.010 | 0.167 | 0.050 | −0.155 |
| | AMSR-E | 0.005 | 0.146 | 0.050 | −0.130 |
| | NLDAS | 0.006 | 0.103 | 0.055 | −0.074 |
| August 18, 2005 | 1 km downscaled | 0.225 | 0.166 | 0.058 | −0.158 |
| | AMSR-E | 0.387 | 0.160 | 0.053 | −0.154 |
| | NLDAS | 0.114 | 0.106 | 0.064 | −0.075 |

* <0.001.

TABLE 14: Comparison statistics between the 1 km downscaled, NLDAS, and AMSR-E soil moisture compared to the Little Washita soil moisture observations for 6 relatively clear days. RMSE, bias, and standard deviations are in $(m^3/m^3)$ [61].

| Day | Dataset | Number of points | RMSE | Standard deviation | Bias |
|---|---|---|---|---|---|
| May 4, 2004 | 1 km downscaled | 8 | 0.043 | 0.015 | 0.009 |
| | AMSR-E | 3 | — | — | −0.019 |
| | NLDAS | 6 | 0.040 | 0.020 | 0.016 |
| May 6, 2004 | 1 km downscaled | 8 | 0.077 | 0.015 | 0.032 |
| | AMSR-E | 3 | — | — | 0.005 |
| | NLDAS | 6 | 0.055 | 0.017 | 0.035 |
| July 3, 2005 | 1 km downscaled | 6 | 0.066 | 0.070 | 0.006 |
| | AMSR-E | 3 | — | — | 0.010 |
| | NLDAS | 4 | 0.061 | 0.070 | 0.020 |
| July 17, 2005 | 1 km downscaled | 2 | 0.050 | 0.015 | 0.047 |
| | AMSR-E | 2 | — | — | 0.031 |
| | NLDAS | 2 | 0.057 | 0.015 | 0.056 |
| August 2, 2005 | 1 km downscaled | 7 | 0.031 | 0.019 | 0.024 |
| | AMSR-E | 3 | — | — | 0.027 |
| | NLDAS | 4 | 0.048 | 0.014 | 0.046 |
| August 4, 2005 | 1 km downscaled | 7 | 0.022 | * | 0.022 |
| | AMSR-E | 3 | — | — | 0.023 |
| | NLDAS | 4 | 0.048 | 0.010 | 0.047 |

* <0.001.

such as July in the Upper Midwest, rapid changes in NDVI due to crop growth could occur and many NDVI curves ranging from NDVI of 1.0 to 5.0 would be created. The curves were fitted to these points—930 points corresponding to the AMSR-E am overpass time (30 years of July data × 31 days) and 930 points corresponding to AMSR-E pm overpass time.

A simple linear regression model between the daily average soil moisture $\theta^{av}(i, j)$ and daily temperature difference $\Delta T_s(i, j)$ for all 30 years for each month was developed as follows:

$$\theta^{av}(i, j) = a_0 + a_1 \Delta T_s(i, j), \qquad (22)$$

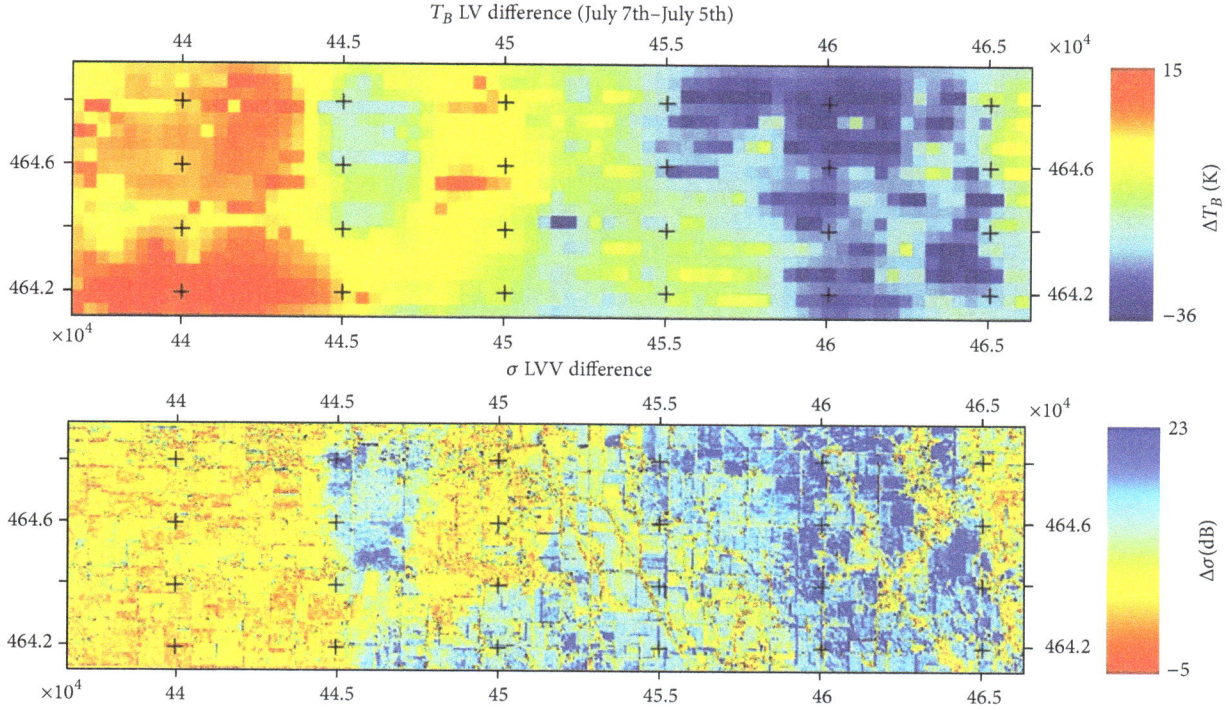

FIGURE 11: Difference images for change in PALS LV brightness temperatures at 400 m resolution and change in AIRSAR LVV backscatter at 30 m resolution for the period July 5 to July 7 (July 5–July 7). The spatial patterns corresponding to wetting or drying are strikingly consistent in both images indicating that the AIRSAR LVV channel is sensitive to near-surface soil moisture [55].

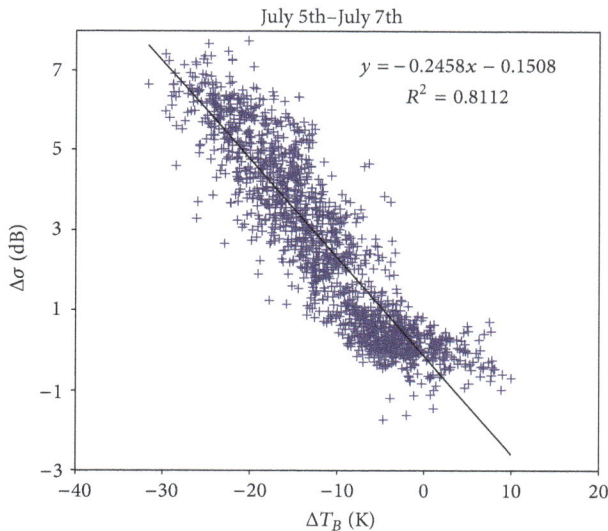

FIGURE 12: Chance in PALS L band V pol. brightness temperature plotted versus change in AIRSAR LVV channel backscattering coefficients. AIRSAR data has been aggregated to the PALS resolution of 400 m. Change is computed for the days July 5 to July 7 [55].

where $i$, $j$ represent the pixel location. $a_0$ and $a_1$ are regression model coefficients which correspond to several different NDVI intervals. The growing season between May–September was examined in this study and the NDVI was subdivided to three intervals: 0~0.3, 0.3~0.6, and 0.6~1. For each month, three NLDAS-based look-up curves of each pixel, which corresponded to the three NDVI intervals were built and the regression coefficients, were obtained.

(4) *Correction of 1 km Downscaled Soil Moisture.* On a daily basis, we used the curves corresponding to the NLDAS-2 data closest to the MODIS pixel to calculate the 1 km $\theta^{av}(i, j)$ from the $\Delta T_s(i, j)$ of each 1 km MODIS pixel. We then averaged $\theta^{av}(i, j)$ from all the 1 km MODIS pixels and compared the values to daily average AMSR-E soil moisture $(\Theta^a + \Theta^p)/2$ and then corrected each $\theta^{av}(i, j)$ with the difference between $(\Theta^a + \Theta^p)/2$ and $\theta^{av}(i, j)$. The corrected soil moisture $\theta^{avc}(i, j)$ is given by

$$\theta^{avc}(i, j) = \theta^{av}(i, j) + \left[ \left( \frac{\Theta^a + \Theta^p}{2} \right) - \frac{1}{N} \sum_{i,j} \theta^{av}(i, j) \right]. \tag{23}$$

We subsequently generated daily values of $\theta^{avc}(i, j)$ at 1 km. This satisfies the following conditions: (a) the average of the disaggregated soil moistures over the AMSR-E pixel is the same as that recorded by AMSR-E, (b) the MODIS 1 km vegetation modulates the distribution of the disaggregated soil moisture through its influence in the change in surface temperature to morning and evening averaged soil moisture computed by NLDAS, and (c) the 1 km scale changes in surface temperature reflect on soil moisture distribution as evidenced in the disaggregated soil moisture. In addition, this

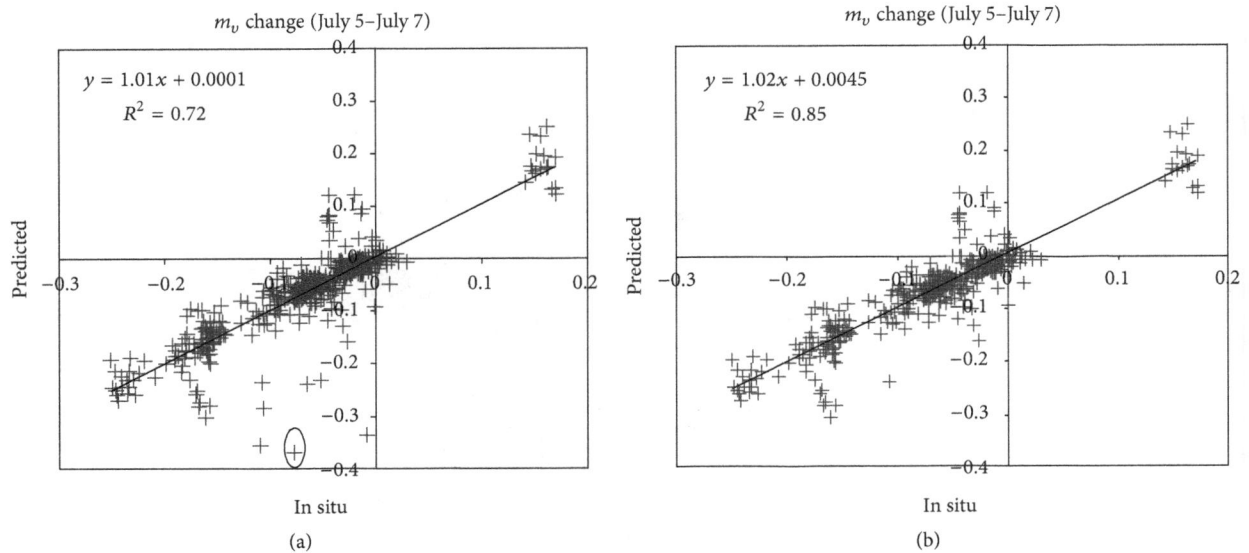

FIGURE 13: 100 m in situ soil moisture change ($x$-axis) compared with the 100 m resolution estimates of soil moisture change derived from the algorithm for the period July 5 to July 7. Plot (a) has all the points with RMSE = 0.046, and plot (b) has 7 outliers removed with RMSE = 0.032 [55].

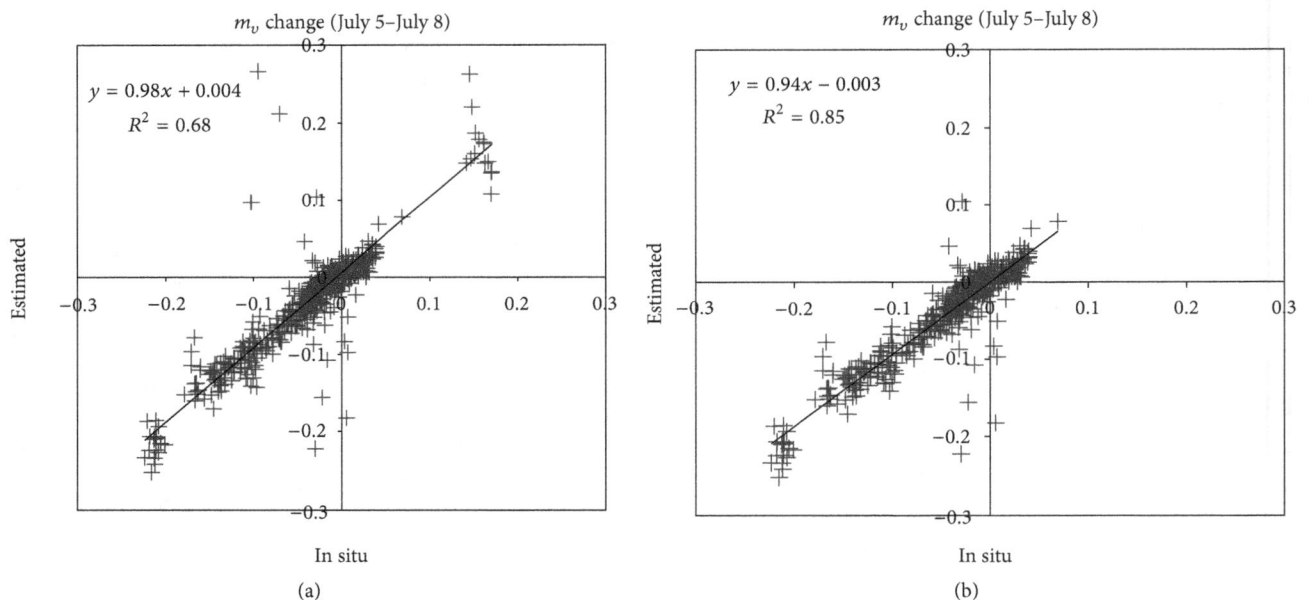

FIGURE 14: 100 m in situ soil moisture change ($x$-axis) compared with the 100 m resolution estimates of soil moisture change derived from the algorithm for the period July 5 to July 8. Plot (a) has all the points with RMSE = 0.046, and plot (b) has the data points from fields WC30 and WC31 removed resulting in an RMSE of 0.024 [55].

methodology can only be applied over areas with no cloud cover.

### 4.2.4. Results

(1) $\theta^{av} - \Delta T_s$ *Look-Up Curves.* Figure 18 shows the regression fitting results between NLDAS-derived daily temperature difference and daily average soil moisture of a pixel (latitude: 101.875°W~102°W; longitude: 35.125°N~35.625°N) for the three growing months: May, July, and August. It can be noticed that the daily average soil moisture values for all the months are in the range from 0.05~0.3. The points that correspond to each NDVI interval (0–0.3, 0.3–0.6, and 0.6–1.0) yield nearly parallel lines and the $R^2$ value of the fit for July are 0.54, 0.56, and 0.40, respectively, for each NDVI

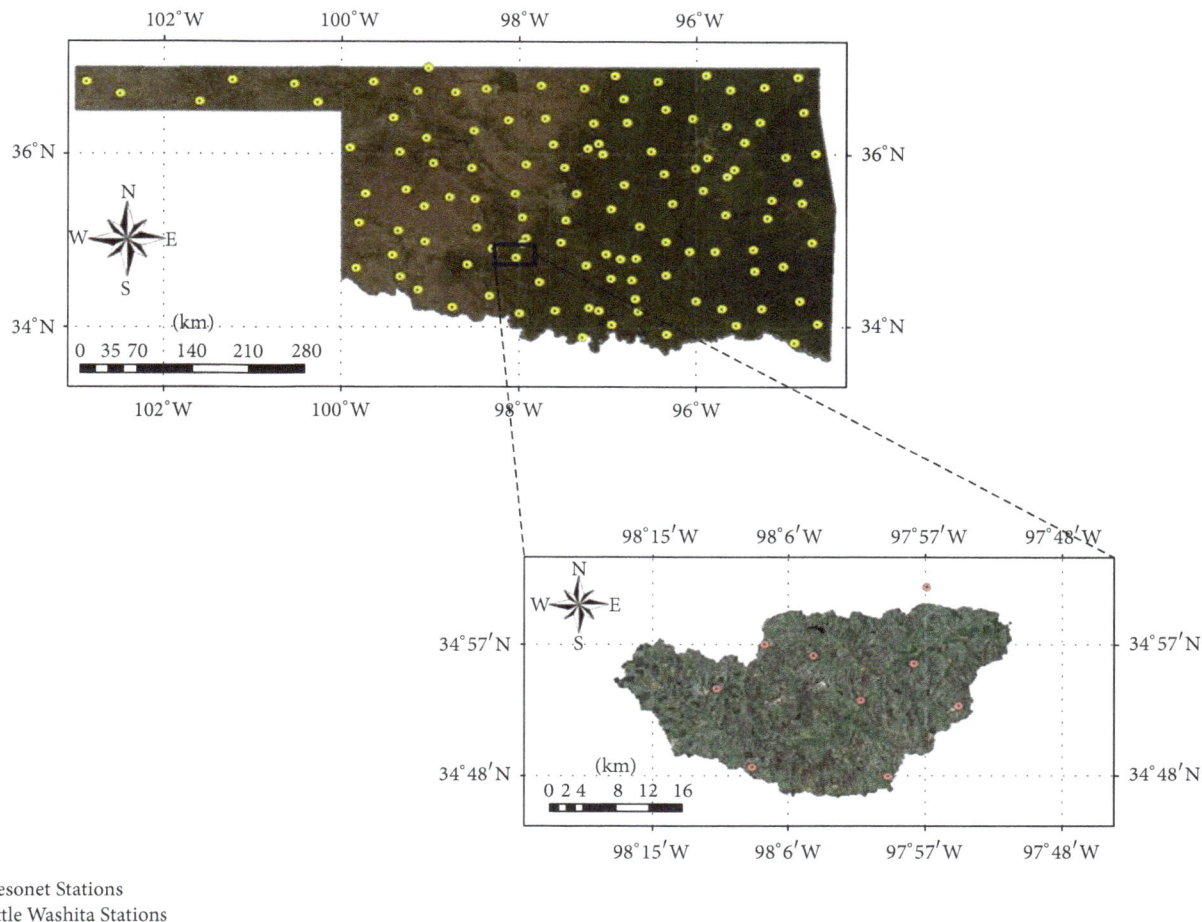

FIGURE 15: Imagery maps of study region of Oklahoma and the Little Washita Watershed. The locations of the Mesonet Stations are denoted in open yellow circles and the soil moisture sites for Little Washita are noted in red dots [61].

interval. Further, the daily average soil moisture has a negative relationship with the daily temperature change which is consistent with the assumptions that (a) the temperature change between morning and night is determined by the wetness of pixel and (b) the vegetation modulates the change of surface temperature and the pixel with higher vegetation is less sensitive to the temperature change.

(2) *1 km Downscaled Soil Moisture Analysis.* Three maps of daily 1 km downscaled soil moisture are shown as examples: May 22, 2004, July 17, 2005, and August 9, 2005 (Figure 19). The 1 km downscaled soil moistures in the lower Mideast part of Oklahoma are missing, which may be due to two reasons: (a) precipitation and heavy cloud cover often dominate this area especially in growing season which may result in missing MODIS surface temperature data and (b) this area corresponds to a gap between AMSR-E sensor swaths. These downscaled maps illustrate the pattern of soil moisture distribution whereby the soil moisture content gradually increases from west to east, which roughly corresponds to the NDVI variation in Oklahoma. In addition, the 1 km

downscaled soil moisture maps also exhibit similar patterns as those of AMSR-E and NLDAS.

For each of the days depicted in Figures 20(a)–20(c), the comparison of the 1 km downscaled soil moisture is shown on the top panel, the AMSR-E 1/4° soil moisture in the central panel, and the NLDAS 1/8° soil moisture in the bottom panel. The NLDAS soil moisture always has complete coverage because it is not impacted by cloud cover or missing data due to swaths not overlapping with each other. On May 22, 2004, a wet area existed in the northeast corner of Oklahoma and this was not captured by the AMSR-E or the 1 km downscaled soil moisture. Even so, in general the spatial patterns of the three estimates resemble each other for the May 22, 2004, case. On July 17, 2005, the western half of Oklahoma was very dry with soil moisture close to 0.02 with larger values in the east. The spatial structure shown by the 1 km soil moisture shows variability even in the dry western part of the state, which cannot be observed using the 1/4° AMSR-E estimates alone. In addition, the 1 km soil moisture captures the wet area in the east central part of the state. A similar west-to-east dry-

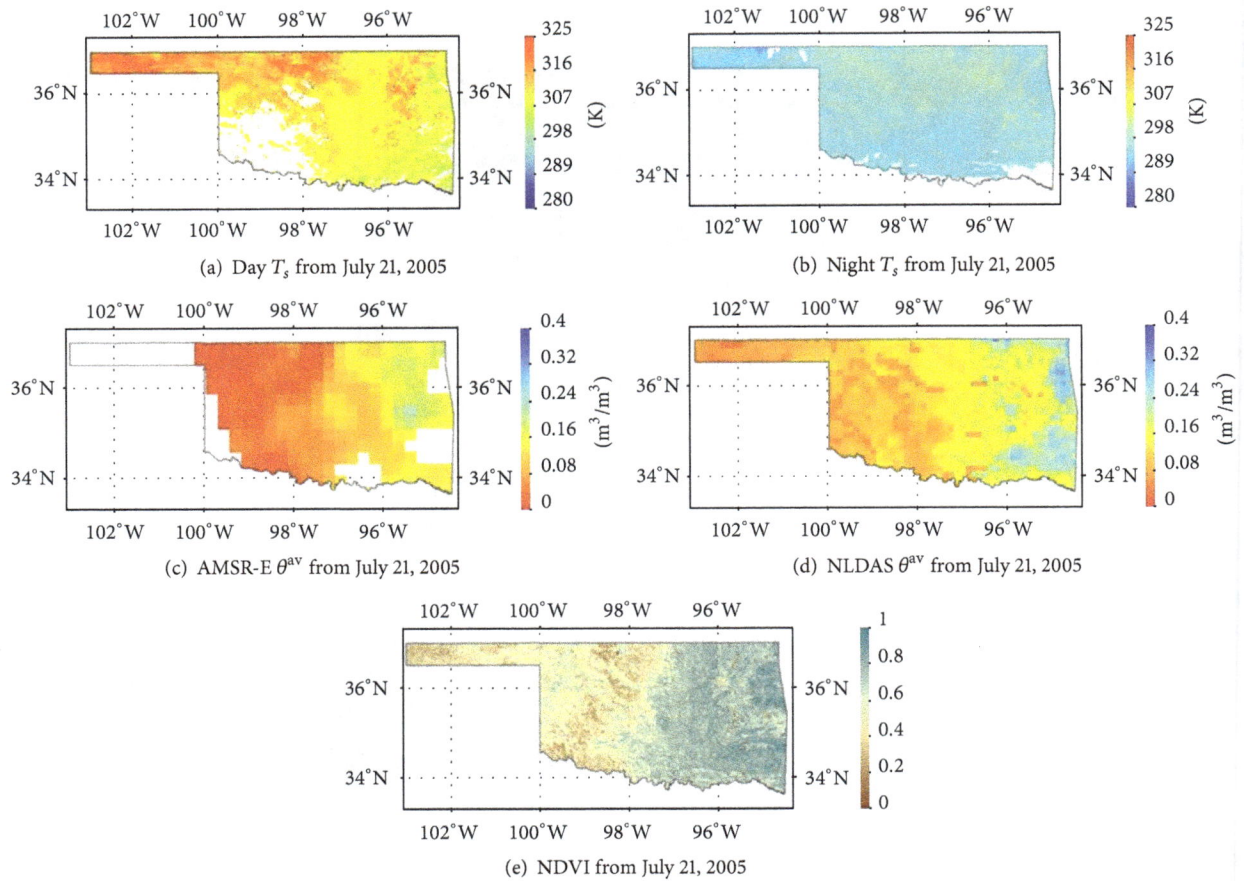

FIGURE 16: Maps of Variables used in the soil moisture downscaling algorithm from July 21, 2005, over Oklahoma (a) MODIS Aqua 1 km land surface temperature during the day. (b) MODIS Aqua 1 km land surface temperature at night. (c) 1/4° spatial resolution AMSR-E soil moisture. (d) 1/8° spatial resolution NLDAS soil moisture. (e) MODIS Aqua 1 km NDVI [61].

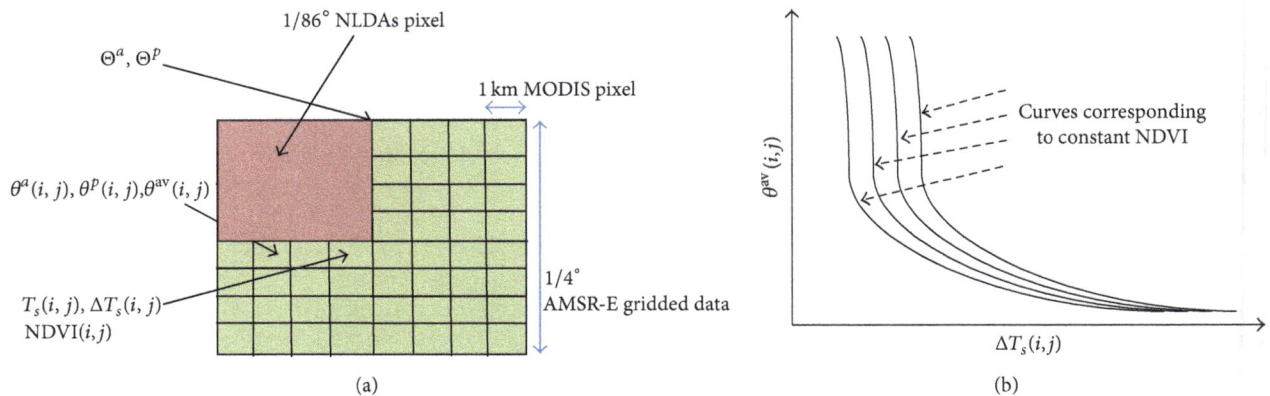

FIGURE 17: (a) Shows the various elements in the disaggregation procedure and (b) shows construction of the curves corresponding to constant NDVI between average soil moisture and change in surface temperature [61].

to-wet pattern was observed in all the estimates of the soil moisture for August 9, 2005.

(3) *Validation by Oklahoma Mesonet Soil Moisture Data.* Table 13 shows that the $R^2$ values of the 1 km downscaled soil moistures are better than AMSR-E and NLDAS, which range from 0.01~0.36, RMSE (range from 0.119~0.168 m³/m³), standard deviation (range from 0.043~0.058 m³/m³), and bias (ranges from −0.158~−0.112 m³/m³). The $R^2$ values for NLDAS ranges from 0.002 to 0.264 and those for AMSR-E

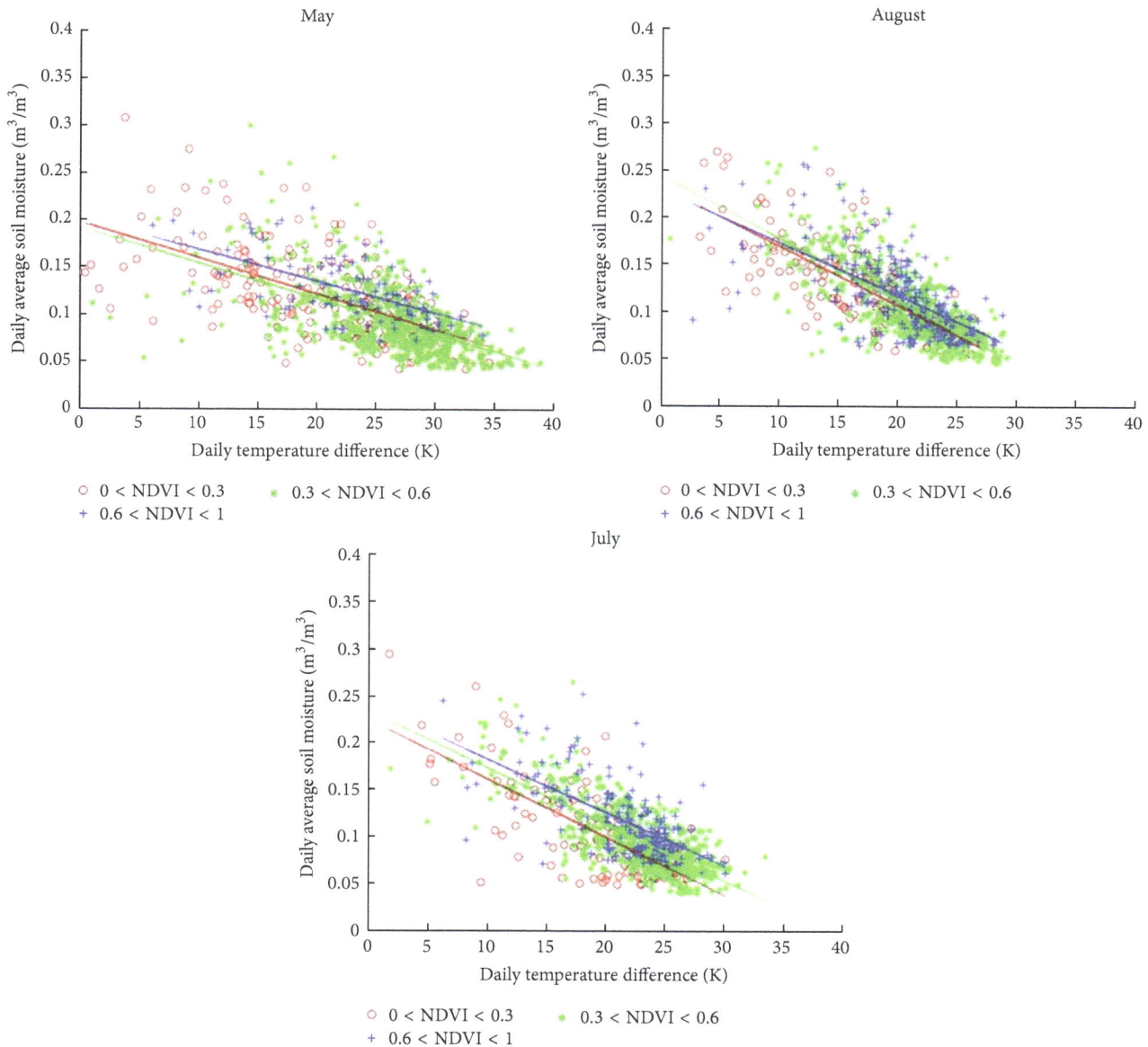

FIGURE 18: Daily temperature difference versus daily average soil moisture corresponding to latitude: 101.875°W~102°W; longitude: 35.125°N~35.625°N and different NDVI values for May, June, and July [61].

from <0.001 to 0.387. These values are considerably lower than those corresponding to the 1 km downscaled soil moistures. Besides, the RMSE values for NLDAS and AMSR-E range from 0.099~0.108 m³/m³ and 0.114~0.160 m³/m³, respectively, which are similar to the range of 1 km downscaled soil moistures. Comparing the correlation scatter plots of the three datasets versus the Mesonet data (Figures 20(a)–20(c)), it can be seen that the 1 km downscaled soil moisture has fewer points than AMSR-E and NLDAS. The reason for this is due to cloudiness and the lack of availability of the surface temperature. Thus it was not possible to downscale the AMSR-E soil moisture to produce the 1 km soil moisture.

It should be noted that the soil moisture values of all three datasets are biased, which indicate that the simulated soil moisture values are lower than the in situ Mesonet observation values. This could be attributed to the following. (a) The accuracy of AMSR-E soil moisture is limited and approximately 0.10. This methodology is based on preserving the 25 km mean soil moisture same as the AMSR-E soil moisture estimates. So, any overall day-to-day bias present in the AMSR-E soil moisture retrievals will be present in the disaggregated 1 km estimates. (b) The MODIS-retrieved daytime surface temperature is higher than the NLDAS-2 land surface model output, particularly during the growing season, which may be the cause of the daily temperature difference being greater than NLDAS and consequently the downscaled soil moisture being lower than NLDAS. (c) The Oklahoma Mesonet consists of Campbell Scientific

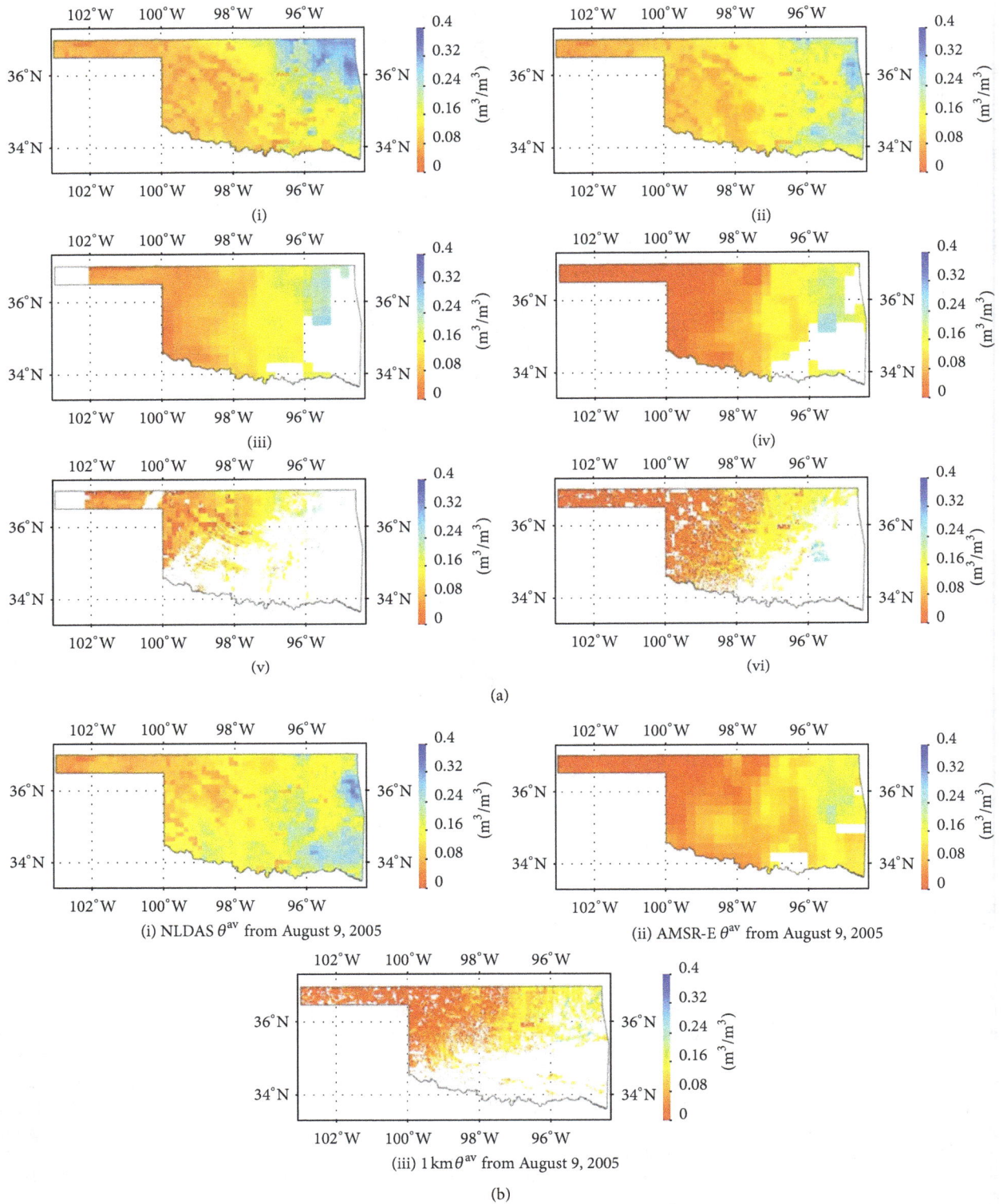

FIGURE 19: (a) Maps of the NLDAS, AMSR-E, and 1 km soil moisture (m³/m³) from May 22, 2004 ((i)–(iii)) and July 17, 2005 ((iv)–(vi)); (b) maps of the NLDAS, AMSR-E, and 1 km soil moisture (m³/m³) from August 9, 2005 [61].

229-L sensors which are soil matric potential sensors (then converted to volumetric soil moisture) which may have biases when compared to the gravimetric and neutron probe samples, of which RMSE is between 0.006~0.052 [45].

(4) *Validation Using Little Washita Watershed Soil Moisture Data.* Table 14 shows the statistical results for six days of Little Washita Watershed soil moisture comparisons with 1 km downscaled, AMSR-E, and NLDAS. AMSR-E statistics are not shown because these were less than three points of AMSR-E soil moisture over this period. Since the comparisons require high coverage of valid 1 km downscaled soil moisture values within the Little Washita region, the days used for the Little Washita comparisons are different from those used for the Mesonet comparisons. Two clear days from each month, (May 4, 2004, May 6, 2004, July 3, 2005, July 17, 2005, August 2, 2005, and August 4, 2005) were selected. In addition, the correlation plots including four clear days and the total 15 clear days of soil moisture in May in the Little Washita region versus the 1 km, AMSR-E, and NLDAS soil moisture are presented in Figures 21 and 22.

For the 1 km downscaled soil moisture, the RMSE values range from 0.022~0.077, while the standard deviations range from <0.001~0.07 and bias ranges from −0.047~0.032. These statistical results demonstrate that the 1 km downscaled soil moisture values are equivalent to the NLDAS and better than the accuracy of AMSR-E soil moisture values. In addition, the downscaled soil moisture values have a higher spatial resolution than the NLDAS soil moisture values since they are at 1 km as opposed to 12.5 km for NLDAS. This will be particularly important in small watershed studies when one pixel of NLDAS might cover the whole catchment and not provide information on spatial variability. Moreover, the results also indicate that the 1 km downscaled soil moisture has a better agreement with Little Washita than Mesonet, although the variables of July 17, 2005, do not perform as well as the other days.

By analyzing the single days correlation plots for May (Figures 21–22), it can be seen that the 1 km downscaled soil moistures match up well with the AMSR-E and NLDAS soil moistures. The correlation for the May 2004 1 km downscaled results versus the Little Washita soil moisture was $R^2 = 0.4$ with an RMSE of 0.018. The statistical results are also better than the comparison with Mesonet soil moistures. A small catchment such as the Little Washita might be covered by only a single pixel of AMSR-E pixel or a few NLDAS pixels; the downscaled soil moisture offers the distinct advantage of having many more pixels (900 1 km pixels versus 6 NLDAS pixels for Little Washita Watershed) and provides spatial variability information.

The frequency distribution of the differences between Little Washita soil moisture values versus 1 km downscaled, AMSR-E, and NLDAS soil moisture values are presented in Figures 23(a)–23(c), respectively. For the Little Washita soil moisture values within a particular AMSR-E or NLDAS are averaged; their correlation plots have fewer points than the 1 km downscaled soil moisture values. The results indicate that the differences between the 1 km downscaled soil

moistures and Little Washita results generally distribute range from −0.05–0.1 and the differences of downscaled soil moisture is around 7%–36% of the total, which is similar to the NLDAS.

## 5. Conclusions and Discussions

Since this paper has discussed three major studies the conclusions from each of them will be highlighted separately in the subsections below. In Section 5.1 the results from the field experiment study of SMEX02 conducted in Ames, Iowa, in summer 2002 will be discussed. The major findings from the study on disaggregation of passive microwave data using active radar observations will be dealt with in Sections 5.2 and 5.3 will explain the major findings from the use of visible near infrared data in disaggregation of AMSR-E data over Oklahoma.

*5.1. Use of PALS in the SMEX02 Field Experiment.* This section applied existing algorithms to previously untested conditions of vegetation cover. The sensitivity of PALS radiometer and radar to soil moisture in these conditions has been studied. Soil moisture retrievals were performed using statistical as well as physical modeling techniques. The root mean square error associated with soil moisture retrieval by statistical regression was around 0.051 g/g while soil moisture retrieval using the zero order incoherent radiative transfer model gave retrieval errors of around 0.036 g/g gravimetric soil moisture. Lower retrieval error associated with using multiple channels as compared to a single channel in soil moisture retrieval by statistical regression has been demonstrated. Existing algorithms for passive radiative transfer were shown to perform satisfactorily even though the vegetation cover was considerable. Vegetation plays a role in reducing the sensitivity of the PALS radar and radiometer and thereby increasing soil moisture retrieval error has been analyzed.

We observed good agreement between the radar- and radiometer-predicted soil moisture. The retrieved values of soil moisture tend to be overestimated which demonstrates the limitations of the change detection algorithm employed in this section, wherein the assumption that vegetation and roughness effects are present only as a bias in PALS active and passive observations are shown to be sources of error.

*5.2. Use of Radar for Disaggregation of Passive Microwave Soil Moisture.* In this section a simple algorithm for estimation of change in soil moisture at the spatial resolution of radar using low-resolution estimate of soil moisture from radiometer and copolarized backscattering coefficients has been proposed. The subpixel scale surface roughness variability does not play an important role in radar sensitivity to soil moisture at the L-band. Observations from combined radar/radiometer data from SMEX02, results from previous studies, and IEM simulations have been presented in support of this argument. Radar sensitivity to soil moisture at the L-band has been assumed to be a function of vegetation opacity only and further a simple soil moisture change estimation algorithm has been developed. Application of the algorithm to data

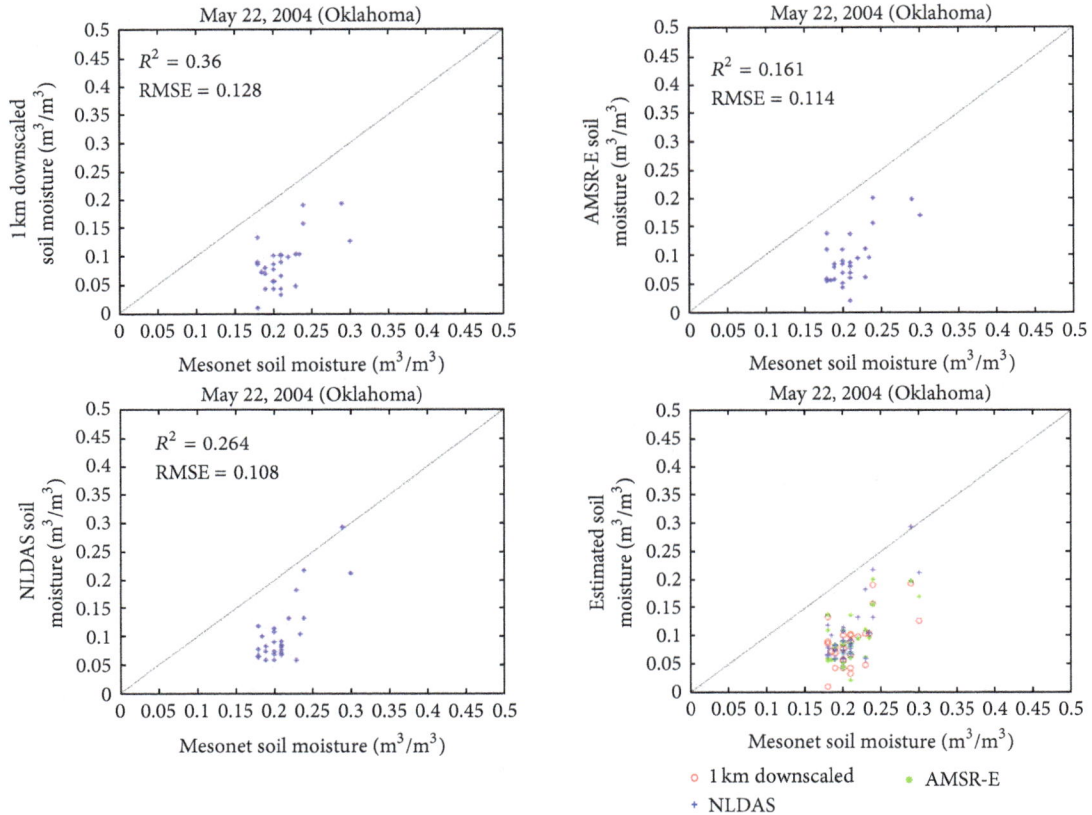

(a)

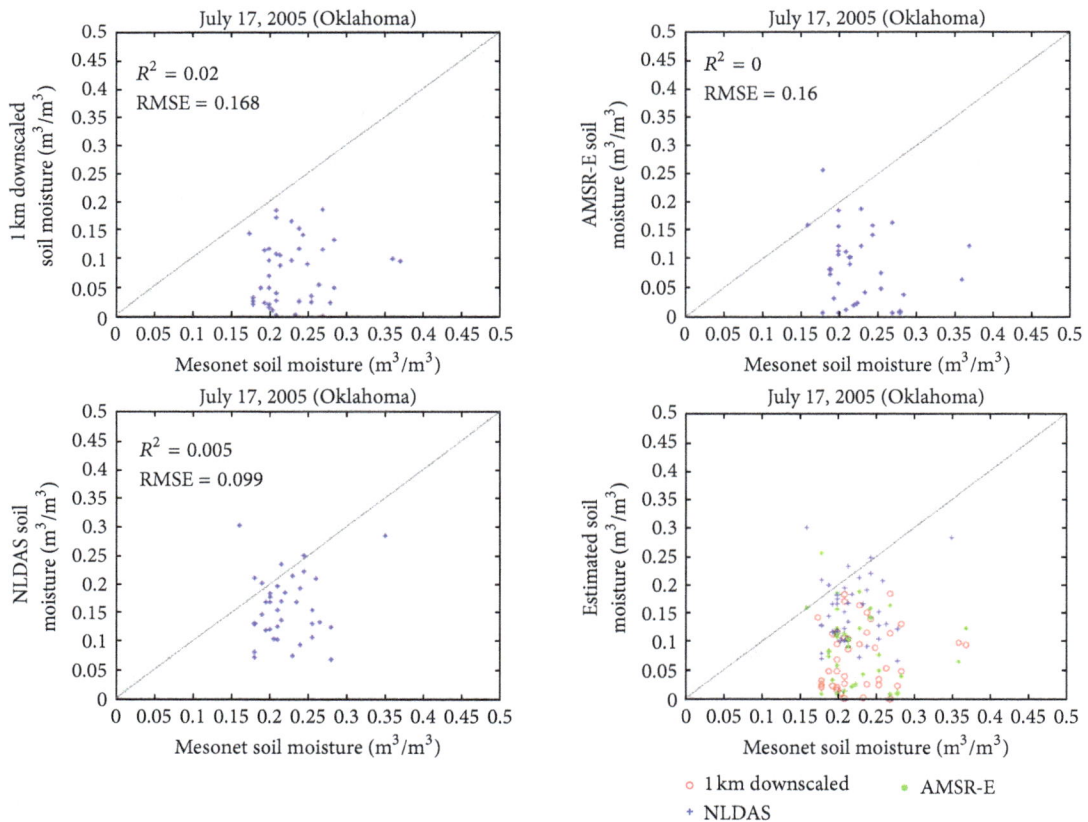

(b)

FIGURE 20: Continued.

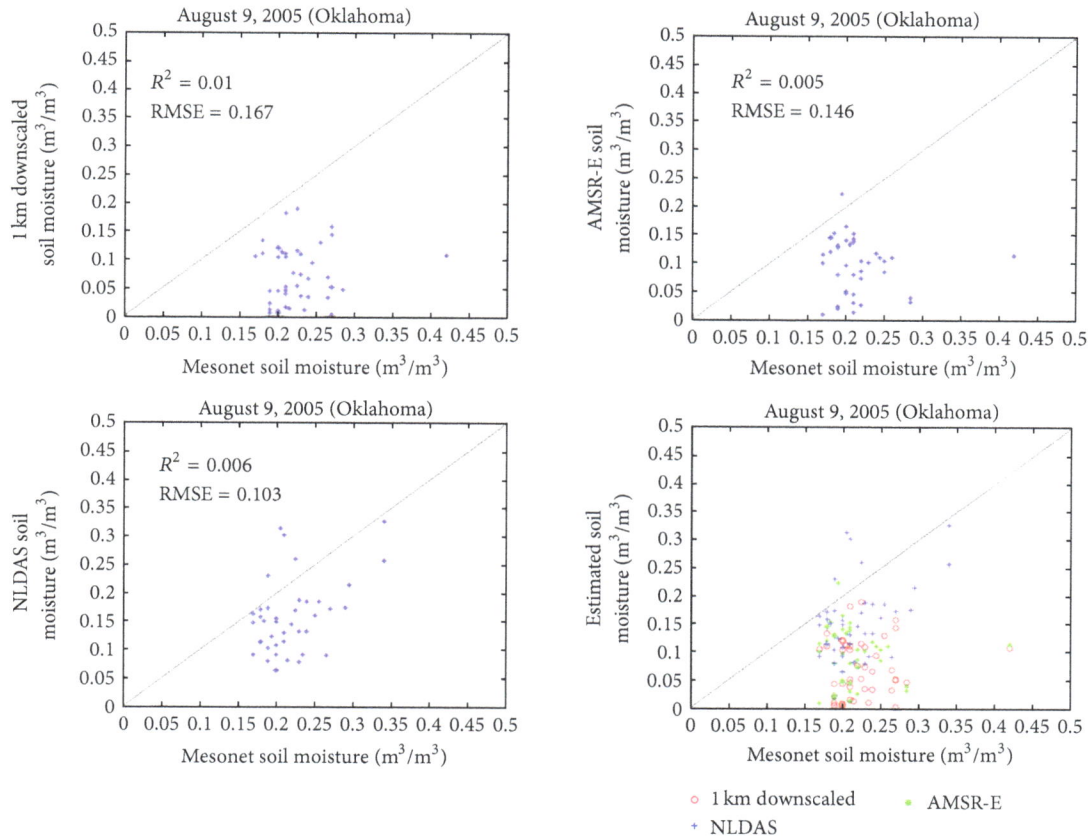

(c)

FIGURE 20: ((a)–(c)) Scatter plots of the 1 km, AMSR-E, and NLDAS soil moisture versus the Oklahoma Mesonet soil moisture observations for May 22, 2004, July 17, 2005, and August 9, 2005 [61].

obtained from the SMEX02 experiments results provided good results with root mean square error of prediction of 0.03 and 0.02 (error for estimated versus measured volumetric soil moisture, both at 100 m resolution) for two periods—July 5 to July 7 and July 7 to July 8. The $R^2$ value for both cases was 0.85.

The originality of the approach presented here lies in using radiometer to estimate soil moisture change at a lower spatial resolution (but with lower ancillary data requirements as compared to radar estimation of soil moisture) and then using change in radar backscatter to estimate the change in soil moisture at higher spatial resolution. The estimated change in soil moisture is a hydrologic variable of significant interest. A simple calibration methodology based on least error between modeled and measured soil moisture change values will allow a better estimation of parameters such as the hydraulic conductivity of soil layers. Mattikalli et al. reported that the 2-day soil moisture change was closely related to the saturated hydraulic conductivity ($K_{sat}$) profile [71]. It will be possible to relate $K_{sat}$ from the radar/radiometer-algorithm-derived change in soil moisture with the added advantage of higher spatial resolution that will lead to more accurate estimation of water and energy fluxes. Future studies on the direction of radar/radiometer combination will have to aim at a better parameterization of the sensitivity relationship

with vegetation opacity allowing the effect of vegetation heterogeneity to be addressed through the $f(\tau)$ parameter in the algorithm. Du et al., 2000, [117] have explored the behavior of soybean and grass canopies over medium rough surfaces. Their results indicate that it should be possible to develop simple parametric relationships between vegetation opacity and relative sensitivity. Estimation of vegetation opacity has been done in the past using optical sensors by estimation of vegetation water content using a proxy such as NDWI [83] and then using the empirical parameter $b$ to relate vegetation opacity and water content [118]. This simple parameterization however has been shown to be too simple for canopies such as corn that are electrically thick scatterers and further research is required to find relationships between vegetation parameters and relative sensitivity over canopies such as corn. The approach presented in the paper should be applicable to data from the Hydros mission, at least over areas of low vegetation water content variability. The algorithm will have to be modified to account for vegetation variability within the radiometer footprint using the $f(\tau)$ parameter before it can be made operational.

5.3. Use of Near Infrared and Visible Data for Disaggregation of AMSR-E Soil Moisture. A soil moisture downscaling

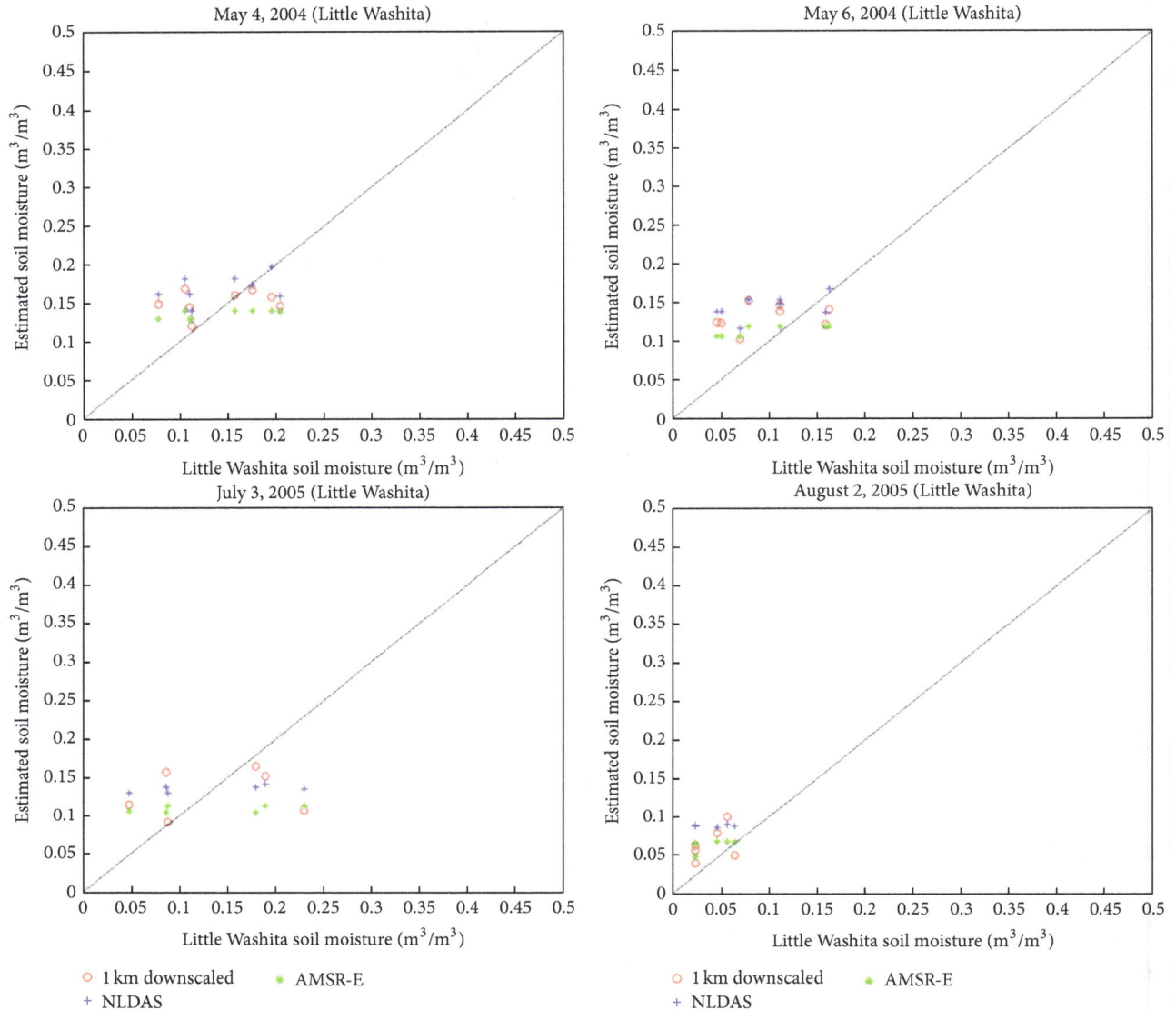

FIGURE 21: Scatter plot of the 1 km, AMSR-E, and NLDAS soil moisture versus the Oklahoma Mesonet soil moisture observations for May 4, 2004, May 6, 2004, July 3, 2005, and August 2, 2005 [61].

algorithm based on NLDAS-derived look-up curves that related daily surface temperature change and average daily soil moisture was developed. This algorithm was applied using MODIS products of clear days during crop growing seasons (May, July, and August of 2004~2005) in Oklahoma. Two sets of validation data, namely, Oklahoma Mesonet and Little Washita soil moisture observations, have been used to compare with the three estimates: 1 km downscaled soil moisture values, AMSR-E soil moisture values, and NLDAS soil moisture values. Statistical analyses and plots were used for analyzing accuracy of the downscaling algorithm.

Our results indicate the following. (a) The look-up regression curves support our assumption that the surface temperature change depends on the wetness of the land surface and that the vegetation modulates this relationship. (b) The 1 km

downscaled maps provide details on the soil moisture spatial distribution patterns in Oklahoma as opposed to the AMSR-E maps. They also compare well with Oklahoma Mesonet soil moisture values. (c) The comparisons of all three datasets with the Oklahoma Mesonet soil moisture observations show an $R^2$ for the 1 km downscaled soil moisture ranging from 0.01~0.36, RMSE values ranging from 0.119~0.168 m$^3$/m$^3$, and standard deviation ranging from 0.043~0.058 m$^3$/m$^3$. The statistical comparisons show that the 1 km downscaled results correlate better to the Oklahoma Mesonet soil moisture observations than do AMSR-E and NLDAS soil moistures; (d) The statistical results for the 1 km downscaled soil moisture comparing with Little Washita Watershed soil moisture show good accuracy for the downscaling algorithm. Single-day comparisons showed RMSE values from 0.022~0.077 m$^3$/m$^3$,

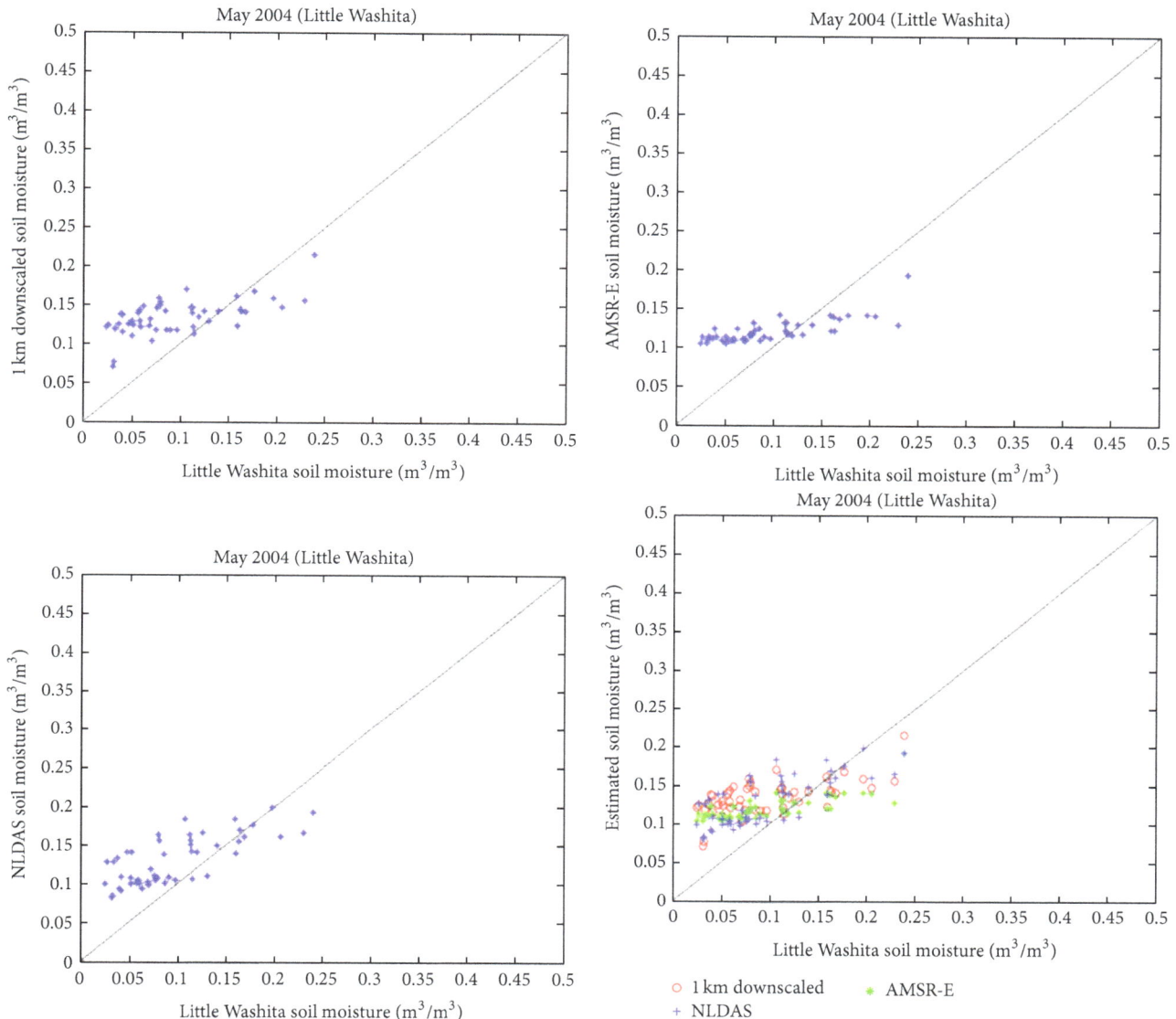

FIGURE 22: Scatter plot comparing AMSR-E, NLDAS, and 1 km downscaled soil moisture to the Little Washita soil moisture observations for all days of May 2004 [61].

standard deviations from <0.001~0.07 m³/m³, and bias values from −0.047~0.032 m³/m³. Taken as a whole for all of the clear days in May 2004, the $R^2$ and RMSE are 0.4 and 0.018 m³/m³, respectively. The errors between estimated and ground data used for validation for 1 km downscaled data are generally from −0.07~0.1 m³/m³, so the downscaled soil moisture has an error of 7%~36% of the total.

However, there are still several limitations that exist in this algorithm: (a) the MODIS temperature and NDVI products are often influenced by the cloud coverage; therefore this method is not an all-weather algorithm for downscaling, (b) the accuracy of the AMSR-E and NLDAS soil moisture determines the accuracy of the 1 km downscaled soil moisture, and (c) Only vegetation and temperature were used in developing this downscaling algorithm, and the high spatial resolution data of these variables would be required. This methodology

is based on preserving the 25 km mean soil moisture same as the AMSR-E soil moisture estimates. So, any overall day-to-day bias present in the AMSR-E soil moisture retrievals will be present in the disaggregated 1 km estimates. But if we compare our results with those reported in the literature and presented in Section 1 and Table 1 we note that this analysis included a large area (the entire state of Oklahoma) and a moderate period of time as opposed to some previous studies, which only included shorter-term field experiments or smaller catchments. However, our correlations values for $R^2$ do compare well with the values noted in these studies [50] of 0.14–0.21; the RMSE shows similar favorable comparisons.

Future work that will combine this approach with our previous active-passive downscaling approach [55] is being pursued and this will offer much better progress in downscaling of soil moisture for catchment studies.

May 2004 (1 km downscaled)

May 2004 (AMSR-E)

May 2004 (NLDAS)

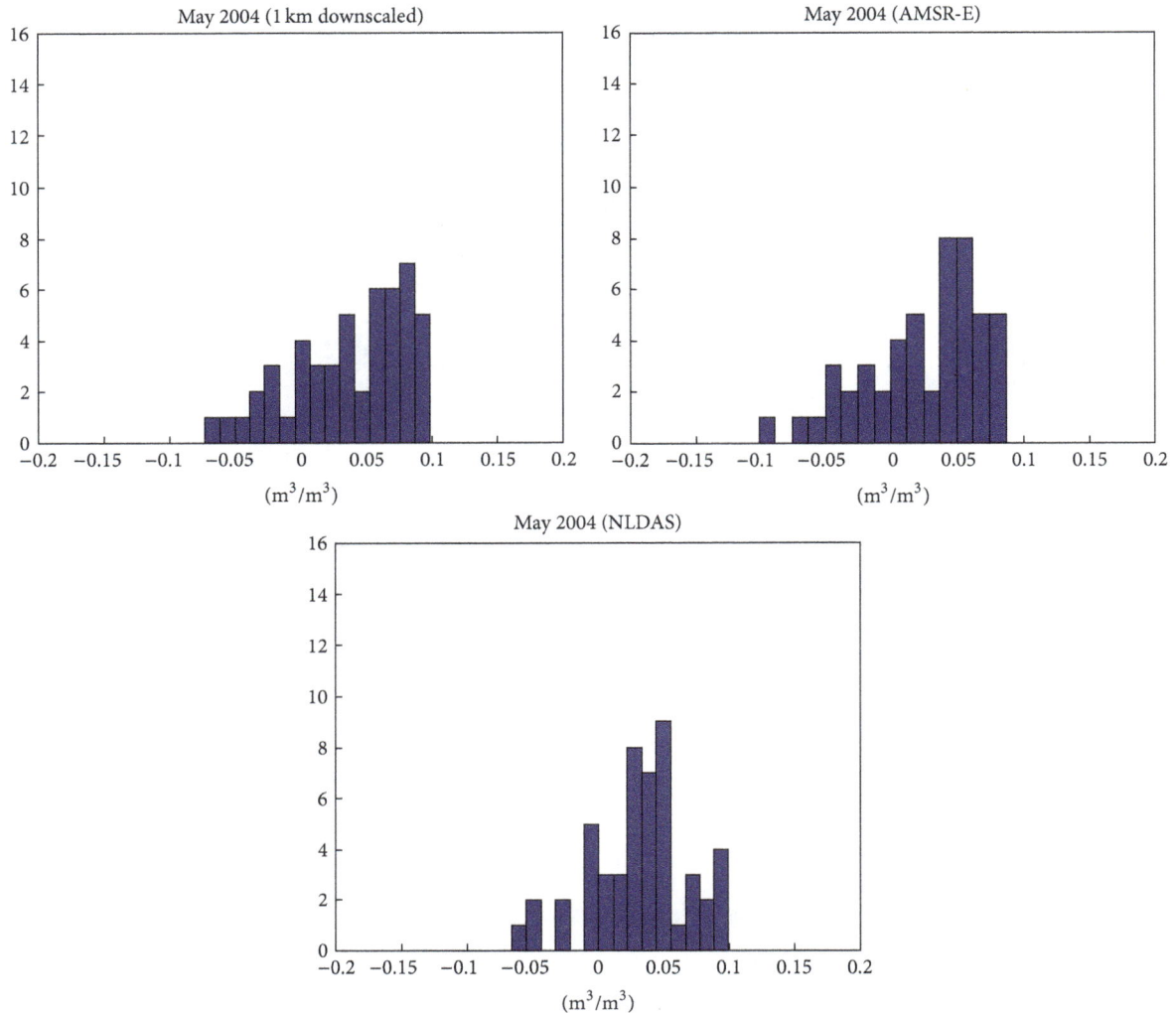

FIGURE 23: Frequency distribution of difference between the 1 km, AMSR-E, and NLDAS estimates of soil moisture and the Little Washita soil moisture observations for May 2004 [61].

## References

[1] K. L. Brubaker and D. Entekhabi, "Analysis of feedback mechanisms in land-atmosphere interaction," *Water Resources Research*, vol. 32, no. 5, pp. 1343–1357, 1996.

[2] T. Delworth and S. Manabe, "The influence of soil wetness on near-surface atmospheric variability," *Journal of Climate*, vol. 2, pp. 1447–1462, 1989.

[3] R. A. Pielke, "Influence of the spatial distribution of vegetation and soils on the prediction of cumulus convective rainfall," *Reviews of Geophysics*, vol. 39, no. 2, pp. 151–177, 2001.

[4] J. Shukla and Y. Mintz, "Influence of land-surface evapotranspiration on the earth's climate," *Science*, vol. 215, no. 4539, pp. 1498–1501, 1982.

[5] Jy-Tai Chang and P. J. Wetzel, "Effects of spatial variations of soil moisture and vegetation on the evolution of a prestorm environment: a numerical case study," *Monthly Weather Review*, vol. 119, no. 6, pp. 1368–1390, 1991.

[6] E. Engman, "Soil moisture, the hydrologic interface between surface and ground waters," in *Remote Sensing and Geographic Information Systems for Design and Operation of Water*

*Resources Systems*, International Association of Hydrological Sciences no. 242, 1997.

[7] J. Mahfouf, E. Richard, and P. Mascarat, "The influence of soil and vegetation on Mesoscale circulations," *Journal of Climate and Applied Climatology*, vol. 26, pp. 1483–1495, 1987.

[8] J. Laccini, T. Carlson, and T. Warner, "Sensitivity of the Great Plains severe storm environment to soil moisture distribution," *Monthly Weather Review*, vol. 115, pp. 2660–2673, 1987.

[9] D. Zhang and R. A. Anthes, "A high-resolution model of the planetary boundary layer—sensitivity tests and comparisons with SESAME-79 data," *Journal of Applied Meteorology*, vol. 21, no. 11, pp. 1594–1609, 1982.

[10] P. R. Houser, W. J. Shuttleworth, J. S. Famiglietti, H. V. Gupta, K. H. Syed, and D. C. Goodrich, "Integration of soil moisture remote sensing and hydrologic modeling using data assimilation," *Water Resources Research*, vol. 34, no. 12, pp. 3405–3420, 1998.

[11] S. Zhang, H. Li, W. Zhang, C. Qiu, and X. Li, "Estimating the soil moisture profile by assimilating near-surface observations with the ensemble Kalman Filter (EnKF)," *Advances in Atmospheric Sciences*, vol. 22, no. 6, pp. 936–945, 2005.

[12] W. Ni-Meister, P. R. Houser, and J. P. Walker, "Soil moisture initialization for climate prediction: assimilation of scanning multifrequency microwave radiometer soil moisture data into a land surface model," *Journal of Geophysical Research D*, vol. 111, no. 20, Article ID D20102, 2006.

[13] J. P. Hollinger, J. L. Pierce, and G. A. Poe, "SSM/I instrument evaluation," *IEEE Transactions on Geoscience and Remote Sensing*, vol. 28, no. 5, pp. 781–790, 1990.

[14] E. Njoku, B. Rague, and K. Fleming, *The Nimbus-7 SMMR Pathfinder Brightness Data Set*, Jet Propulsion Lab, Pasadena, Calif, USA, 1998.

[15] S. Paloscia, G. Macelloni, E. Santi, and T. Koike, "A multifrequency algorithm for the retrieval of soil moisture on a large scale using microwave data from SMMR and SSM/I satellites," *IEEE Transactions on Geoscience and Remote Sensing*, vol. 39, no. 8, pp. 1655–1661, 2001.

[16] A. Guha and V. Lakshmi, "Sensitivity, spatial heterogeneity, and scaling of C-band microwave brightness temperatures for land hydrology studies," *IEEE Transactions on Geoscience and Remote Sensing*, vol. 40, no. 12, pp. 2626–2635, 2002.

[17] A. Guha and V. Lakshmi, "Use of the Scanning Multichannel Microwave Radiometer (SMMR) to retrieve soil moisture and surface temperature over the central United States," *IEEE Transactions on Geoscience and Remote Sensing*, vol. 42, no. 7, pp. 1482–1494, 2004.

[18] J. Wen, T. J. Jackson, R. Bindlish, A. Y. Hsu, and Z. B. Su, "Retrieval of soil moisture and vegetation water content using SSM/I data over a corn and soybean region," *Journal of Hydrometeorology*, vol. 6, no. 6, pp. 854–863, 2005.

[19] V. Lakshmi, "Special sensor microwave imager data in field experiments: FIFE-1987," *International Journal of Remote Sensing*, vol. 19, no. 3, pp. 481–505, 1998.

[20] V. Lakshmi, E. F. Wood, and B. J. Choudhury, "A soil-canopy-atmosphere model for use in satellite microwave remote sensing," *Journal of Geophysical Research D*, vol. 102, no. 6, pp. 6911–6927, 1997.

[21] V. Lakshmi, E. F. Wood, and B. J. Choudhury, "Investigation of effect of heterogeneities in vegetation and rainfall on simulated SSM/I brightness temperatures," *International Journal of Remote Sensing*, vol. 18, no. 13, pp. 2763–2784, 1997.

[22] V. Lakshmi, E. F. Wood, and B. J. Choudhury, "Evaluation of Special Sensor Microwave/Imager satellite data for regional soil moisture estimation over the Red River basin," *Journal of Applied Meteorology*, vol. 36, no. 10, pp. 1309–1328, 1997.

[23] L. Li, P. Gaiser, T. J. Jackson, R. Bindlish, and J. Du, "WindSat soil moisture algorithm and validation," in *Proceedings of the International Geoscience and Remote Sensing Symposium*, pp. 1188–1191, Barcelona, Spain, July 2007.

[24] H. Gao, E. F. Wood, T. J. Jackson, M. Drusch, and R. Bindlish, "Using TRMM/TMI to retrieve surface soil moisture over the southern United States from 1998 to 2002," *Journal of Hydrometeorology*, vol. 7, no. 1, pp. 23–38, 2006.

[25] M. Owe, R. de Jeu, and T. Holmes, "Multisensor historical climatology of satellite-derived global land surface moisture," *Journal of Geophysical Research F*, vol. 113, no. 1, Article ID F01002, 2008.

[26] W. Wagner, V. Naeimi, K. Scipal, R. Jeu, and J. Martínez-Fernández, "Soil moisture from operational meteorological satellites," *Hydrogeology Journal*, vol. 15, no. 1, pp. 121–131, 2007.

[27] C. Rüdiger, J. C. Calvet, C. Gruhier, T. R. H. Holmes, R. A. M. de Jeu, and W. Wagner, "An intercomparison of ERS-Scat and AMSR-E soil moisture observations with model simulations over France," *Journal of Hydrometeorology*, vol. 10, no. 2, pp. 431–447, 2009.

[28] C. S. Draper, J. P. Walker, P. J. Steinle, R. A. M. de Jeu, and T. R. H. Holmes, "An evaluation of AMSR-E derived soil moisture over Australia," *Remote Sensing of Environment*, vol. 113, no. 4, pp. 703–710, 2009.

[29] R. A. M. Jeu, W. Wagner, T. R. H. Holmes, A. J. Dolman, N. C. Giesen, and J. Friesen, "Global soil moisture patterns observed by space borne microwave radiometers and scatterometers," *Surveys in Geophysics*, vol. 29, no. 4-5, pp. 399–420, 2008.

[30] K. Imaoka, M. Kachi, H. Fujii et al., "Global change observation mission (GCOM) for monitoring carbon, water cycles, and climate change," *Proceedings of the IEEE*, vol. 98, no. 5, pp. 717–734, 2010.

[31] E. G. Njoku and D. Entekhabi, "Passive microwave remote sensing of soil moisture," *Journal of Hydrology*, vol. 184, no. 1-2, pp. 101–129, 1996.

[32] F. T. Ulaby, P. C. Dubois, and J. Van Zyl, "Radar mapping of surface soil moisture," *Journal of Hydrology*, vol. 184, no. 1-2, pp. 57–84, 1996.

[33] T. J. Schmugge, W. P. Kustas, J. C. Ritchie, T. J. Jackson, and A. Rango, "Remote sensing in hydrology," *Advances in Water Resources*, vol. 25, no. 8-12, pp. 1367–1385, 2002.

[34] F. T. Ulaby, M. Razani, and M. C. Dobson, "Effect of vegetation cover on the microwave radiometric sensitivity to soil moisture," *IEEE Transactions on Geoscience and Remote Sensing*, vol. 21, no. 1, pp. 51–61, 1983.

[35] B. J. Choudhury, T. J. Schmugge, R. W. Newton, and A. Chang, "Effect of surface roughness on the microwave emission from soils," *Journal of Geophysical Research*, vol. 89, no. 9, pp. 5699–5706, 1979.

[36] F. Ulaby, R. K. Moore, and A. K. Fung, *Microwave Remote Sensing: Active and Passive, Vol. III—Volume Scattering and Emission Theory, Advanced Systems and Applications*, Artech House, Dedham, Mass, USA, 1986.

[37] J. R. Wang, J. C. Shiue, T. J. Schmugge, and E. T. Engman, "The L-band PBMR measurements of surface soil moisture in FIFE," *IEEE Transactions on Geoscience and Remote Sensing*, vol. 28, no. 5, pp. 906–914, 1990.

[38] T. Schmugge, T. J. Jackson, W. P. Kustas, and J. R. Wang, "Passive microwave remote sensing of soil moisture: results from HAPEX, FIFE and MONSOON' 90," *ISPRS Journal of Photogrammetry and Remote Sensing*, vol. 47, no. 2-3, pp. 127–143, 1992.

[39] P. E. O'Neill, N. S. Chauhan, and T. J. Jackson, "Use of active and passive microwave remote sensing for soil moisture estimation through corn," *International Journal of Remote Sensing*, vol. 17, no. 10, pp. 1851–1865, 1996.

[40] J. D. Bolten, V. Lakshmi, and E. G. Njoku, "A passive-active retrieval of soil moisture for the Southern great plains 1999 experiment," *IEEE Transactions on Geoscience and Remote Sensing*, vol. 41, no. 12, pp. 2792–2801, 2003.

[41] U. Narayan, V. Lakshmi, and E. G. Njoku, "Retrieval of soil moisture from passive and active L/S band sensor (PALS) observations during the Soil Moisture Experiment in 2002 (SMEX02)," *Remote Sensing of Environment*, vol. 92, no. 4, pp. 483–496, 2004.

[42] E. G. Njoku, W. J. Wilson, S. H. Yueh et al., "Observations of soil moisture using a passive and active low-frequency microwave airborne sensor during SGP99," *IEEE Transactions on Geoscience and Remote Sensing*, vol. 40, no. 12, pp. 2659–2673, 2002.

[43] F. Brock, K. Crawford, R. L. Elliott et al., "The oklahoma mesonet—a technical overview," *Journal of Atmospheric and Oceanic Technology*, vol. 12, no. 1, pp. 5–19, 1995.

[44] B. G. Illston, J. B. Basara, and K. C. Crawford, "Seasonal to interannual variations of soil moisture measured in Oklahoma," *International Journal of Climatology*, vol. 24, no. 15, pp. 1883–1896, 2004.

[45] B. G. Illston, J. B. Basara, D. K. Fisher et al., "Mesoscale monitoring of soil moisture across a statewide network," *Journal of Atmospheric and Oceanic Technology*, vol. 25, no. 2, pp. 167–182, 2008.

[46] S. E. Hollinger and S. A. Isard, "A soil moisture climatology of Illinois," *Journal of Climate*, vol. 7, no. 5, pp. 822–833, 1994.

[47] G. L. Schaefer, M. H. Cosh, and T. J. Jackson, "The USDA natural resources conservation service soil climate analysis network (SCAN)," *Journal of Atmospheric and Oceanic Technology*, vol. 24, no. 12, pp. 2073–2077, 2007.

[48] G. C. Heathman, M. H. Cosh, E. Han, T. J. Jackson, L. G. McKee, and S. McAfee, "Field scale spatiotemporal analysis of surface soil moisture for evaluating point-scale in situ networks," *Geoderma*, vol. 170, pp. 195–205, 2012.

[49] O. Merlin, A. Al Bitar, J. P. Walker, and Y. Kerr, "An improved algorithm for disaggregating microwave-derived soil moisture based on red, near-infrared and thermal-infrared data," *Remote Sensing of Environment*, vol. 114, no. 10, pp. 2305–2316, 2010.

[50] M. Piles, A. Camps, M. Vall-llossera et al., "Downscaling SMOS-derived soil moisture using MODIS visible/infrared data," *IEEE Transactions on Geoscience and Remote Sensing*, vol. 49, no. 9, pp. 3156–3166, 2011.

[51] O. Merlin, A. Chehbouni, J. P. Walker, R. Panciera, and Y. H. Kerr, "A simple method to disaggregate passive microwave-based soil moisture," *IEEE Transactions on Geoscience and Remote Sensing*, vol. 46, no. 3, pp. 786–796, 2008.

[52] O. Merlin, A. Al Bitar, J. P. Walker, and Y. Kerr, "A sequential model for disaggregating near-surface soil moisture observations using multi-resolution thermal sensors," *Remote Sensing of Environment*, vol. 113, no. 10, pp. 2275–2284, 2009.

[53] D. Entekhabi, E. G. Njoku, P. E. O'Neill et al., "The soil moisture active passive (SMAP) mission," *Proceedings of the IEEE*, vol. 98, no. 5, pp. 704–716, 2010.

[54] T. Schmugge, P. Gloersen, T. Wilheit, and F. Geiger, "Remote sensing of soil moisture with microwave radiometers," *Journal of Geophysical Research*, vol. 79, no. 2, pp. 317–323, 1974.

[55] U. Narayan, V. Lakshmi, and T. J. Jackson, "High-resolution change estimation of soil moisture using L-band radiometer and radar observations made during the SMEX02 experiments," *IEEE Transactions on Geoscience and Remote Sensing*, vol. 44, no. 6, pp. 1545–1554, 2006.

[56] U. Narayan and V. Lakshmi, "Characterizing subpixel variability of low resolution radiometer derived soil moisture using high resolution radar data," *Water Resources Research*, vol. 44, no. 6, Article ID W06425, 2008.

[57] N. N. Das, D. Entekhabi, and E. G. Njoku, "An algorithm for merging SMAP radiometer and radar data for high-resolution soil-moisture retrieval," *IEEE Transactions on Geoscience and Remote Sensing*, vol. 49, no. 5, pp. 1504–1512, 2011.

[58] O. Merlin, A. Al Bitar, V. Rivalland, P. Béziat, E. Ceschia, and G. Dedieu, "An analytical model of evaporation efficiency for unsaturated soil surfaces with an arbitrary thickness," *Journal of Applied Meteorology and Climatology*, vol. 50, no. 2, pp. 457–471, 2011.

[59] O. Merlin, F. Jacob, J.-P. Wigneron et al., "Multidimensional disaggregation of land surface temperature using high-resolution red, near-infrared, shortwave-infrared, and microwave-L bands," *IEEE Transactions on Geoscience and Remote Sensing*, vol. 50, no. 5, pp. 1864–1880, 2012.

[60] O. Merlin, C. Rudiger, A. Al Bitar et al., "Disaggregation of SMOS soil moisture in Southeastern Australia," *IEEE Transactions on Geoscience and Remote Sensing*, vol. 50, no. 5, pp. 1556–1571, 2012.

[61] B. Fang, V. Lakshmi, R. Bindlish, T. Jackson, M. Cosh, and J. Basara, "Passive microwave soil moisture downscaling using vegetation and surface temperature," *Vadose Zone Journal*. In review.

[62] M. A. Friedl, D. K. McIver, J. C. F. Hodges et al., "Global land cover mapping from MODIS: algorithms and early results," *Remote Sensing of Environment*, vol. 83, no. 1-2, pp. 287–302, 2002.

[63] J. R. Wang and T. J. Schmugge, "An empirical model for the complex dielectric permittivity of soils as a function of water content," *IEEE Transactions on Geoscience and Remote Sensing*, vol. GE-18, no. 4, pp. 288–295, 1980.

[64] M. C. Dobson, F. T. Ulaby, M. T. Hallikainen, and M. A. El-Rayes, "Microwave dielectric behavior of wet soil—part 2: dielectric mixing models," *IEEE Transactions on Geoscience and Remote Sensing*, vol. GE-23, no. 1, pp. 35–46, 1985.

[65] A. K. Fung, Z. Li, and K. S. Chen, "Backscattering from a randomly rough dielectric surface," *IEEE Transactions on Geoscience and Remote Sensing*, vol. 30, no. 2, pp. 356–369, 1992.

[66] T. Schmugge, "Remote sensing of soil moisture," in *Hydrological Forecasting*, M. G. Anderson and T. P. Burt, Eds., John Wiley & Sons, New York, NY, USA, 1985.

[67] R. R. Gillies and T. N. Carlson, "Thermal remote sensing of surface soil water content with partial vegetation cover for incorporation into climate models," *Journal of Applied Meteorology*, vol. 34, no. 4, pp. 745–756, 1995.

[68] S. J. Goetz, "Multi-sensor analysis of NDVI, surface temperature and biophysical variables at a mixed grassland site," *International Journal of Remote Sensing*, vol. 18, no. 1, pp. 71–94, 1997.

[69] T. Mo, B. J. Choudhury, T. J. Schmugge, J. R. Wang, and T. J. Jackson, "A model for the microwave emission from vegetation covered fields," *Journal of Geophysical Research*, vol. 87, pp. 11 229–11 237, 1982.

[70] E. T. Engman and N. Chauhan, "Status of microwave soil moisture measurements with remote sensing," *Remote Sensing of Environment*, vol. 51, no. 1, pp. 189–198, 1995.

[71] N. M. Mattikalli, E. T. Engman, L. R. Ahuja, and T. J. Jackson, "Microwave remote sensing of soil moisture for estimation of profile soil property," *International Journal of Remote Sensing*, vol. 19, no. 9, pp. 1751–1767, 1998.

[72] C. A. Laymon, W. L. Crosson, V. V. Soman et al., "Huntsville '98: an expirement in ground-based microwave remote sensing of soil moisture," *International Journal of Remote Sensing*, vol. 20, no. 2–4, pp. 823–828, 1999.

[73] E. G. Njoku and L. Li, "Retrieval of land surface parameters using passive microwave measurements at 6-18 GHz," *IEEE Transactions on Geoscience and Remote Sensing*, vol. 37, no. 1, pp. 79–93, 1999.

[74] Y. H. Kerr and E. G. Njoku, "Semiempirical model for interpreting microwave emission from semiarid land surfaces as seen from space," *IEEE Transactions on Geoscience and Remote Sensing*, vol. 28, no. 3, pp. 384–393, 1990.

[75] Y. Oh, K. Sarabandi, and F. T. Ulaby, "An empirical model and an inversion technique for radar scattering from bare soil surfaces," *IEEE Transactions on Geoscience and Remote Sensing*, vol. 30, no. 2, pp. 370–381, 1992.

[76] P. C. Dubois, J. van Zyl, and T. Engman, "Measuring soil moisture with imaging radars," *IEEE Transactions on Geoscience and Remote Sensing*, vol. 33, no. 4, pp. 915–926, 1995.

[77] J. Shi, J. Wang, A. Y. Hsu, P. E. O'Neill, and E. T. Engman, "Estimation of bare surface soil moisture and surface roughness parameter using L-band SAR image data," *IEEE Transactions on Geoscience and Remote Sensing*, vol. 35, no. 5, pp. 1254–1266, 1997.

[78] T. J. Jackson, J. Schmugge, and E. T. Engman, "Remote sensing applications to hydrology: soil moisture," *Hydrological Sciences Journal*, vol. 41, no. 4, pp. 517–530, 1996.

[79] M. Owe, A. A. Van De Griend, R. De Jeu, J. J. De Vries, E. Seyhan, and E. T. Engman, "Estimating soil moisture from satellite microwave observations: past and ongoing projects, and relevance to GCIP," *Journal of Geophysical Research D*, vol. 104, no. 16, pp. 19735–19742, 1999.

[80] J. D. Villasenor, D. R. Fatland, and L. D. Hinzman, "Change detection on Alaska's North Slope using repeat-pass ERS-1 SAR images," *IEEE Transactions on Geoscience and Remote Sensing*, vol. 31, no. 1, pp. 227–236, 1993.

[81] M. C. Dobson and F. T. Ulaby, "Active microwave soil-moisture research," *IEEE Transactions on Geoscience and Remote Sensing*, vol. 24, no. 1, pp. 23–36, 1986.

[82] M. C. Dobson and F. T. Ulaby, "Preliminary evaluation of the Sir-B response to soil-moisture, surface-roughness, and crop canopy cover," *IEEE Transactions on Geoscience and Remote Sensing*, vol. 24, no. 4, pp. 517–526, 1986.

[83] T. J. Jackson, D. Chen, M. Cosh et al., "Vegetation water content mapping using Landsat data derived normalized difference water index for corn and soybeans," *Remote Sensing of Environment*, vol. 92, no. 4, pp. 475–482, 2004.

[84] M. H. Cosh, T. J. Jackson, R. Bindlish, and J. H. Prueger, "Watershed scale temporal and spatial stability of soil moisture and its role in validating satellite estimates," *Remote Sensing of Environment*, vol. 92, no. 4, pp. 427–435, 2004.

[85] A. S. Limaye, W. L. Crosson, C. A. Laymon, and E. G. Njoku, "Land cover-based optimal deconvolution of PALS L-band microwave brightness temperatures," *Remote Sensing of Environment*, vol. 92, no. 4, pp. 497–506, 2004.

[86] L. Yunling, K. Yunjin, and J. van Zyl, "The NASA/JPL airborne synthetic aperture radar system," 2002, http://airsar.jpl .nasa.gov/documents/genairsar/airsar_paper1.pdf.

[87] K. Mallick, B. K. Bhattacharya, and N. K. Patel, "Estimating volumetric surface moisture content for cropped soils using a soil wetness index based on surface temperature and NDVI," *Agricultural and Forest Meteorology*, vol. 149, no. 8, pp. 1327–1342, 2009.

[88] I. Sandholt, K. Rasmussen, and J. Andersen, "A simple interpretation of the surface temperature/vegetation index space for assessment of surface moisture status," *Remote Sensing of Environment*, vol. 79, no. 2-3, pp. 213–224, 2002.

[89] T. N. Carlson, R. R. Gillies, and T. J. Schmugge, "An interpretation of methodologies for indirect measurement of soil water content," *Agricultural and Forest Meteorology*, vol. 77, no. 3-4, pp. 191–205, 1995.

[90] T. N. Carlson and D. A. Ripley, "On the relation between NDVI, fractional vegetation cover, and leaf area index," *Remote Sensing of Environment*, vol. 62, no. 3, pp. 241–252, 1997.

[91] M. Minacapilli, M. Iovino, and F. Blanda, "High resolution remote estimation of soil surface water content by a thermal inertia approach," *Journal of Hydrology*, vol. 379, no. 3-4, pp. 229–238, 2009.

[92] T. R. Karl, G. Kukla, and J. Gavin, "Decreasing diurnal temperature range in the United States and Canada from 1941 through 1980," *Journal of Climate & Applied Meteorology*, vol. 23, no. 11, pp. 1489–1504, 1984.

[93] G. J. Collatz, L. Bounoua, S. O. Los, D. A. Randall, I. Y. Fung, and P. J. Sellers, "A mechanism for the influence of vegetation on the response of the diurnal temperature range to changing climate," *Geophysical Research Letters*, vol. 27, no. 20, pp. 3381–3384, 2000.

[94] A. Dai and C. Deser, "Diurnal and semidiurnal variations in global surface wind and divergence fields," *Journal of Geophysical Research D*, vol. 104, no. 24, pp. 31109–31125, 1999.

[95] T. J. Jackson, D. M. Le Vine, A. Y. Hsu et al., "Soil moisture mapping at regional scales using microwave radiometry: the Southern great plains hydrology experiment," *IEEE Transactions on Geoscience and Remote Sensing*, vol. 37, no. 5, pp. 2136–2151, 1999.

[96] T. J. Jackson, A. J. Gasiewski, A. Oldak et al., "Soil moisture retrieval using the C-band polarimetric scanning radiometer during the Southern Great Plains 1999 Experiment," *IEEE Transactions on Geoscience and Remote Sensing*, vol. 40, no. 10, pp. 2151–2161, 2002.

[97] T. J. Jackson, M. H. Cosh, R. Bindlish et al., "Validation of advanced microwave scanning radiometer soil moisture products," *IEEE Transactions on Geoscience and Remote Sensing*, vol. 48, no. 12, pp. 4256–4272, 2010.

[98] I. Mladenova, V. Lakshmi, T. J. Jackson, J. P. Walker, O. Merlin, and R. A. M. de Jeu, "Validation of AMSR-E soil moisture using L-band airborne radiometer data from National Airborne Field Experiment 2006," *Remote Sensing of Environment*, vol. 115, no. 8, pp. 2096–2103, 2011.

[99] K. E. Mitchell, D. Lohmann, P. R. Houser et al., "The multi-institution North American Land Data Assimilation System (NLDAS): utilizing multiple GCIP products and partners in a continental distributed hydrological modeling system," *Journal of Geophysical Research D*, vol. 109, no. D7, article D07, 2004.

[100] B. A. Cosgrove, D. Lohmann, K. E. Mitchell et al., "Real-time and retrospective forcing in the North American Land Data Assimilation System (NLDAS) project," *Journal of Geophysical Research D*, vol. 108, no. 22, pp. 1–12, 2003.

[101] D. Lohmann, K. E. Mitchell, P. R. Houser et al., "Streamflow and water balance intercomparisons of four land surface models in the North American Land Data Assimilation Project," *Journal of Geophysical Research*, vol. 109, Article ID D07S91, 2004.

[102] D. Lohmann, K. E. Mitchell, P. R. Houser et al., "Streamflow and water balance intercomparisons of four land surface models in the North American Land Data Assimilation System project," *Journal of Geophysical Research D*, vol. 109, no. 7, Article ID D07S91, 2004.

[103] A. Robock, L. Luo, E. F. Wood et al., "Evaluation of the North American Land Data Assimilation System over the Southern Great Pkains during the warm season," *Journal of Geophysical Research*, vol. 108, no. D22, article 8846, 2003.

[104] J. C. Schaake, Q. Duan, V. Koren et al., "An intercomparison of soil moisture fields in the North American Land Data Assimilation System (NLDAS)," *Journal of Geophysical Research D*, vol. 109, no. 1, Article ID D01S90, 2004.

[105] E. G. Njoku, T. J. Jackson, V. Lakshmi, T. K. Chan, and S. V. Nghiem, "Soil moisture retrieval from AMSR-E," *IEEE*

*Transactions on Geoscience and Remote Sensing*, vol. 41, no. 2, pp. 215–229, 2003.

[106] E. G. Njoku and S. K. Chan, "Vegetation and surface roughness effects on AMSR-E land observations," *Remote Sensing of Environment*, vol. 100, no. 2, pp. 190–199, 2006.

[107] T. J. Jackson, "III. Measuring surface soil moisture using passive microwave remote sensing," *Hydrological Processes*, vol. 7, no. 2, pp. 139–152, 1993.

[108] C. O. Justice, J. R. G. Townshend, E. F. Vermote et al., "An overview of MODIS Land data processing and product status," *Remote Sensing of Environment*, vol. 83, no. 1-2, pp. 3–15, 2002.

[109] C. J. Tucker, "Red and photographic infrared linear combinations for monitoring vegetation," *Remote Sensing of Environment*, vol. 8, no. 2, pp. 127–150, 1979.

[110] R. B. Myneni, F. G. Hall, P. J. Sellers, and A. L. Marshak, "Interpretation of spectral vegetation indexes," *IEEE Transactions on Geoscience and Remote Sensing*, vol. 33, no. 2, pp. 481–486, 1995.

[111] R. B. Myneni, S. Hoffman, Y. Knyazikhin et al., "Global products of vegetation leaf area and fraction absorbed PAR from year one of MODIS data," *Remote Sensing of Environment*, vol. 83, no. 1-2, pp. 214–231, 2002.

[112] R. S. De Fries, M. Hansen, J. R. G. Townshend, and R. Sohlberg, "Global land cover classifications at 8 km spatial resolution: the use of training data derived from Landsat imagery in decision tree classifiers," *International Journal of Remote Sensing*, vol. 19, no. 16, pp. 3141–3168, 1998.

[113] Z. Wan and Z. L. Li, "A physics-based algorithm for retrieving land-surface emissivity and temperature from EOS/MODIS data," *IEEE Transactions on Geoscience and Remote Sensing*, vol. 35, no. 4, pp. 980–996, 1997.

[114] R. A. McPherson, C. A. Fiebrich, K. C. Crawford et al., "Statewide monitoring of the mesoscale environment: a technical update on the Oklahoma Mesonet," *Journal of Atmospheric and Oceanic Technology*, vol. 24, no. 3, pp. 301–321, 2007.

[115] J. L. Heilman and D. G. Moore, "Evaluating near-surface soil moisture using heat capacity mapping mission data," *Remote Sensing of Environment*, vol. 12, no. 2, pp. 117–121, 1982.

[116] S. Hong, V. Lakshmi, E. E. Small, and F. Chen, "The influence of the land surface on hydrometeorology and ecology: new advances from modeling and satellite remote sensing," *Hydrology Research*, vol. 42, no. 2-3, pp. 95–112, 2011.

[117] Y. Du, F. T. Ulaby, and M. C. Dobson, "Sensitivity to soil moisture by active and passive microwave sensors," *IEEE Transactions on Geoscience and Remote Sensing*, vol. 38, no. 1, pp. 105–114, 2000.

[118] T. J. Jackson and T. J. Schmugge, "Vegetation effects on the microwave emission of soils," *Remote Sensing of Environment*, vol. 36, no. 3, pp. 203–212, 1991.

# 6

# Response of Cold-Tolerant *Aspergillus* spp. to Solubilization of Fe and Al Phosphate in Presence of Different Nutritional Sources

**K. Rinu,[1] Mukesh Kumar Malviya,[1] Priyanka Sati,[1] S. C. Tiwari,[2] and Anita Pandey[1]**

[1] Biotechnological Applications, G. B. Pant Institute of Himalayan Environment and Development, Kosi-Katarmal, Almora, Uttarakhand 263 643, India

[2] Department of Botany and Microbiology, HNB Garhwal University, Srinagar, Uttarakhand 246 174, India

Correspondence should be addressed to Anita Pandey; anita@gbpihed.nic.in

Academic Editors: L. A. Dawson and R. R. Dupont

Three species of *Aspergillus*, namely, *A. niger*, *A. glaucus* and *A. sydowii*, isolated from soil samples collected from the Indian Himalayan Region (IHR), have been investigated for solubilization of aluminium phosphate and iron phosphate in the presence of different carbon and nitrogen sources. Preference of each fungal species varied for nitrogen and carbon sources, in terms of phosphate-solubilization. Among three species, *Aspergillus niger* gave the best results; it solubilized 32% and 8% of the supplemented aluminium phosphate and iron phosphate, respectively. The results indicated that the effect of carbon and nitrogen sources can influence the phosphate solubilizing efficiency of all the three *Aspergillus* spp. tested. All the three species were found to be plant-growth promoters in bioassays conducted under greenhouse conditions. The Al and Fe phosphate solubilization efficiency, investigated in the present study, is at the lower end of their previously reported tricalcium phosphate solubilization efficiency. The cultures are likely to have better field applications in agrobiotechnology, due to their potential towards solubilization of Al and Fe phosphates, which are known to have lower solubility through microbial activity.

## 1. Introduction

Microorganisms are known to play a fundamental role in biogeochemical cycling of P in natural and agricultural ecosystems. Gerretsen [1] initially demonstrated that microbiological activity in the rhizosphere can dissolve sparingly soluble inorganic P and increase plant growth. Several species of *Aspergillus* and *Penicillium* are known for solubilizing insoluble phosphates [2–4]. Usually, in vitro P solubilization activity is known to be associated with decline in pH [5]. Organic acids can greatly increase the rate of soluble P through chelation and ion exchange reactions [6]. In addition to organic acid production [7], phosphate solubilization can also occur due to proton extrusion [8, 9].

The widespread use of phosphate-solubilizing microbial inoculants remains limited due to inconsistent results under field conditions. This is mainly due to the functional efficiency of microorganisms, which is largely governed by environmental conditions. The forest and cropped soils in

the Indian Himalayan Region (IHR) are generally acidic [10, 11]. In acidic soils, free oxides and hydroxides of Al and Fe are known to fix phosphorus [12, 13]. Inoculation with phosphate-solubilizing microorganisms has been considered as a feasible approach to increase the amount of available P in soil. Fungal species, in this context, have been recognized as more appropriate compared to bacterial species [14]. In a recent study, tricalcium phosphate solubilization efficiency of ten cold- and pH-tolerant species of *Aspergillus* have been reported, with *A. niger* followed by *A. glaucus* and *A. sydowii*, respectively, being the best performers [15]. The suboptimal conditions (temperatures) were found to be optimal for solubilization of phosphate in the cited study. Al and Fe phosphates are known for their low solubility through microbial activity (bacteria and fungi), as compared to tricalcium phosphate. In the present study, response of these three cold-tolerant *Aspergillus* spp. (*A. niger*, *A. glaucus*, and *A. sydowii*) has been investigated with respect to solubilization of Al and Fe phosphates, in the presence of different nutritional sources. The plant-growth promoting ability of these fungal species has also been carried out under greenhouse conditions, using maize and wheat as test species.

## 2. Materials and Methods

*2.1. Fungal Cultures.* The fungal cultures were originally isolated from soil samples collected from various forest locations in higher altitudes (1800–3610 m above sea level) of the IHR, and maintained on agar slants at 4°C in the culture collection developed in the microbiology laboratory of the GB Pant Institute of Himalayan Environment and Development, Almora. The fungal cultures have also been deposited and accessioned at the Indian Type Culture Collection (ITCC), Indian Agricultural Research Institute, New Delhi (Accession numbers ITCC2546- *A. niger* and ITCC4210- *A. sydowii*), and the Agharkar Research Institute Fungal Culture Collection (ARIFCC), Pune, India (Accession number ARIFCC771- *A. glaucus*), respectively [15].

### 2.2. Estimation of AlPO₄ and FePO₄ Solubilization

*AlPO₄.* 0.5 g/100 mL of $AlPO_4$ (HiMedia, India) was added as a P source in Pikovskaya's broth (yeast extract, 0.50 g; dextrose, 10.00 g; ammonium sulphate, 0.50 g; potassium chloride, 0.20 g; magnesium sulphate, 0.10 g; manganese sulphate, 0.0001 g; and ferrous sulphate, 0.0001 g) [16]. For estimation of the effect of different carbon and nitrogen sources on solubilization of $AlPO_4$, dextrose was replaced with D-fructose, starch, and sucrose as the carbon source (separate treatments). Similarly, ammonium sulphate was replaced with ammonium chloride, sodium nitrate, and potassium nitrate (separate treatments) as the nitrogen source.

*FePO₄.* 0.5 g/100 mL of $FePO_4$ (HiMedia, India) was added as the P source in Pikovskaya's broth. For estimation of the effect of different carbon and nitrogen sources on solubilization of $FePO_4$, dextrose was replaced with D-fructose, starch, and sucrose as the carbon source (separate treatments). Similarly, ammonium sulphate was replaced with ammonium chloride, sodium nitrate, and potassium nitrate as the nitrogen source (separate treatments).

The initial pH of the media, before autoclaving, was 7.50. The autoclaved medium was then inoculated with 5 mm disc cultures of the respective fungi and incubated at 21°C for 42 days. The culture filtrate was withdrawn from selected flasks on every 7th day of incubation by vacuum filtration through Whatman numbers 42 filter paper. The filtrate was then analyzed for $P_2O_5$ production following the chlorostannous reduced molybdophosphoric acid blue method [12]. The absorbance was recorded at 700 nm using Uvikon spectrophotometer (Kontron Instruments, UK). The pH (Systronics, India) of the culture filtrate was also recorded upon each sampling event. The biomass of the cultures was estimated every 7th day of incubation following drying at 65°C for 72 h, from the same flasks that were used for the above experiments.

*2.3. Microcosm Studies.* Plant-growth-promoting abilities of the three fungal species were performed following

a pot-based assay under greenhouse conditions (temp.: 25 ± 0.5°C; humidity: 75–85%). Maize (*Zea mays*) and wheat (*Triticum aestivum*) were used as test crops. The seeds were obtained from local farmers at a nearby village, Kosi-Katarmal. Plastic pots (10 × 15 cm) for each treatment were filled with 300 g soil (sandy loam with $pH_{H_2O}$ 6.7, 40% (w/w) moisture content). The treatments were arranged in a randomized block design with 25 replicates.

For taking measurements on growth parameters (shoot and root length and dry weight of shoot, root, and seed), plants were harvested at peak flowering time (approx. 3 months after sowing, $n = 25$). The plants were randomly uprooted from treated and control pots and washed several times with running water. First, the shoot and root length were measured then the roots were cutoff, and following drying at 65°C for 72 h, the measurements of biomass of root and shoot parts were performed, separately. For yield parameters (biological and economic yield), ten plants were collected randomly from control and inoculated plots, at the time of senescence. The plants were dried at 70°C for 72 h for calculating the biological yield. Ten seeds from each plant were used for measurement of seed weight (an estimate of economic yield). Harvest index was calculated following the formula: harvest index = economic yield/biological yield × 100.

*2.4. Statistical Analysis.* The data were analyzed with the computer programme Excel (Microsoft Corp.) for graphical representation, and the mean values and variance among the means (fixed effects or model I one-way ANOVA), SPSS/PC [17], were used to explore the correlations between phosphate solubilization, biomass production, and pH.

## 3. Results

### 3.1. Aluminium Phosphate Solubilization

*3.1.1. A. glaucus.* A. glaucus solubilized 17% of the aluminium phosphate on day 21, which was found to be significantly correlated to decline in pH ($r = -0.987$, $P \leq 0.01$) and production of biomass ($r = 0.993$, $P \leq 0.01$). Less solubilization ($P \leq 0.01$) was recorded in the case of fructose (15.34%)-containing medium as compared to the normal Pikovskaya's broth medium (Figures 1 and 3(a)). Among nitrogen sources (Figure 3(b)), sodium nitrate gave the best results for solubilization of aluminium phosphate (11.56% on day 21). A statistically significant ($P \leq 0.01$) negative correlation was developed between solubilization of aluminium phosphate and a decline in pH. Significant positive correlation ($P \leq 0.01$) between production of biomass and solubilization of aluminium phosphate was also developed in the case of both media.

*3.1.2. A. niger.* A. niger exhibited maximum efficiency for solubilization of aluminium phosphate on day 35 (32%) with maximum decline in pH (2.39) and production of fungal biomass (Figure 1). A significant correlation was developed between the decline in pH of the medium ($r = -0.870$, $P \leq 0.01$) and the production of biomass ($r = 0.928$). With all

Response of Cold-Tolerant Aspergillus spp. to Solubilization of Fe and Al Phosphate in Presence of Different Nutritional Sources

77

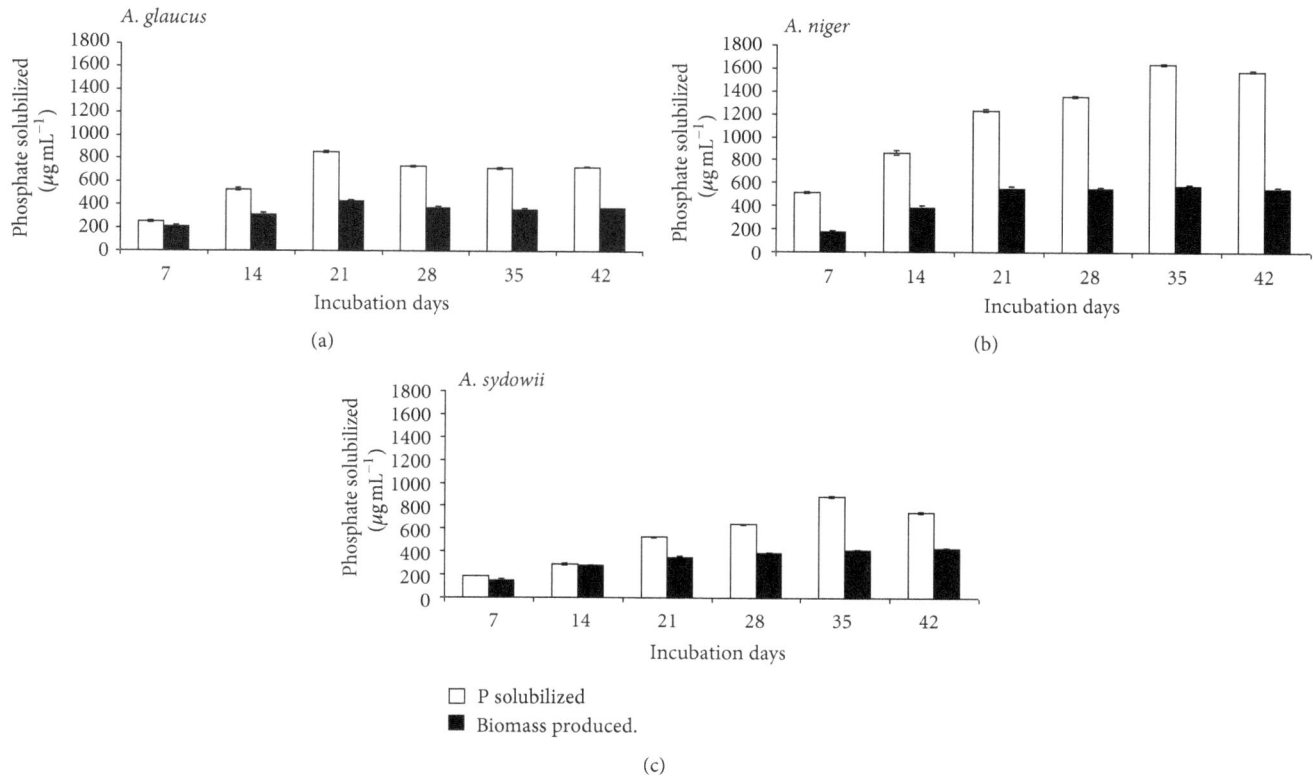

FIGURE 1: Solubilization of aluminium phosphate and biomass production by *Aspergillus* spp.

media compositions, except the medium supplemented with ammonium chloride, the solubilization of aluminium phosphate persisted even after day 28 of incubation (Figure 1). Significantly higher aluminium phosphate solubilization ($P \leq 0.01$) occurred in the medium supplemented with ammonium chloride (Figure 3(d), 37.78%) as compared to the normal Pikovskaya's broth (32%, Figure 1). Fructose was found to be the best among the carbon sources (Figure 3(c), 32.90% on day 28) in terms of enhancing phosphate solubilization. In the presence of starch, the solubilization was the lowest, and the decline in pH (up to 2.74) was also found to be comparatively low persisting even after day 42 of incubation. The decline in pH in the presence of fructose, ammonium chloride, sodium nitrate, and potassium nitrate was found to be equivalent to the decline in pH in the normal Pikovskaya's medium supplemented with aluminium phosphate. The correlations between phosphate solubilization, the decline in pH, and the production of biomass were found to be significant ($P \leq 0.01$) for all the media tested.

### 3.1.3. A. sydowii.
Maximum aluminium phosphate solubilization (18%, Figure 1), pH decline (2.84), and biomass production were recorded on day 35 of incubation. The decline in pH and biomass production were found to be significantly correlated with solubilization ($r = -0.989, P \leq 0.01; r = 0.929, P \leq 0.01$, resp.). Solubilization in the medium supplemented with ammonium chloride was (17.66% on day 35, Figure 3(f)) equivalent to the maximum solubilization found in the Pikovskaya's medium supplemented with

aluminium phosphate. The decline in pH was also found to be 2.92 on day 42 in this medium. Among the carbon sources, fructose was the least efficient (solubilized 11.6% of supplied aluminium phosphate on day 45, Figure 3(e)), while the media supplemented with starch and sucrose solubilized up to 17% of the added aluminium phosphate. The production of biomass and reduction in pH were found to be minimum in the medium supplemented with fructose; however, the fungal growth and the solubilization activity persisted even after day 42. The correlations between phosphate solubilization, the decline in pH, and the production of biomass were found to be statistically significant ($P \leq 0.01$) for all of the media tested.

### 3.2. Iron Phosphate Solubilization

#### 3.2.1. A. glaucus.
Maximum (6%) solubilization of iron phosphate, decline in pH, and production of biomass (522 mg) were recorded on day 28 (Figure 2). Solubilization of iron phosphate exhibited a significant correlation with a reduction in pH ($r = -0.866, P \leq 0.01$) and production of biomass ($r = 0.848, P \leq 0.01$). Medium supplemented with starch performed best in terms of solubilization of iron phosphate (Figure 4(a)), resulting in significantly higher values ($P \leq 0.05$), as compared to the normal Pikovskaya's broth (Figure 2) containing iron phosphate. The maximum decline in pH (3.07) and production of biomass were also recorded in this medium. Slightly higher solubilization (statistically not significant) was also recorded in the medium

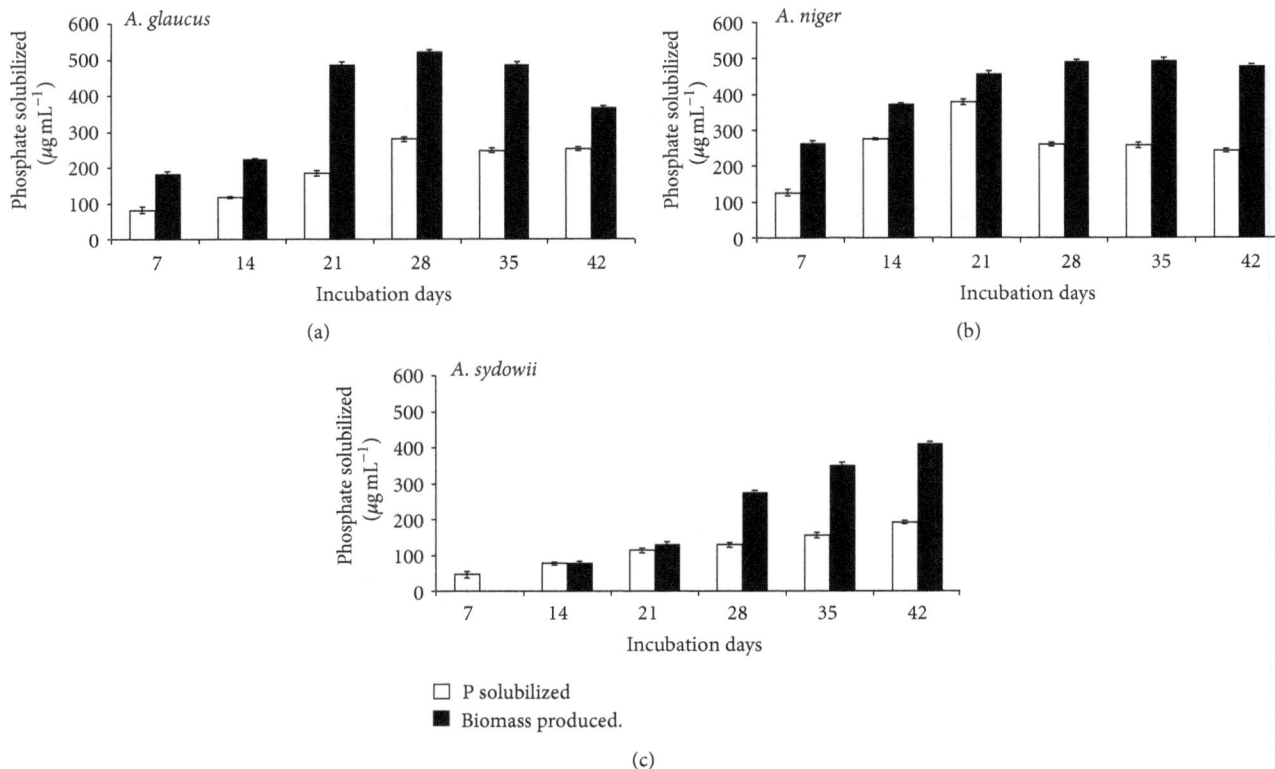

FIGURE 2: Solubilization of iron phosphate and biomass production by *Aspergillus* spp.

supplemented with ammonium chloride (Figure 4(b)) as compared to normal Pikovskaya's medium containing iron phosphate.

### 3.2.2. A. niger.

Maximum solubilization of iron phosphate (8%), reduction in pH (4.03), and production of biomass were recorded on day 21 (Figure 2). Correlation between phosphate solubilization and acidification was not found to be statistically significant ($r = -0.870$), while the production of biomass exhibited significant ($P \leq 0.001$) correlation with phosphate solubilization. Significantly higher ($P \leq 0.009$) phosphate solubilization (8.32%) was found in the medium supplemented with sodium nitrate (Figure 4(d)) as compared to the normal iron-phosphate-containing Pikovskaya's broth (Figure 2). This was recorded as the best among the tested compositions where pH was reduced to 3.21. Among the carbon sources, fructose was found to be the best (6.58% on day 35), followed by sucrose (6.5% on day 28), and starch (5.6% on day 28) (Figure 4(c)).

### 3.2.3. A. sydowii.

Maximum iron phosphate solubilization (4%), the minimum pH (4.73) and the maximum production in biomass were recorded on day 42 (Figure 2). Reduction in pH and biomass production were found to be significantly correlated with iron phosphate solubilization ($r = -0.993$, $P \leq 0.01$; $r = 0.974$, $P \leq 0.01$, resp.). Among the modified media, maximum solubilization (4.36%) was recorded in the case of the medium supplemented with starch on day 28 (Figure 4(e)). The minimum decline in pH (4.23 on day 35)

was also recorded in the same medium. In this case, biomass production was observed to be persisting even after day 42 of incubation. The persistence of solubilization was found to be correlated to the biomass production (significant at $P \leq 0.01$). The solubilization of iron phosphate by *A. sydowii*, in both B2 and B5, was significantly ($P \leq 0.0003$, and $P \leq 0.05$, resp.) higher, than the iron phosphate in the normal Pikovskaya's broth containing iron phosphate.

### 3.3. Effect of Fungal Inoculation on Plant Growth of Maize and Wheat.

Positive response of inoculation with three species of *Aspergillus* in two test crops, maize and wheat, were recorded under greenhouse conditions. In case of maize, in *A. glaucus* inoculated plants, the increase in root and shoot length, root, shoot, and seed weight, and harvest index ranged between 1.20 to 1.35 times, while, in the case of *A. niger*-inoculated treatments, the increases ranged between 1.18 to 3.0 times. Except root weight, all the parameters tested were significantly higher due to inoculation with *A. glaucus* and *A. niger*. *A. sydowii* inoculation resulted in the increases in various growth parameters, ranging between 1.21 and 2.50 times, with all the parameters being significantly higher, except shoot length (Table 1).

In wheat-based experiments, inoculation with *A. niger* and *A. sydowii* resulted in increases in root and shoot length, root, shoot, and seed weight, and harvest index ranging between 1.22 to 2.11 times, and 1.14 with 2.07 times, respectively, all parameters being statistically increased over the controls. These increases ranged between 1.11 to 1.82 times

Response of Cold-Tolerant Aspergillus spp. to Solubilization of Fe and Al Phosphate in Presence of Different
Nutritional Sources

79

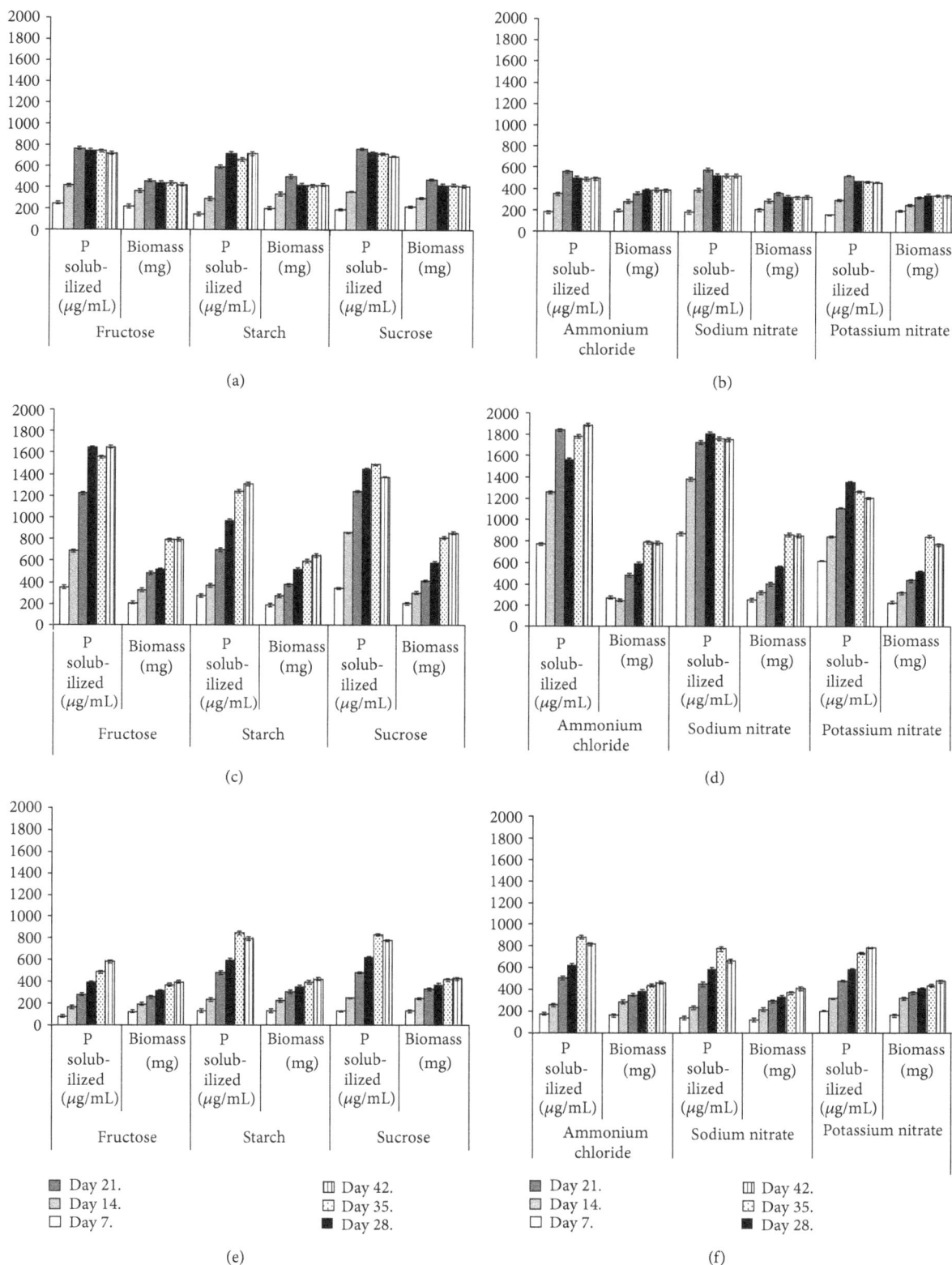

FIGURE 3: Effect of carbon ((a), (c), and (e)) and nitrogen ((b), (d), and (f)) sources on the aluminum phosphate solubilization efficiency and biomass production of *A. glaucus* ((a) and (b)), *A. niger* ((c) and (d)), and *A. sydowii* ((e) and (f)). *y*-axis shows both phosphate solubilized and biomass produced (separately mentioned on *x*-axis).

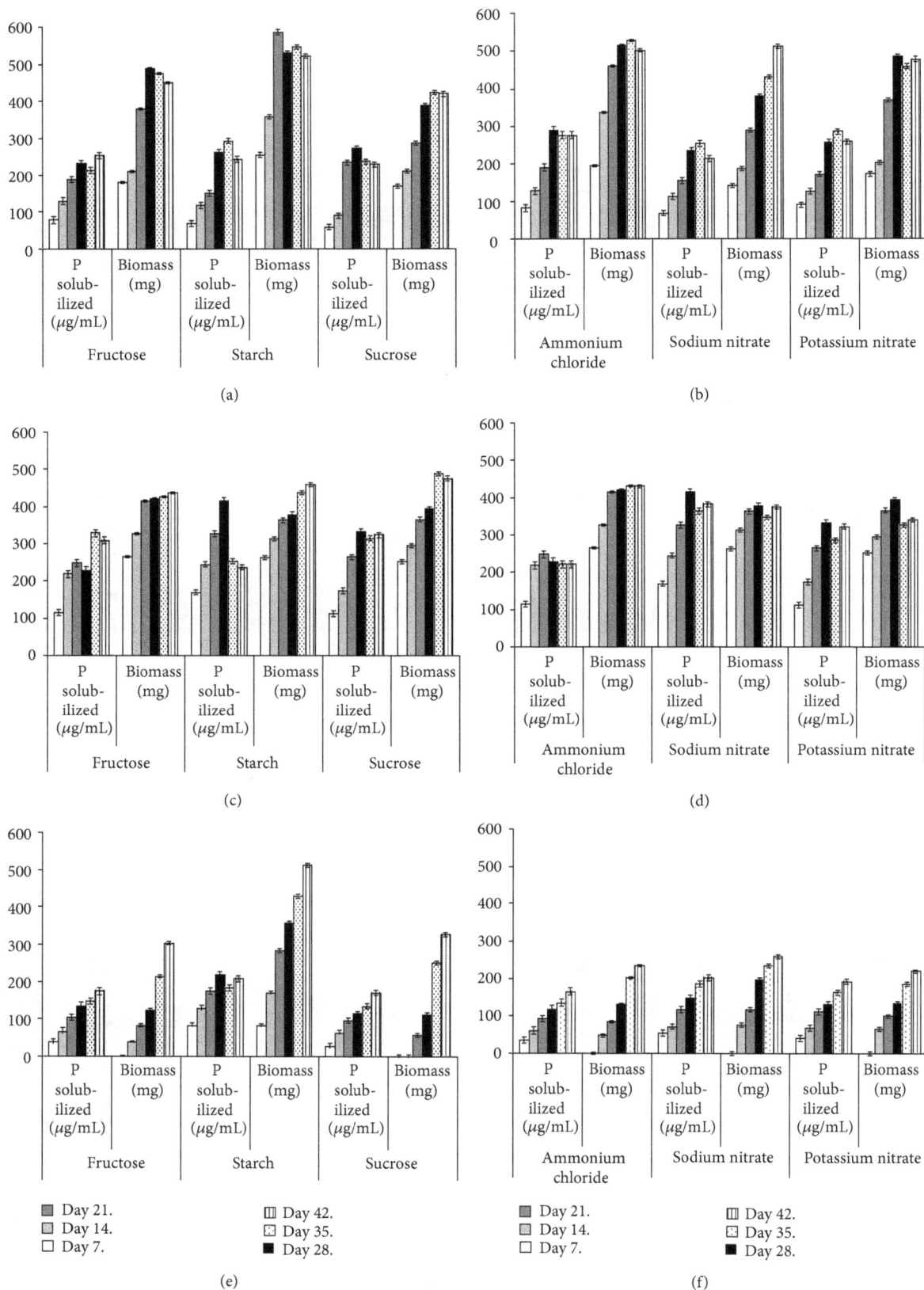

FIGURE 4: Effect of carbon ((a), (c), and (e)) and nitrogen ((b), (d), and (f)) sources on the iron phosphate solubilization efficiency and biomass production of *A. glaucus* ((a) and (b)), *A. niger* ((c) and (d)), and *A. sydowii* ((e) and (f)). *y*-axis shows both phosphate solubilized and biomass produced (separately mentioned on *x*-axis).

Response of Cold-Tolerant Aspergillus spp. to Solubilization of Fe and Al Phosphate in Presence of Different Nutritional Sources

81

TABLE 1: Effect of phosphate-solubilizing *Aspergillus* species on the growth of maize and wheat under greenhouse conditions.

| Treatment | Shoot length (cm) | Root length (cm) | Biomass production and yield (g dry weight) | | | Harvest index |
|---|---|---|---|---|---|---|
| | | | Shoot | Root | Seed | |
| **Maize** | | | | | | |
| Control | $56.70 \pm 8.37$ | $11.50 \pm 1.90$ | $0.28 \pm 0.16$ | $0.04 \pm 0.22$ | $0.84 \pm 0.02$ | 0.32 |
| *A. glaucus* | $68.20 \pm 3.20^*$ | $15.30 \pm 2.51^*$ | $0.38 \pm 0.06^*$ | $0.05 \pm 0.03$ | $0.93 \pm 0.06^*$ | 0.39 |
| ANOVA | $F_{1,18} = 15.53$ $P = 0.001$ | $F_{1,18} = 25.73$ $P = 0.05$ | $F_{1,18} = 5.39$ $P = 0.03$ | $F_{1,18} = 2.14$ $P = 0.16$ | $F_{1,18} = 3.16$ $P = 0.01$ | NA |
| *A. niger* | $67.31 \pm 2.16^*$ | $15.23 \pm 1.25^*$ | $0.61 \pm 0.14^*$ | $0.12 \pm 0.02$ | $1.12 \pm 0.06^*$ | 0.69 |
| ANOVA | $F_{1,18} = 11.93$ $P = 0.002$ | $F_{1,18} = 22.83$ $P = 0.0002$ | $F_{1,18} = 5.39$ $P = 0.03$ | $F_{1,18} = 2.14$ $P = 0.16$ | $F_{1,18} = 6.23$ $P = 0.006$ | NA |
| *A. sydowii* | $74.81 \pm 5.32$ | $17.21 \pm 3.26^*$ | $0.58 \pm 0.16^*$ | $0.10 \pm 0.03^*$ | $1.02 \pm 0.04^*$ | 0.68 |
| ANOVA | $F_{1,18} = 25.21$ $P = 0.87$ | $F_{1,18} = 9.46$ $P = 0.006$ | $F_{1,18} = 12.83$ $P = 0.002$ | $F_{1,18} = 11.36$ $P = 0.003$ | $F_{1,18} = 4.56$ $P = 0.004$ | NA |
| **Wheat** | | | | | | |
| Control | $29.80 \pm 4.86$ | $16.85 \pm 2.68$ | $0.09 \pm 0.03$ | $0.13 \pm 0.02$ | $0.27 \pm 0.06$ | 68.08 |
| *A. glaucus* | $33.20 \pm 4.10$ | $20.40 \pm 3.63^*$ | $0.13 \pm 0.05^*$ | $0.16 \pm 0.05$ | $0.36 \pm 0.06^*$ | 124.14 |
| ANOVA | $F_{1,18} = 2.86$ $P = 0.10$ | $F_{1,18} = 6.18$ $P = 0.02$ | $F_{1,18} = 4.62$ $P = 0.04$ | $F_{1,18} = 2.41$ $P = 0.13$ | $F_{1,18} = 11.58$ $P = 0.003$ | NA |
| *A. niger* | $39.15 \pm 6.03^*$ | $20.60 \pm 3.84^*$ | $0.19 \pm 0.03^*$ | $0.17 \pm 0.05^*$ | $0.37 \pm 0.10^*$ | 102.78 |
| ANOVA | $F_{1,18} = 14.58$ $P = 0.001$ | $F_{1,18} = 6.41$ $P = 0.02$ | $F_{1,18} = 43.74$ $P = 0.0001$ | $F_{1,18} = 6.40$ $P = 0.02$ | $F_{1,18} = 7.11$ $P = 0.015$ | NA |
| *A. sydowii* | $35.40 \pm 5.32^*$ | $19.20 \pm 2.15^*$ | $0.14 \pm 0.04^*$ | $0.27 \pm 0.17^*$ | $0.34 \pm 0.05^*$ | 82.92 |
| ANOVA | $F_{1,18} = 6.04$ $P = 0.02$ | $F_{1,18} = 4.66$ $P = 0.04$ | $F_{1,18} = 4.97$ $P = 0.04$ | $F_{1,18} = 5.94$ $P = 0.02$ | $F_{1,18} = 6.42$ $P = 0.02$ | NA |

ANOVA: analysis of variance; NA: not detected; *significant increment as compared to control.

higher than the controls and were statistically significant in case of *A. glaucus*.

## 4. Discussion

The phosphate solubilization efficiency by microorganisms depends on the form of insoluble phosphate. The three fungal cultures used in the present study exhibited differential responses towards solubilization of aluminium and iron phosphate. The Al and Fe phosphate solubilization efficiency observed, in the present study, are lower than previously reported tricalcium phosphate solubilization efficiency [15]. The cultures are likely to have better applications due to their potential towards solubilization of Al and Fe phosphates, which are known for lesser solubility through microbial activity. Lopez et al. [18] and Puente et al. [19] have demonstrated the solubilization efficiency of more insoluble forms of phosphates in rock weathering through microbial activities, with reference to desert-plant-based studies. Positive correlation between phosphate solubilization and pH reduction developed in most cases is attributed to the production of organic acids and proton extrusion. Production of organic acids by microbial cultures has been emphasized as one of the main mechanisms involved in phosphate solubilization [2, 20].

The effect of nutritional aspects, such as carbon and nitrogen sources, on microbial phosphate solubilization efficiency has been studied by various workers [8, 21]. In the present study, the nitrogen-source-based modifications in the media brought significant changes in the solubilization efficiency of aluminium phosphate, production of biomass, and changes in pH, as compared to carbon sources. Nitrogen in ammonium form is necessary for P solubilization by production of organic acids through $NH_4^+/H^+$ exchange mechanisms [5, 22]. Solubilization of iron phosphate was found to be the best in starch-containing medium as compared to other carbon sources. Starch is the only polysaccharide used in the present study as a carbon source and is likely to create nutritional stress conditions that, in turn, might support phosphate solubilization by virtue of producing secondary metabolites.

Maximum solubilization of aluminium phosphate, in some of the instances, was recorded with less production of biomass. In case of *A. glaucus*, maximum biomass was produced in presence of starch, while maximum aluminium phosphate was solubilized in the media supplemented with fructose and sucrose. Depending on the nutrient and P sources, the fungus may use alternative metabolic pathways resulting in secretion of different organic acids [3, 23]. Similar results have been reported in previous studies, where species of *Aspergillus* and *Paecilomyces* have been evaluated for solubilization of tricalcium phosphate at different temperatures [15, 24]. Evaluation of the factors responsible for production of lesser fungal biomass in the medium vis-à-vis lethal and sublethal concentration of Al and Fe is an area of important future research. Independent of the excretion of organic acids, the toxic effect of $Al^{3+}$ can affect the release of P from

AlPO$_4$ [2]. The results in the present study are contradictory to the earlier findings of Barroso et al. [21] that reported a correlation between the production of greater biomass and a reduction in pH in the medium supplemented with AlPO$_4$ than that containing calcium phosphate. Decreased solubilization of phosphate in the presence of aluminium in comparison to iron has also been reported by Vyas et al. [25]. Inefficiency of *A. sydowii* growth in the presence of Fe during the initial period of incubation was an interesting observation. However, in later phases of growth, the fungus again started acting on the substrate probably because it was in need of the nutrients, thus releasing P from insoluble forms. In addition, after the initial shock, the cells are likely to utilize the available free P for metabolism first, later solubilizing P to acclimatize to the stressed environment [26].

At suboptimal growth conditions, where phosphate solubilization was found to be maximum, a highly significant negative correlation was observed between a reduction in pH and the production of biomass. This was an indication of the direct relationship between cell proliferation and phosphate solubilization with reference to pH and growth conditions. Metabolic alteration, due to extracellular pH and involvement of different signaling pathways, has been reported [27]. The role of these signaling molecules in view of the mechanisms involved in solubilization of phosphate also requires attention in future research.

Use of P-solubilizing fungi in plant nutrition is well reported [28, 29]. In the present study, the three fungal cultures exhibited their ability to enhance the growth of two crops. Inoculation with *A. niger* produced a maximum harvest index in maize, while *A. glaucus* produced a maximum index in wheat. Yield and harvest index are considered the most important parameters, evaluating the performance of PGP microorganisms. Microorganisms are important in agricultural nutrient cycling to reduce the need for chemical fertilizers [30]. Wahid and Mehana [31] reported the impact of phosphate-solubilizing fungi on wheat and faba bean, *Penicillium pinophilum* being the most efficient. Organic acids help in the mobility of nutrients in soil and hence can have a positive impact on plant growth. There is evidence that organic acids are capable of mobilizing phosphorus [32, 33]. Several factors have been known to improve the quality of soil, phosphorus mobilization, in particular [27]. The efficiency of organic acid production by the *Aspergillus* spp., used in the present study, has been evaluated [34]. These cultures were also found to produce phosphatases [15], which are likely one of the factors responsible for P mobilization [35].

The fixation of P depends on the soil acidity. In acidic soils, free oxides and hydroxides of Al and Fe fix P, while in alkaline soils P is fixed with Ca and Mg. Aluminium and iron phosphate-solubilizing microorganisms play a major role in P mobilization in low temperature environments of Himalayan soils, which are mostly acidic. Effectiveness of the *Aspergillus* spp., used in the present study, has already been demonstrated for their tricalcium phosphate solubilization efficiency [15]. The cited report also revealed the tolerance of these fungi to a wide range of pH. The present study documented the ability of the best three tricalcium phosphate-solubilizing fungi for their aluminium and iron phosphate solubilization

in the presence of different carbon and nitrogen sources. Based on the previous [15] and present investigations, it is concluded that these fungi can be used as phosphate solubilizers in different environmental conditions and hence are ecologically important in soil nutrient management in mountain ecosystems. Further, due to their cold-tolerant properties, these fungi also provide an opportunity to develop bioformulations for field application in low-temperature environments.

Documentation of microbial diversity, including mycorrhizae, in various ecological niches of the Indian Himalayan Region (IHR) with particular reference to biotechnological applications, such as development of microbial inoculants for colder regions, has received attention in recent times [36–39]. Cold-tolerant bacterial inoculants, possessing plant growth promoting and biocontrol properties, have already been developed in suitable formulations with particular reference to field applications under mountain ecosystems [40–46]. Phosphate solubilization, biocontrol properties, and tolerance to low temperature have been the preferred characteristics for screening suitable inoculants for biotechnological applications. These investigations are important as the distribution of microorganisms is largely governed by environmental specificities.

## Acknowledgments

Director, G. B. Pant Institute of Himalayan Environment and Development, Almora, is thanked for encouragement and extending the facilities. Department of Science and Technology and Ministry of Environment and Forests, Government of India, New Delhi, are acknowledged for their financial support.

## References

[1] F. C. Gerretsen, "The influence of microorganisms on the phosphate intake by the plant," *Plant and Soil*, vol. 1, no. 1, pp. 51–81, 1948.

[2] P. Illmer, A. Barbato, and F. Schinner, "Solubilization of hardly-soluble AlPO$_4$ with P-solubilizing microorganisms," *Soil Biology and Biochemistry*, vol. 27, no. 3, pp. 265–270, 1995.

[3] C. B. Barroso and E. Nahas, "Solubilization of iron phosphate by free or immobilized spores and pellets of *Aspergillus niger*," *Research Journal of Microbiology*, vol. 1, pp. 210–219, 2006.

[4] H. Singh and M. S. Reddy, "Effect of inoculation with phosphate solubilizing fungus on growth and nutrient uptake of wheat and maize plants fertilized with rock phosphate in alkaline soils," *European Journal of Soil Biology*, vol. 47, no. 1, pp. 30–34, 2011.

[5] P. E. A. Asea, R. M. N. Kucey, and J. W. B. Stewart, "Inorganic phosphate solubilization by two *Penicillium* species in solution culture and soil," *Soil Biology and Biochemistry*, vol. 20, no. 4, pp. 459–464, 1988.

[6] G. M. Gadd, "Bioremedial potential of microbial mechanisms of metal mobilization and immobilization," *Current Opinion in Biotechnology*, vol. 11, no. 3, pp. 271–279, 2000.

[7] J. E. Cunningham and C. Kuiack, "Production of citric and oxalic acids and solubilization of calcium phosphate by *Penicillium bilaii*," *Applied and Environmental Microbiology*, vol. 58, no. 5, pp. 1451–1458, 1992.

Response of Cold-Tolerant Aspergillus spp. to Solubilization of Fe and Al Phosphate in Presence of Different
Nutritional Sources

83

[8] I. Reyes, L. Bernier, R. R. Simard, P. Tanguay, and H. Antoun, "Characteristics of phosphate solubilization by an isolate of a tropical *Penicillium rugulosum* and two UV-induced mutants," *FEMS Microbiology Ecology*, vol. 28, no. 3, pp. 291–295, 1999.

[9] A. E. Carrillo, C. Y. Li, and Y. Bashan, "Increased acidification in the rhizosphere of cactus seedlings induced by *Azospirillum brasilense*," *Naturwissenschaften*, vol. 89, no. 9, pp. 428–432, 2002.

[10] S. S. Pal, "Interactions of an acid tolerant strain of phosphate solubilizing bacteria with a few acid tolerant crops," *Plant and Soil*, vol. 198, no. 2, pp. 169–177, 1998.

[11] A. Pandey and L. M. S. Palni, "The rhizosphere effect in trees of the Indian Central Himalaya with special reference to altitude," *Applied Ecology and Environmental Research*, vol. 5, no. 1, pp. 93–102, 2007.

[12] M. L. Jackson, *Soil Chemical Analysis*, Prentice-Hall, New Delhi, India, 1967.

[13] J. R. McLaughlin, J. C. Ryden, and J. K. Syers, "Sorption of inorganic phosphate by iron- and aluminium- containing components," *Journal of Soil Science*, vol. 32, no. 3, pp. 365–377, 1981.

[14] P. N. Rajankar, D. H. Tambekar, and S. R. Wate, "Study of phosphate solubilization efficiencies of fungi and bacteria isolated from saline belt of Purna river basin," *Research Journal of Agricultural and Biological Sciences*, vol. 3, pp. 701–703, 2007.

[15] K. Rinu and A. Pandey, "Temperature-dependent phosphate solubilization by cold- and pH-tolerant species of *Aspergillus* isolated from Himalayan soil," *Mycoscience*, vol. 51, no. 4, pp. 263–271, 2010.

[16] R. I. Pikovskaya, "Mobilization of phosphorus in soil in connection with vital activity of some microbial species," *Mykrobiologiya*, vol. 17, pp. 362–370, 1948.

[17] SPSS/PC, "SPSS/PC for the IBM PC/XT/AT," SPSS, Chicago, Ill, USA, 1986.

[18] B. R. Lopez, Y. Bashan, and M. Bacilio, "Endophytic bacteria of *Mammillaria fraileana*, an endemic rock-colonizing cactus of the southern Sonoran Desert," *Archives of Microbiology*, vol. 193, no. 7, pp. 527–541, 2011.

[19] M. E. Puente, C. Y. Li, and Y. Bashan, "Rock-degrading endophytic bacteria in cacti," *Environmental and Experimental Botany*, vol. 66, no. 3, pp. 389–401, 2009.

[20] A. Pandey, L. M. S. Palni, P. Mulkalwar, and M. Nadeem, "Effect of temperature on solubilization of tricalcium phosphate by *Pseudomonas corrugata*," *Journal of Scientific and Industrial Research*, vol. 61, no. 6, pp. 457–460, 2002.

[21] C. B. Barroso, G. T. Pereira, and E. Nahas, "Solubilization of CAHPO$_4$ and ALPO$_4$ by *Aspergillus niger* in culture media with different carbon and nitrogen sources," *Brazilian Journal of Microbiology*, vol. 37, no. 4, pp. 434–438, 2006.

[22] S. A. Omar, "The role of rock-phosphate-solubilizing fungi and vesicular-arbusular-mycorrhiza (VAM) in growth of wheat plants fertilized with rock phosphate," *World Journal of Microbiology and Biotechnology*, vol. 14, no. 2, pp. 211–218, 1998.

[23] A. G. Moat and J. W. Foster, *Microbial Physiology*, John Wiley & Sons, New York, NY, USA, 2nd edition, 1988.

[24] K. Rinu and A. Pandey, "Slow and steady phosphate solubilization by a Psychrotolerant strain of *Paecilomyces hepiali* (MTCC 9621)," *World Journal of Microbiology and Biotechnology*, vol. 27, pp. 1055–1062, 2011.

[25] P. Vyas, P. Rahi, A. Chauhan, and A. Gulati, "Phosphate solubilization potential and stress tolerance of *Eupenicillium parvum* from tea soil," *Mycological Research*, vol. 111, no. 8, pp. 931–938, 2007.

[26] S. Seshadri, R. Muthukumarasamy, C. Lakshminarasimhan, and S. Ignacimuthu, "Solubilization of inorganic phosphates by *Azospirillum halopraeferans*," *Current Science*, vol. 79, pp. 565–567, 2000.

[27] L. Palomo, N. Claassen, and D. L. Jones, "Differential mobilization of P in the maize rhizosphere by citric acid and potassium citrate," *Soil Biology and Biochemistry*, vol. 38, no. 4, pp. 683–692, 2006.

[28] M. A. Whitelaw, T. J. Harden, and G. L. Bender, "Plant growth promotion of wheat inoculated with *Penicillium radicum* sp. nov," *Australian Journal of Soil Research*, vol. 35, no. 2, pp. 291–300, 1997.

[29] I. Reyes, L. Bernier, and H. Antoun, "Rock phosphate solubilization and colonization of maize rhizosphere by wild and genetically modified strains of *Penicillium rugulosum*," *Microbial Ecology*, vol. 44, no. 1, pp. 39–48, 2002.

[30] R. Çakmakçi, F. Dönmez, A. Aydin, and F. Şahin, "Growth promotion of plants by plant growth-promoting rhizobacteria under greenhouse and two different field soil conditions," *Soil Biology and Biochemistry*, vol. 38, no. 6, pp. 1482–1487, 2006.

[31] O. A. A. Wahid and T. A. Mehana, "Impact of phosphate-solubilizing fungi on the yield and phosphorus-uptake by wheat and faba bean plants," *Microbiological Research*, vol. 155, no. 3, pp. 221–227, 2000.

[32] C. P. Vance, C. Uhde-Stone, and D. L. Allan, "Phosphorus acquisition and use: critical adaptations by plants for securing a nonrenewable resource," *New Phytologist*, vol. 157, no. 3, pp. 423–447, 2003.

[33] L. Wei, C. Chen, and Z. Xu, "Citric acid enhances the mobilization of organic phosphorus in subtropical and tropical forest soils," *Biology and Fertility of Soils*, vol. 46, no. 7, pp. 765–769, 2010.

[34] K. Rinu, A. Pandey, and L. M. S. Palni, "Utilization of Psychrotolerant phosphate solubilizing fungi under low temperature conditions of the mountain ecosystem," in *Environment Management and Biotechnology Springer Science + Business Media*, T. Satyanarayana, B. N. Johri, and A. Prakash, Eds., pp. 77–90, 2012.

[35] P. Vazquez, G. Holguin, M. E. Puente, A. Lopez-Cortes, and Y. Bashan, "Phosphate-solubilizing microorganisms associated with the rhizosphere of mangroves in a semiarid coastal lagoon," *Biology and Fertility of Soils*, vol. 30, no. 5-6, pp. 460–468, 2000.

[36] A. Pandey, P. Trivedi, B. Kumar, B. Chaurasia, and L. M. S. Palni, "Soil Microbial diversity from the Himalaya: need for documentation and conservation," NBA Scientific Bulletin 5, National Biodiversity Authority, Tamil Nadu, India, 2006.

[37] B. Chaurasia, A. Pandey, and L. M. S. Palni, "Distribution, colonization and diversity of arbuscular mycorrhizal fungi associated with central Himalayan rhododendrons," *Forest Ecology and Management*, vol. 207, no. 3, pp. 315–324, 2005.

[38] S. Singh, A. Pandey, B. Kumar, and L. M. S. Palni, "Enhancement in growth and quality parameters of tea [*Camellia sinensis* (L.) O. Kuntze] through inoculation with arbuscular mycorrhizal fungi in an acid soil," *Biology and Fertility of Soils*, vol. 46, no. 5, pp. 427–433, 2010.

[39] A. Kumar, S. Singh, and A. Pandey, "General microflora, arbuscular mycorrhizal colonization and occurrence of endophytes in the rhizosphere of two age groups of *Ginkgo biloba* L. of Indian Central Himalaya," *Indian Journal of Microbiology*, vol. 49, no. 2, pp. 134–141, 2009.

[40] A. Pandey, P. Trivedi, B. Kumar, and L. M. S. Palni, "Characterization of a phosphate solubilizing and antagonistic strain of *Pseudomonas putida* (B0) isolated from a sub-alpine location in the Indian Central Himalaya," *Current Microbiology*, vol. 53, no. 2, pp. 102–107, 2006.

[41] A. Pandey, L. M. S. Palni, and K. P. Hebbar, "Suppression of damping-off in maize seedlings by *Pseudomonas corrugata*," *Microbiological Research*, vol. 156, no. 2, pp. 191–194, 2001.

[42] A. Pandey, A. Durgapal, M. Joshi, and L. M. S. Palni, "Influence of *Pseudomonas corrugata* inoculation on root colonization and growth promotion of two important hill crops," *Microbiological Research*, vol. 154, no. 3, pp. 259–266, 1999.

[43] B. Kumar, P. Trivedi, and A. Pandey, "*Pseudomonas corrugata*: a suitable bacterial inoculant for maize grown under rainfed conditions of Himalayan region," *Soil Biology and Biochemistry*, vol. 39, no. 12, pp. 3093–3100, 2007.

[44] P. Trivedi, B. Kumar, A. Pandey, and L. M. S. Palni, "Growth promotion of rice by phosphate solubilizing bioinoculants in a Himalayan location," in *Proceedings of the 1st International Meeting on Microbial Phosphate Solubilization*, E. Velazqez and C. Rodriguez-Barrueco, Eds., vol. 102 of *Plant and Soil, Developments in Plant and Soil Sciences*, pp. 291–299, Springer, 2007.

[45] P. Trivedi and A. Pandey, "Plant growth promotion abilities and formulation of *Bacillus megaterium* strain B 388 (MTCC6521) isolated from a temperate Himalayan location," *Indian Journal of Microbiology*, vol. 48, no. 3, pp. 342–347, 2008.

[46] K. Rinu and A. Pandey, "*Bacillus subtilis* NRRL B-30408 inoculation enhances the symbiotic efficiency of *Lens esculenta* Moench at a Himalayan location," *Journal of Plant Nutrition and Soil Science*, vol. 172, no. 1, pp. 134–139, 2009.

# 7

# Maya and WRB Soil Classification in Yucatan, Mexico: Differences and Similarities

**Héctor Estrada-Medina,[1] Francisco Bautista,[2] Juan José María Jiménez-Osornio,[1] José Antonio González-Iturbe,[3] and Wilian de Jesús Aguilar Cordero[1]**

[1] *Departamento de Manejo y Conservación de Recursos Naturales Tropicales (PROTROPICO), Campus de Ciencias Biológicas y Agropecuarias (CCBA), Universidad Autónoma de Yucatán (UADY), Km 15.5 Carretera Mérida - Xmatkuil, Mérida, Yucatán 97315, Mexico*
[2] *Centro de Investigaciones en Geografía Ambiental (CIGA), Universidad Nacional Autónoma de México (UNAM), Antigua Carretera a Pátzcuaro No. 8701, Col. Ex-Hacienda de San José de La Huerta, Morelia, Michoacán 58190, Mexico*
[3] *Facultad de Arquitectura (UADY), Calle 50 S/N x 57 y 59 Ex-Convento de La Mejorada, Mérida, Yucatán 97000, Mexico*

Correspondence should be addressed to Héctor Estrada-Medina; hector.estrada@uady.mx

Academic Editors: M. B. Adams and C. Martius

Soils of the municipality of Hocabá, Yucatán, México, were identified according to both Mayan farmers' knowledge and the World Reference Base for Soil Resources (WRB). To identify Maya soil classes, field descriptions made by farmers and semistructured interviews were utilized. WRB soils were identified by describing soil profiles and analyzing samples in the laboratory. Mayan farmers identified soils based on topographic position and surface properties such as colour and amount of rock fragments and outcrops. Farmers distinguished two main groups of soils: *K'ankab* or soils of plains and *Boxlu'um* or soils of mounds. *K'ankab* is a group of red soils with two variants (*K'ankab* and *Haylu'um*), whereas *Boxlu'um* is a group of dark soils with five variants (*Tsek'el, Ch'ich'lu'um, Chaltun, Puslu'um,* and *Ch'och'ol*). Soils on the plains were identified as Leptosoils, Cambisols, and Luvisols. Soils identified in mounds were Leptosols and Calcisols. Many soils identified by farmers could be more than one WRB unit of soil and *vice versa*; in these cases no direct relationship between both classification systems was possible. Mayan and WRB soil types are complementary; they should be used together to improve regional soil classifications, help transference of agricultural technologies, and make soil management decisions.

## 1. Introduction

Local soil classification systems play an important role in many agricultural sites throughout the world but they have not considered to construct scientific classification systems [1]. Opportunities to use traditional systems to improve scientific soil classifications, mapping, and environmental impact monitoring are not fully exploited [2]. In countries like Mexico, indigenous soil knowledge of ancestral groups [3–7] need to be understood to facilitate planning, transmission, and implementation of new agricultural technologies [3, 8].

Local knowledge is restricted geographically, dynamic, collective, diachronic, and holistic; it is the product of a long observation history, analysis, and management of the natural resources, transmitted orally from generation to generation [9]. Traditional soil classification systems, created by the users, have a local importance and are based on properties easily affected by management [10]. This knowledge is enough to understand and manage the soil in a local way to solve short term specific problems [2, 11, 12]. On the other hand, scientific soil classification systems are based in measurable and observable soil characteristics defined in terms of diagnostic properties, materials, and horizons related to the soil morphology [13]. Traditional knowledge is being lost because these new regionally applied scientific schemes do not consider it. Incorporation of both types of knowledge into a more useful scheme requires the development of a

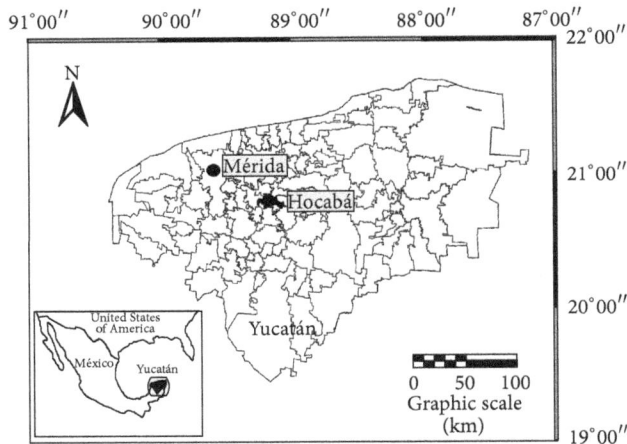

FIGURE 1: Study zone.

TABLE 1: Characterization of interviewed farmers from Hocabá, Yucatán, México.

| Farmer age | | Years doing *milpa* | | Number of *mecates*[*] | |
|---|---|---|---|---|---|
| Range (years) | n | Range | n | Range | N |
| 20–29 | 3 | 1–9 | 6 | 1–9 | 3 |
| 30–39 | 1 | 10–19 | 7 | 10–19 | 6 |
| 40–49 | 8 | 20–29 | 6 | 20–29 | 12 |
| 50–59 | 7 | 30–39 | 4 | 30–39 | 6 |
| 60–69 | 10 | 40–49 | 4 | 40–49 | 5 |
| 70–80 | 8 | 50–59 | 9 | 50–59 | 3 |
| >80 | 3 | >60 | 4 | >60 | 5 |

[*]1 *mecate* = 400 m$^2$.

common language among farmers, extensionists, technicians, and researchers.

In Yucatán, farmers descended from the old Mayan culture still have a great quantity of knowledge about soils, which they continue using for their agricultural practices [5, 8, 14]; studies concerning this matter are descriptive and only a few have attempted to systematize this knowledge and relate it to scientific soil classification systems [15–17]. In this study the soils of the municipality of Hocabá, Yucatán, México, were identified according to the Mayan farmers' knowledge and the WRB system. The differences and similarities between the two systems were analyzed in order to identify the best correspondence between them.

## 2. Materials and Methods

*2.1. Study Zone.* The municipality of Hocabá is located in the central region of the state of Yucatan at 20° 49′N and 89° 15′W within the geomorphologic landscape defined by Lugo [18] as a "structural plain almost horizontal marginal to the coast" with up to 10 m of altitude (Figure 1). Hocabá occupies an area of 81.75 km$^2$ that represents 0.18% of the state territory [19]. The climate is subhumid tropical with a summer rain season Aw$_1$(i′)g [20, 21]. The dominant vegetation is low deciduous forest [22] and the main crops of the land are sisal (*Agave fourcroydes* Lem.) and corn [23]. Two geologic zones converge in this area: a 58 million years ago limestone zone, with fine grain silicated and scarce presence of fossils in the majority of the municipality, and a 13 to 25 million years ago limestone zone, in the southeast part of the municipality, with cream and brownish microcrystalline grey rocks with great amount of fossils [24]. Intercalated zones of plains and mounds compound the topography. Mounds reach diameters of 3 to 10 m and heights up to 3 m; the plains usually have a diameter of 10 to 30 m [14].

Forty semistructured interviews with Mayan farmers were carried out in order to obtain information about the Mayan soil knowledge of the municipality of Hocabá. Interviewed farmers were "milperos" (farmers who grow *milpa*—association of corn, bean, and pumpkin) because they are the ones who have more contact and experience using the soil resource. Interviews were conducted directly on the parcel of each farmer where they were asked to mention and show the types of soils they knew, their properties or ways to recognize them, and their abundance and distribution. Farmers were also asked about types of crops they prefer to grow on each kind of soil, type of management, fertilizing, main weeds, and typical problems. The only criterion to select a farmer to be interviewed was the occurrence of their parcel in any of the two main areas of corn production within the municipality [23]. Farmers that only spoke Maya were interviewed with the help of a translator. Based mainly on the predominant responses obtained during the interviews as well as the observations made on the field, the scheme of the Mayan soil classification for this area was built. Once the Mayan soils types were recognized, representative pits for each Mayan soil identified were excavated and profiles were described [25], sampled, analysed, and classified using the WRB classification system [13]. A comparative approach was used to establish similarities and differences between Mayan knowledge and WRB system.

## 3. Results

The 40 interviewed farmers (4% of the milperos of the municipality) recognized 11 different classes of soils. Most of the interviewed farmers of the study area were older than 40 years, with variable experience on making *milpa* and worked an average area of 1 ha per year (Table 1). Farmers from 60 to 69 years old provided the majority of information about the recognition of the soils, identifying eight classes. Farmers of the three older ranges of age recognized all the types of soils found on the municipality (Table 2). There were only 2 out of 11 classes of soils recognized by all the farmers (*Boxlu'um* and *K'ankab*); the other 9 classes were recognized only by 25% or less of the farmers. The soil properties that Mayan farmers considered to classify their soils are very easy to be observed, these properties included topographic position and colour followed by amount of rock fragments, outcrops, and water retention (Table 3).

Farmers also recognized differences between soils according to the crops they prefer to grow on each class of soil

TABLE 2: Number of soils recognized by the interviewed farmers from Hocabá, Yucatán, México.

| Name | Age range | | | | | | | % |
|---|---|---|---|---|---|---|---|---|
| | 20–29 | 30–39 | 40–49 | 50–59 | 60–69 | 70–80 | >80 | |
| | $n = 3$ | $n = 1$ | $n = 8$ | $n = 7$ | $n = 10$ | $n = 8$ | $n = 3$ | $N = 40$ |
| K'ankab | 3 | 1 | 8 | 7 | 10 | 8 | 3 | 100 |
| Boxlu'um | 3 | 1 | 8 | 7 | 10 | 8 | 3 | 100 |
| Puslu'um | | | | 4 | 3 | 3 | | 25.0 |
| Ch'ich'lu'um | 1 | | 1 | 2 | 1 | | | 12.5 |
| Muluch buk'tun | | | 2 | | 2 | | 1 | 12.5 |
| Ek'lu'um | 1 | | 1 | | 1 | | | 7.5 |
| Ch'och'ol | | | | | | 1 | | 2.5 |
| Tsek'el | | | | | | 1 | | 2.5 |
| Chaltun | | | | | 1 | | | 2.5 |
| Chaklu'um | | | | | 1 | | | 2.5 |
| Haylu'um | | | | | | | 1 | 2.5 |

TABLE 3: Characteristics of the Mayan soils of the municipality of Hocabá, Yucatán, México.

| Name | Visual characteristics | | | | |
|---|---|---|---|---|---|
| | Soil color | Topographic position | Superficial rock fragments | Outcrops | Water retention |
| K'ankab, Chaklu'um | Red | Only on plains | Low or none | Low or none | Good |
| Haylu'um | Brown or reddish-brown (dark colors) | Base of the mounds | Low amount of fine gravels or none | Hard rock within first 10 cm. | Bad |
| Chichlu'um | Clear brown or black | On the flat top of the mounds | A lot of fine gravels | Low | Good |
| Puslu'um | Black | Mounds | Low amount of gravels or none | Hard rock within first 10 cm. | Very bad |
| Ch'och'ol | Black | Base of the mounds | Piles of cobbles | None | Good |
| Tsek'el, Yan yan tunichi', Muluch buk'tun | Black | Mounds | High in gravels, stones, and cobbles | High | Bad |
| Chaltun | Black | Mounds | Low or none | Very high | Bad |

and the agronomic problems they perceive (Table 4). They do *milpa* in any type of soil without discriminating between mound or plain soils. Specifically, they usually prefer sowing varieties of local chilli (*Capsicum* spp.) in the mound soils called *Ch'ich'lu'um*. Similarly, vegetable crops and other great diversity of crops are usually sowed in the plain soils free of rock fragments and outcrops called *K'ankab*. Weeds develop quicker on plain soils because there are more seeds there than in the mound soils and they can germinate at any moment when conditions become favourable. Among the plants that are exclusive or develop quicker on mound soils are *Chichibé* (*Sida acuta* Burm.) and *Sac kaatzim* (*Mimosa bahamensis* Benth.), while on plain soils *Sacchiu* (*Abutilon permolle* (willd.) Sweet) and other grasses, *Habín* (*Piscidia piscipula* Sarg.), *Tzalam* (*Lysiloma latisiliquum* (L.) Benth), *Tsotsk'ab* (*Mentzelia aspera* L.), *Kiintal* (*Desmodium purpureum* (Mill.) Fawe), and *Tajonal* (*Viguiera dentata* (Cav.)) were also mentioned.

Farmers pointed out the following problems, remarking that they are present with different intensity in each soil class. Generally, mound soils have lower water retention and

incidence of gophers, raccoons, and weeds than red soils. The sum of all these factors results, according to farmers, in low yields.

Farmers also use the type of rock associated with soils to classify them. Even more, farmers classify and use those different types of rocks according to their properties and use (Table 5). Farmers recognized five types of rocks; from those, two of them have relevant properties to agriculture, as they appear to have good water retention.

According to the WRB, soil units identified on the plains were Chromic Luvisols (LVcr), characterized by the presence of a Bt horizon and CEC > 24 cmol kg$^{-1}$ through the whole profile; Epileptic Cambisols (lep-CM), Endoleptic Cambisols (len-CM), and Endoskeletic Cambisols (skn-CM) having a Bw horizon but varying in depth and amount of rock fragments; and Lithic Leptosols (li-LP), which are soils up to 10 cm depth (Table 6).

On the mounds, the soil groups were Leptosols (LP) and Calcisols (CL). Both are dark colored (chroma less than 3) and have high organic matter contents from 23 to 50% (Table 7). Both groups have minimal amounts of fine earth due to

TABLE 4: Crops and agronomic problems of the soils of Hocabá, Yucatán, México.

| Maya soil name | Preferred crops* | Detected problems |
|---|---|---|
| K'ankab | Jamaica (*Hibiscus* sp.), macal (*Xanthosoma yucatanense*), jicama (*Pachyrhizus erosus*), yuca (*Manihot esculenta*), and sweet potato (*Ipomoea batatas*). | Weeds grow faster. *Tuzas* (*Dasyprocta mexicana*) and raccoons (*Procycon* spp.) are more frequents. |
| Haylu'um | Maize (*Zea mays*), beans (*Phaseolus* spp., *Vigna* spp.), and pumpkins (*Cucurbita* spp.). | As they are shallow soils, maize falls down easily. |
| Ch'ich'lu'um | Chili pepper (*Capsicum* spp.) and sometimes sweet potato. | *Tuzas* (less frequent). |
| Puslu'um | Maize, beans, and pumpkins | Shallow soils, low water retention. |
| Ch'och'ol | None | Little surface for planting (too many rock fragments) |
| Tsek'el, Yan yan tunichi', Mulu'ch buk'tun | Maize, beans, and pumpkins | Little surface for planting (many rock fragments). Presence of weeds. |
| Chaltun | Maize, beans, and pumpkins | Little surface for planting (too much rock). Very shallow soils. |

*Farmers do not have any preference to where to grow *Sisal* (*Agave fourcroydes*); they all agreed that the more rock fragments and outcrops in the soil the better the growth of *Sisal*.

TABLE 5: Types of rock and their characteristics according to the farmers of Hocabá, Yucatán, México.

| Maya name | Spanish name* | Use | Characteristics |
|---|---|---|---|
| Saktunich o Sascab | Creta | To build roads | It converts in powder and absorbs much water |
| Xuxtunich | Roca desgranable | Like sandpaper to cleaning animals, to complete albarradas** | It breaks easily even by hand or when it is burned and absorbs water |
| Toktunich | Roca fracturable | To build albarradas | Very hard, when it is buried is not broken only turns black and does not absorb water |
| Sakalbox | Roca soluble | To make hand grinders and albarradas | It is the hardest one and does not absorb water |
| Haysaltunich | Laja | Pib***, to complete albarradas | Does not absorb water |

*Only *laja* is a common name among farmers, the other 4 names were derived from observations of their properties; **Albarrada* is a wall made of rocks; ***Pib* means cooking in pits.

the high content of rock fragments. There are three different types of Leptosols: (1) Hyperskeletic Leptosol (LPhsk), having more than 80% by weight of rock fragments; (2) Nudilithic and Lithic Leptosol (LPli), having a depth less than 5 and 10 cm, respectively, and; (3) Calcaric Humic Leptosol (LPca-hu), more than 10 cm in depth, high organic matter content, and calcium carbonate content less than 40%. Two types of Calcisols were recognized (CL): (1) Epipetric Skeletic Calcisol (CLptpsk) and (2) Epileptic Skeletic Calcisol (CLlepsk) both of them differing in their depth.

## 4. Discussion

*4.1. Soils Identification.* In Hocabá, Yucatán, soil knowledge is being lost because there are less young people interested in making *milpa*, the main activity that relates farmers to soil. Whit each generation, fewer young people engage in this activity because most of them prefer a salaried work or studying. Moreover, most of the adults younger than 50 years old perform *milpa* in an intermittent way combining it with a salaried work [26]. The reduction of the available forest area

to make *milpa* is also a factor in the abandonment of this activity [23]. All these causes are promoting the loss of the traditional soil knowledge; this is supported in this study by the observed relationship between farmers' age and number of soils they recognize. Loss of traditional soil knowledge is occurring similar to other parts of the world [2].

No classification system is static [12] and the Mayan soil classification is not an exception. Synonymies and differences in the descriptions given by the interviewed farmers confirmed this situation. In this study, four cases of possible synonymies were found: *Puslu'um* and *Ch'ich'lu'um*, *K'ankab* and *Chaklu'um*, *Boxlu'um* and *Eklu'um*, and *Muluch buk'tun* and *Tsek'el*. The first three cases are reported by [8, 15] as different soil classes.

Soil names and descriptions provided by the farmers were contrasted with those of previous works [15, 17]; all of them presented a similar number of soil classes and the descriptions were highly consistent, although some names varied (Table 8). In those works done at state and regional levels, only three additional soils were reported for the study area (*Ya'axhom*, *Ak'alche*, and *Kacab*), suggesting that Mayan

TABLE 6: Chemical and physical properties of soils located in plains at Hocabá, Yucatán, México.

| Soil horizon | Depth cm | Dry color | Structure AS, ASi | PSD Sand % | PSD Clay % | PSD Silt % | TC | Fgr % | Cgr % | CO3 % | pH | OM % | CEC cmol+ kg-1 | Ca | Mg | Na | K | BS % |
|---|---|---|---|---|---|---|---|---|---|---|---|---|---|---|---|---|---|---|
| **Lithic Leptosol Rhodic** | | | | | | | | | | | | | | | | | | |
| A | 0–9 | 5YR 4/4 | SBK, VF-M | 43.0 | 19.0 | 38.0 | L | 0 | 0 | 0.4 | 7.4 | 14.0 | 47.2 | 23.6 | 16.6 | 0.1 | 2.8 | 100 |
| **Leptic Cambisol** | | | | | | | | | | | | | | | | | | |
| A | 0–11 | 5YR 3/3 | SBK, F-H | 48.8 | 20.6 | 31.4 | L | 0 | 0 | 0.1 | 7.6 | 15.8 | 39.7 | 31.5 | 2.7 | 0.3 | 0.8 | 89 |
| Bw1 | 11–23 | 5YR 3/3 | SBK, F-H | 48.0 | 19.6 | 32.4 | L | 0 | 0 | 0.1 | 7.6 | 13.8 | 47.8 | 31.0 | 2.9 | 0.1 | 4.4 | 81 |
| Bw2 | 23–38 | 5YR 3/2 | ABK, VF-L | 46.1 | 24.5 | 29.4 | L | 0 | 0 | 0.1 | 7.5 | 11.7 | 29.2 | 25.8 | 3.2 | 0.2 | 0.4 | 100 |
| **Endoskeletal Cambisol** | | | | | | | | | | | | | | | | | | |
| A1 | 0–9 | 5YR 3/3 | SBK, VF-M | 42 | 26 | 32 | L | 0 | 0 | 0.1 | 7.2 | 17.2 | 51.2 | 32.0 | 21.4 | 0.1 | 0.8 | 100 |
| A2 | 9–17 | 5YR 4/4 | SBK, VF-L | 50 | 22 | 28 | L | 0 | 0 | 0.1 | 7.0 | 11.0 | 49.9 | 27.8 | 25.3 | 0.7 | 0.6 | 100 |
| Bw1 | 17–39 | 5YR 4/6 | ABK, VF-L | 51 | 21 | 28 | L | 0 | 0 | 0.1 | 6.9 | 10.0 | 43.2 | 30.5 | 15.9 | 0.1 | 0.3 | 100 |
| Bw2 | 39–56 | 5YR 4/6 | ABK, VF-L | 58 | 16 | 26 | CL | 0 | 20 | 0.1 | 7.0 | 6.9 | 44.1 | 29.2 | 18.8 | 0 | 0.2 | 100 |
| C | 56–100 | 5YR 4/6 | ABK, VF-L | 47 | 19 | 33 | L | 23 | 67 | 0.1 | 7.2 | 7.2 | 29.2 | 20.5 | 12.2 | 0.1 | 0.3 | 100 |
| **Endoleptic Cambisol** | | | | | | | | | | | | | | | | | | |
| A1 | 0–4 | 5YR 4/3 | GR, F-H | 45.1 | 21.6 | 33.3 | L | 5 | 0 | 0.5 | 7.4 | 18.8 | 46.0 | 39.0 | 3.6 | 0.2 | 2.3 | 100 |
| A2 | 4–22 | 5YR 4/4 | SBK, VF-M | 49.0 | 19.6 | 31.4 | L | 0 | 0 | 0.1 | 7.3 | 12.3 | 33.2 | 38.5 | 2.3 | 0.1 | 1.0 | 100 |
| Bw1 | 22–33 | 5YR 3/6 | ABK, VF-M | 54.9 | 15.7 | 25.5 | CL | 5 | 0 | 0.1 | 7.3 | 8.1 | 29.0 | 30.0 | 2.5 | 0.1 | 0.4 | 100 |
| Bw2 | 33–55 | 5YR 4/3 | ABK, VF-L | 61.8 | 16.5 | 22.5 | CL | 3 | 0 | 0.4 | 7.5 | 9.1 | 25.5 | 19.5 | 1.2 | 0.1 | 0.3 | 100 |
| Bw3 | 55–75 | 5YR 4/3 | ABK, VF-L | 51.0 | 17.5 | 32.5 | L | 2 | 0 | 1.5 | 7.5 | 5.5 | 34.8 | 35.5 | 1.0 | 0.1 | 0.2 | 100 |
| **Haplic Luvisol Rhodic** | | | | | | | | | | | | | | | | | | |
| A1 | 0–6 | 5YR 2.5/3 | GR, F-H | 47.1 | 25.5 | 27.5 | L | 0 | 0 | 0.1 | 7.4 | 11.4 | 36.2 | 21.6 | 3.6 | 0.1 | 1.1 | 73 |
| A2 | 6–20 | 5YR 3/3 | SBK, F-M | 48.0 | 26.0 | 26.0 | L | 0 | 0 | 0.1 | 6.6 | 8.1 | 25.5 | 12.6 | 2.5 | 0.3 | 0.3 | 61 |
| Bt1 | 20–45 | 2.5YR 3/6 | ABK, M-L | 40.2 | 37.7 | 22.1 | CL | 0 | 0 | 0.1 | 6.7 | 3.8 | 24.9 | 15.3 | 0.9 | 0.2 | 0.3 | 67 |
| Bt2 | 45–85 | 2.5YR 3/6 | ABK, M-L | 30.4 | 47.7 | 22.5 | C | 0 | 0 | 0.1 | 7.0 | 3.1 | 19.0 | 16.2 | 1.8 | 0.4 | 0.3 | 97 |
| Bt3 | 85–109 | 2.5YR 4/6 | ABK, M-L | 36.8 | 39.2 | 24.0 | CL | 0 | 0 | 0.1 | 7.1 | 3.3 | 15.1 | 16.2 | 1.0 | 0.3 | 0.3 | 100 |
| Bt4 | 109–150 | 2.5YR 4/6 | ABK, M-L | 34.3 | 37.3 | 28.4 | CL | 0 | 0 | 0.1 | 7.2 | 2.7 | 10.4 | 14.8 | 3.2 | 0.3 | 0.3 | 100 |

PSD: particle size distribution. AS: aggregates shape (GR: granular, SBK: subangular blocky, ABK: angular blocky). ASi: aggregates size (VF: very fine, F: fine, M: medium). Stability (L: low, M: moderate, H: high). L: loam; Cl: clay loam; C: clay; TC: textural class. Fgr: fine gravels (<2 cm), Cgr: coarse gravels (>2 cm); OM: organic matter, CEC: cation exchange capacity, BS: base saturation.

TABLE 7: Properties of soils located in mounds at Hocabá, Yucatán, México.

| Soil horizon | Depth cm | Dry color | Structure AS, ASi | PSD Sand % | Clay % | Silt % | TC | Gr % | St % | CO₃ % | pH | OM % | CEC cmol⁺ kg⁻¹ | Exchange Cations cmol⁺ kg⁻ Ca | Mg | Na | K | BS % |
|---|---|---|---|---|---|---|---|---|---|---|---|---|---|---|---|---|---|---|
| **Hyperskeletic Leptosol** | | | | | | | | | | | | | | | | | | |
| A | 0–10 | 7.5YR 2.5/1 | GR, VF-M | 70.6 | 15.7 | 13.7 | SL | 50 | 30 | 12 | 8.0 | 45.0 | 66.2 | 54.0 | 1.8 | 0.1 | 3.3 | 89 |
| Ak/C | 10–45 | 7.5YR 3/1 | GR, VF-L | 58.8 | 17.6 | 23.5 | SL | 67 | 25 | 4 | 8.0 | 36.4 | 19 | 19.2 | 5.4 | 0.4 | 3.1 | 100 |
| **Rendzic Hyperskeletic Leptosol** | | | | | | | | | | | | | | | | | | |
| A | 0–7 | 10YR 2.5/1 | GR, F-M | 63.7 | 15.7 | 20.6 | SL | 51 | 40 | 31 | 7.8 | 34.5 | 54.1 | 38.4 | 12.6 | 0.2 | 0.9 | 100 |
| A/C | 7–23 | 10YR 2.5/1 | GR, VF-M | 71.6 | 13.7 | 14.7 | SL | 51 | 45 | 43 | 7.7 | 28.6 | 24.4 | 39.0 | 27.0 | 0.2 | 1.0 | 100 |
| **Leptic Skeletal Calcisol** | | | | | | | | | | | | | | | | | | |
| A | 0–1 | 7.5YR 2.5/1 | GR, VF-L | 55.0 | 20.0 | 25.0 | SCL | 30 | 10 | 41 | 8.0 | 30.6 | 40.7 | 38.6 | 0.7 | 0.2 | 2.8 | 100 |
| Ak/Ck | 1–15 | 7.5YR 4/3 | GR, VF-M | 62.7 | 13.7 | 23.5 | SL | 36 | 17 | 34 | 8.1 | 19.4 | 32.2 | 35.3 | 1.1 | ND | 0.6 | 100 |
| Ck/Ak | 15–50 | 7.5YR 4/3 | SBK, VF-L | 62.7 | 15.7 | 21.6 | SL | 34 | 20 | 42 | 8.0 | 19.5 | 26.7 | 35.3 | 0.9 | 0.5 | 0.6 | 100 |
| **Epipetric Skeletic Calcisol** | | | | | | | | | | | | | | | | | | |
| Ak | 0–4 | 10YR 3/3 | GR, F-H | 49.0 | 19.6 | 31.4 | L | 22 | 0 | 30 | 8.0 | 21.3 | 35.5 | 25.4 | 4.0 | 0.3 | 1.8 | 100 |
| Bk1 | 4–20 | 10YR 4/3 | SBK, F-M | 53.9 | 31.4 | 14.7 | SCL | 25 | 15 | 35 | 8.2 | 11.3 | 16.7 | 14.9 | 1.1 | 0.3 | 1.0 | 100 |
| Bk2 | 20–35 | 10YR 4/1 | SBK, VF-L | 52.9 | 21.6 | 25.6 | SCL | 21 | 20 | 36 | 8.3 | 12.9 | 18.2 | 16.5 | 1.1 | 0.3 | 0.8 | 100 |
| Ckm | 35–40 | | | | | | | | | >50 | | | | | | | | |
| IIAk | 40–60 | 10YR 5/1 | GR, VF-L | 59.8 | 19.6 | 20.6 | SL | 22 | 15 | 37 | 8.6 | 10.6 | 14.4 | 12.1 | 0.4 | 0.2 | 0.5 | 100 |
| **Humic-Hyperskeletal Leptosol** | | | | | | | | | | | | | | | | | | |
| Ak1 | 0–2 | 7.5YR 2.5/1 | GR, VF-M | 69.0 | 11.0 | 20.0 | SL | 31 | 0 | 36 | 7.5 | 49.9 | 59.9 | 32.8 | 34.1 | 0.1 | 1.1 | 100 |
| Ak2 | 2–22 | 7.5YR 3.5/1 | GR, VF-L | 67.0 | 13.0 | 20.0 | SL | 0 | 90 | 44 | 7.7 | 42.6 | 37.6 | 41.8 | 22.3 | 0.1 | 1.2 | 100 |
| C/A | 22–80 | 7.5YR 4/1 | SBK, VF-L | 71.0 | 12.0 | 17.0 | SL | 0 | 95 | 47 | 7.8 | 18.0 | 28.6 | 21.1 | 8.4 | 0.1 | 0.2 | 100 |

PSD: particle size distribution. AS: aggregates shape (GR: granular, SBK: subangular blocky, ABK: angular blocky). ASi: aggregates size (VF: very fine, F: fine, M: medium). Stability (L: low, M: moderate, H: high). L: loam; CL: clay loam; C: clay; TC: textural class. Gr: gravels, St: stones; OM: organic matter; CEC: cation exchange capacity; BS: base saturation.

TABLE 8: Comparison between the descriptions of the soils found in the municipality of Hocabá, Yucatán, México, and those presented in others studies.

| Maya name* | Aguilera (1958) [31] | Duch (1988, 1991) [15, 16] | This study |
|---|---|---|---|
| K'ankab | Light red, deep soils | Reddish brown color. Found it in plain terrains | Red soil |
| Boxlu'um | — | Black color | Black or dark soil |
| Puslu'um | — | Remarkable less amount of rock fragments than Boxlu'um | Black soil, soft, with no rock fragments. It dries out quickly |
| Ch'ich'lu'um | — | Very hard and gravel aggregates | Soil with abundant amount of gravels |
| Muluch buk'tun | — | — | Soil with a lot of rock fragments |
| Ek'lu'um | Organic soil on calcareous rock (Ek'lu'um Tsek'el) | — | Black soil |
| Ch'och'ol | Soil with calcareous rocks along profile | Soil with abundant rock fragments on the surface | Soil under piled rock fragments |
| Tsek'el | Calcareous rock with a thin layer of soil | Shallow soil with abundant rock fragments | Soil with abundant rock fragments |
| Chaltun | Soil over laja rock | Soil with calcareous armour exposed | Black shallow soil. with cracked or holed rock |
| Chaklu'um | — | Soil more red than K'ankab | Dark red soil |
| Haylu'um | — | — | Very shallow soil. Less than 10 cm depth |

*Writing of the Maya names is according to the Porrúa dictionary [32].

knowledge of soils is similar in the whole state. Typically, farmer classifications are highly variable or they have little consistency from region to region [8, 12, 27]; however, it seems that the Mayan soil knowledge is quite homogeneous, even at regional level [8]. This homogeneity can help to facilitate its systematization.

On the other hand, this apparent homogeneity could also indicate a loss of the soil knowledge. This statement is supported by the results of this work in which only two classes of soils were recognized by all the interviewed farmers (Boxlu'um and K'ankab) and these two terms were used to refer to the rest of the soils when they did not know them.

Most of the classification systems reflect the priorities of who propose them [28]. Characteristics that Mayan farmers use to classify soils are mainly visual and intimately related to their agricultural activities.

Farmers recognized colour differences among soils; however, it was observed that there are different tones that farmers do not consider as distinctive elements. A particular case is the Mayan term Box that means black or dark. Many farmers referred the soil called Chichlu'um that is usually light brown to dark brown as simply Boxlu'um. However, apparently this is related to the absence of a Mayan word to designate brown color, although some farmers used the Spanish term "achocolatado" (colored as chocolate) to refer to this colour. Another special case is the soil called K'ankab whose common translation is "yellow place at the bottom" in attention to subsuperficial soils horizons [8]. The Mayan word Chak means red; thus, Chaklu'um are red soils. These soils are darker than other red top soils such as K'ankab.

Soil depth was only an important characteristic for farmers to differentiate between shallow (soils < 10 cm) and deep soils (soils > 10 cm). Farmers recognized these soils

empirically during sowing when they insert their sowing stick, by observation of aerial roots on maize plants, or when maize plants fall down due to the wind action because roots lack deep anchorage. Farmers affirmed that mound soils are not very deep but they pointed out that roots always find cracks in the rocks to continue growing. Shallow soils are called Haylu'um in both plains and mounds and correspond to Nudilithic Leptosols. Farmers judge the depth of K'ankab by its surface color, the darker the soil is the shallower it is, and the lighter the soil is the deeper it is. Bautista and Zinck [8] reported these differences in deepness for K'ankab soils.

The microtopographic position, superficial amount of rock fragments, and outcrops are three characteristics that the farmers always consider together. We found five rock types with different uses and it is possible that they influence soil characteristics and soil genesis [14, 26]. For example, soils called Ch'och'ol can only be found at the base of the mounds under stone accumulations (Hyperskeletic Leptosols), while soils designated Chaltun are almost always near the mounds but present bedrock very near to the surface (Nudilithic Leptosols).

Water retention is a characteristic that many farmers recognized but their comments relative to this property were inconsistent. For many farmers soils that do not retain water (they dried out first) were those on the mounds, while others assured that it was the soils on the plains. This disagreement can be due to the farmers not considering the variability in the amount of rock fragments or the depth of the soils as factors that determine the water retention. In fact, this characteristic was only relevant to recognize the soil called Puslu'um, which farmers consistently designated as the soil that dried out first.

Other studies have found that soil texture is an important property for local classifications [11, 29], but it seems that this is not the case of the Mayan classification, because none of their agricultural practices requires a direct physical contact with the deeper layers of soil.

*4.2. Soils Uses.* Preferences for growing crops were mainly linked to the more availability of workable surface for sowing in plain soils. Farmers like to grow most of their crops on the *K'ankab* soils, because the absence of rock fragments and outcrops makes field operations easier and quicker. The one clear exception is the soil called *Ch'ich'lu'um*, in which farmers prefer to sow peppers and sweet potato, arguing that they only grow well in that class of soil. It is possible that a nutritional reason exists to explain the best development of those crops in that class of soil, but this remains to be confirmed with soil fertility analyses.

Although some farmers said that with good rain the mound soils give better production, most of them assured that crops on mound soils as well as plain soils grow well if it has rained well and on time. On the other hand, when rain is not good, some farmers said that the mound soils produce higher yields while others argue that the plain soils do. The reason for these inconsistencies could be the depth of the soils and the amount of rock fragments and outcrops. Comparatively, shallow soils can store less water than deep soils, but amount of rocks and stones on mound soils help to conserve the humidity better than in shallow plain soils.

Soil water content is a property associated by farmers with the presence of weeds. In this regard, they said that they have many problems to control weeds because as soon as it rains, weeds appear in the soil. They also pointed out differences in weed composition and abundance between mound and plain soils. A farmer making *milpa* in the same place for five years in a row said that every year he had to use more herbicide and dedicate more hours to remove weeds than the earlier year.

In contrast to what the authors of [16, 17] found, the farmers interviewed in this study did not use any terms to refer to soil fertility. This is perhaps because these authors did their studies using a deeper anthropologic approach. The author of [16] developed the hierarchical classification of the soils of the Puuc region of the state of Yucatan and outlined a classification departed from a linguistic point of view, grouping the soil classes according to the meaning of the Mayan names as well as some management aspects. Such research is important to obtain information that may no longer exist among the contemporary inhabitants of a region.

## 5. Comparison and Systematization

Our results suggest that Mayan soils knowledge is given at three levels (Figure 2). In level one, topographic position mound (*Muluch*) or plain (*K'ankabal*). Mound soils are dark, generally black, grey, or brown, while plain soils are red to red-brown which makes a first division among soil classes resulting in a general designation for soils according to their color. Dark soils are designated as *Box'luum* and red soils as *K'ankab*. The prefix *Box* means literally black but it is used to refer to all classes of dark colours. On the other hand, the prefix *K'an* means literally yellow but is used to designate light red soils on plains. Level two is almost exclusive for

mound soils since in the plains the amount of rock fragments and outcrops are nearly absent. However, in some cases, red mound soils, having abundant amount of rock fragments and outcrops, were observed and recognized by farmers as *Ch'ich-K'ankab* or *Tsek'el-K'ankab*. This was the only case where farmers used a compound name mixing two single names.

In level three, variations of soils from the second level were recognized according to their association with specific topographic position. Here, there were two subdivisions: (1) soil names ending with the Mayan word *lu'um*, which means soil, for example, *Ch'ich'lu'um*—soil with gravels—(Hsk-LP) or *Haylu'um*—very shallow soil—(Nu-LP) and (2) soils that were designated according to the specific microtopographic positions on which they occur, for example, *Ch'och'ol*—soil under piled rock fragments—(Hsk-LP), *Tsek'el*—soil among rock fragments—(Nu-LP), *Chaltun*—soil between outcrops—(Nu-LP). At this level, soils in mounds were recognized as *K'ankab, Haylu'um Ch'ich'lu'um, Tsek'el, Chaltun, Puslu'um, Ch'och'ol*, and so on. In the case of the plain soils, farmers only recognized two variants: *K'ankab* and *Hay'lu'um*. In both cases when farmers were not sure of the specific name of the soil, they designate as *Boxluu'm* to all soils in mounds and *K'ankab* to all soils in plains.

Following this scheme, it can be seen that level one (mound soils and plain soils) is the most studied level so far [5, 14, 15]. It is in the second and third levels that research is needed. It is in those levels where the participation of the farmers is important in order to better understand each one of the elements of the landscape and topography that they recognize and use to identify soils.

Mayan soil types and WRB units cannot be directly related to each other because these systems share few diagnostic properties and assign them different relative importance [8]. Many soils identified by farmers relate with more than one WRB group of soil and vice versa; in these cases, no direct relationship between both classification systems is possible (Table 9).

People's understanding of soils constitutes a complex knowledge system, with some categories similar or complementary to those used by modern soil science [8, 30]. For example, even though hierarchical levels of the WRB system are based on qualitative and quantitative data, they use qualifiers to distinguish soils at secondary levels. Some of those characteristics, that is, gravels or rock fragments percentage, are related to the Mayan approach. For instance, amount of rock fragments is a very important property for building hierarchal levels in the Mayan soil nomenclature (e.g., *Ch'och'ol*) as well as in the WRB classification at the qualifier level (e.g., Hyperskeletic and Skeletic Leptosols).

## 6. Conclusions

In Hocabá, Yucatán farmers distinguished two main groups of soils: *K'ankab* or soils of plains and *Boxlu'um* or soils of mounds. *K'ankab* is a group of red soils with two variants (*K'ankab* and *Haylu'um*), whereas *Boxlu'um* is a group of dark soils with five variants (*Tsek'el, Ch'ich'lu'um, Chaltun, Puslu'um*, and *Ch'och'ol*). Soils on the plains were identified as Leptosols, Cambisols, Cambisols, and Luvisols. Soils

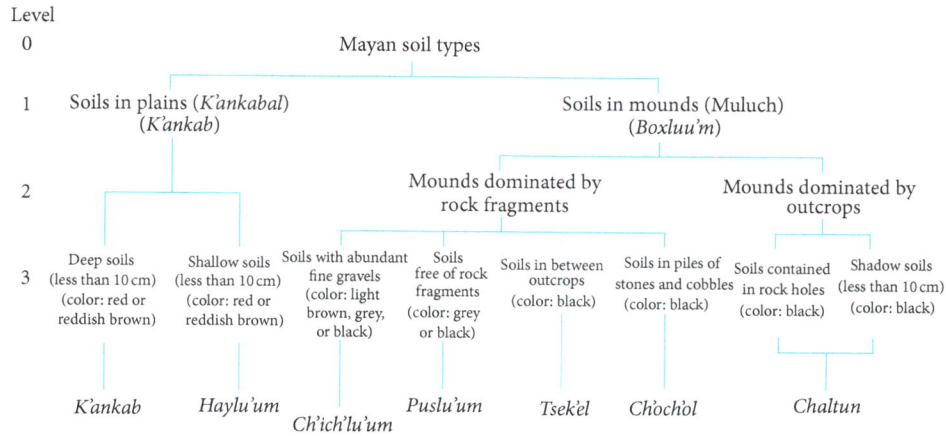

FIGURE 2: Mayan soil types in Hocabá, Yucatán.

TABLE 9: Relationship between WRB soil units and Mayan soil types.

| WRB soil group | Dominant kind of rock | Topographic position | Mayan soil class |
|---|---|---|---|
| Chromic Luvisols<br>Eutric Cambisols<br>Mollic Leptosol | Fracturable | Any part of the plains | K'ankab |
| Nudilithic Leptosol | Fracturable | Base of the mounds | Haylu'um |
| Hyperskeletic Leptosol | | | Ch'ich'lu'um |
| Haplic Calcisol<br>Hyperskeletic Leptosol | Graintable and powderable | Flat tops of the mounds | Puslu'um<br>Ch'ich'lu'um |
| Petrocalcic Calcisol | Fracturable and grainable | Top of the mounds | Puslu'um |
| Nudilithic Leptosol<br>Hyperskeletic Leptosol | Fracturable<br>Fracturable<br>Fracturable | Top of the mounds<br>Sides of the mounds<br>Base of the mounds | Boxlu'um<br>Tsek'el<br>Ch'och'ol |
| Rendzic Leptosol<br>Nudilithic Leptosol | Fracturable | Top of the mounds<br>Sides of the mounds | Boxlu'um<br>Tsek'el |
| Nudilithic Leptosol<br>Lithic Leptosol<br>Nudilithic Leptosol | Soluble<br>Soluble<br>Fracturable | Top of the mounds<br>Sides of the mounds<br>Base of the mounds | Chaltun<br>Boxlu'um<br>Haylu'um |

identified in mounds were Leptosols and Calcisols. Mayan soil types and WRB groups are complementary; they should be used together in order to improve both soil classifications, to help transference of agricultural technologies, and make soil management decisions. Soil characteristics that should be considered for a local soil classification system are topographic position (plain or mound), colour, amount of rock fragments and outcrops, and soil depth or effective rooting depth.

## Acknowledgments

Acknowledgments are due to the Science and Technology National Council (CONACyT) for the scholarship granted to the first author for his master degree studies; to CONA-CyT project: *Base de datos para la península de Yucatán, incluyendo nomenclatura maya y FAO*, Convenio: R31624-B; to the American Institute for Global Change Research (IAI), project: Biogeochemical cycles under land use change in the semiarid Americas; to Dr. Steve Gliessman, Dr. Robert Graham, and M.Sc. Arturo Caamal Maldonado for their comments and corrections to the paper; to Wendy Huchim and Miguel Huicab for their collaboration on some of the tables and figures. This paper is dedicated to the memory of my friend Mr. Pedro Canché farmer from Hocabá who with his wide knowledge of soils, roads, and people helped me as guide and Mayan-Spanish translator and in contacting other farmers.

## References

[1] A. M. G. A. Winklerprins, "Local soil knowledge: a tool for sustainable land management," *Society and Natural Resources*, vol. 12, no. 2, pp. 151–161, 1999.

[2] P. V. Krasilnikov and J. A. Tabor, "Perspectives on utilitarian ethnopedology," *Geoderma*, vol. 111, no. 3-4, pp. 197–215, 2003.

[3] N. Barrera-Bassols and J. A. Zinck, "Ethnopedology: a world-wide view on the soil knowledge of local people," *Geoderma*, vol. 111, no. 3-4, pp. 171–195, 2003.

[4] N. Barrera-Bassols, "Etnoedafología purhépecha," *México Indígena*, vol. 6, no. 24, pp. 47–52, 1988.

[5] F. Bautista, J. Jiménez, J. Navarro, A. Manu, and R. Lozano, "Microrelieve y color del suelo como propiedades de diagnóstico en Leptosoles cársticos," *Terra*, vol. 21, no. 1, pp. 1–12, 2003.

[6] B. J. Williams, "Aztec soil science," *Boletín del Instituto de Geografía*, vol. 7, pp. 115–120, 1975.

[7] B. J. Williams and C. A. Ortiz-Solorio, "Middle American folk soil taxonomy," *Annals, Association of American Geographers*, vol. 71, no. 3, pp. 335–358, 1981.

[8] F. Bautista and J. A. Zinck, "Construction of an Yucatec Maya soil classification and comparison with the WRB framework," *Journal of Ethnobiology and Ethnomedicine*, vol. 6, article 7, 2010.

[9] V. Toledo, "Indigenous knowledge on soils: an ethnoecological conceptualization," in *Ethnopedology in A Worldwide Perspectives: An Annotated Bibliography*, N. Barrera-Bassols and J. A. Zinck, Eds., International Institute for Aerospace Survey and Earth Science, Enschede, The Netherlands, 2000.

[10] A. J. Tabor, "Soil classification systems," in *Soil of Arid Regions of the U.S. and Israel*, 1997, http://cals.arizona.edu/oals/soils/classifsystems.html.

[11] J. A. Sandor and L. Furbee, "Indigenous knowledge and classification of soils in the Andes of Southern Peru," *Soil Science Society of America Journal*, vol. 60, no. 5, pp. 1502–1512, 1996.

[12] A. J. Tabor, "Soil surveys and indigenous soil classification," *Indigenous Knowledge and Development Monitor*, vol. 1, no. 1, 1993.

[13] IUSS-ISRIC-FAO, "2006 World reference base for soil resources 2006: A framework for international classification, correlation and communication," World Soil Resour Report 103, FAO, Rome, Italy.

[14] F. Bautista, H. Estrada-Medina, J. Jiménez-Osornio, and J. González-Iturbe, "Relación entre relieve y suelos en zonas cársticas," *Terra*, vol. 22, no. 3, pp. 243–254, 2004.

[15] G. J. Duch, *La conformación territorial del estado de Yucatán—Los componentes del medio físico*, Centro Regional de la Península de Yucatán CRUPY, Universidad Autónoma de Chapingo, México, Mexico, 1988.

[16] G. J. Duch, *Fisiografía del estado de Yucatán—su relación con la agricultura*, Centro Regional de la Península de Yucatán CRUPY, Universidad Autónoma de Chapingo, México, Mexico, 1991.

[17] P. N. Dunning, "Soils and vegetation," in *Lords of the Hills: Ancient Maya Settlement in the Puuc Region, Yucatán, México*, vol. 15 of *Monographs in World Archaeology*, Prehistory Press, Madison, Wis, USA, 1992.

[18] J. Lugo, "Geomorfología," in *Atlas de Procesos Territoriales de Yucatán*, P. P. Chico, Ed., Universidad Autónoma de Yucatán, Yucatán, México, 1999.

[19] Gobierno Constitucional del Estado de Yucatán, "Programa de desarrollo regional de la zona Henequenera de Yucatán 1992–1994," México, 1992.

[20] INEGI, "Anuario estadístico del estado de Yucatán," Gobierno del estado de Yucatán, 1995.

[21] R. Orellana, "Evaluación Climática (Climatología de la Península de Yucatán)," in *Atlas de Procesos Territoriales de Yucatán*, A. García de Fuentes, C. Y. Ordóñez, and P. Chico Ponce de León, Eds., pp. 162–182, Facultad de Arquitectura, Universidad Autónoma de Yucatán, Mérida, Mexico, 2000.

[22] S. J. Flores and C. I. Espejel, "Tipos de vegetación de la península de Yucatán," Fascículo 3, Sostenibilidad Maya. Universidad Autónoma de Yucatán, México, Mexico, 1994.

[23] L. Cano, *Cambio del uso del suelo en el municipio de Hocabá, Yucatán, México [Tesis de maestría en Manejo y Conservación de Recursos Naturales Tropicales]*, Facultad de Medicina Veterinaria y Zootecnia, Universidad Autónoma de Yucatán, 2000.

[24] INEGI, "Instituto Nacional de Estadística, Geografía e Informática," Carta Geológica 1:250000, Secretaría de Programación y Presupuesto, México, Mexico, 1983.

[25] C. Siebe, R. Janh, and K. Stahr, "Manual para la descripción y evaluación ecológica de suelos en el campo." Publicación Especial 4, Sociedad Mexicana de la Ciencia del Suelo, A. C. Edo. de México, México, Mexico, 1996.

[26] F. Bautista, J. Garcia, and A. Mizrahi, "Diagnóstico campesino de la situación agrícola en Hocabá, Yucatán," *Terra Latinoamericana*, vol. 23, no. 4, 2005.

[27] E. Habarurema and K. G. Steiner, "Soil suitability classification by farmers in southern Rwanda," *Geoderma*, vol. 75, no. 1-2, pp. 75–87, 1997.

[28] M. Corbeels, A. Shiferaw, and M. Haile, "Farmers' knowledge of soil fertility and local managament strategies in Tigray, Ethiopia," Managing Africa's Soils 10, IIED, London, UK, 2000.

[29] L. Pool-Novelo, E. Cervantes-Trejo, and S. Mesa-Díaz, "La clasificación *Tsotsil* de soils en el paisaje cárstico de la subregión San Cristobal de las casas, Chiapas, México," *Terra*, vol. 9, no. 1, pp. 11–23, 1991.

[30] N. Barrera-Bassols and A. J. Zinck, *Ethnopedology in a Worldwide Perspectives: An Annotated Bibliography*, International Institute for Aerospace Survey and Earth Science, Enschede, The Netherlands, 2000.

[31] N. Aguilera Herrera, "Suelos," in *Los recursos naturales del Sureste y su aprovechamiento*, E. Beltrán, Ed., Instituto Mexicano de Recursos Naturales Renovables, México City, Mexico, 1958.

[32] Porrúa, *Dicccionario Maya*, Editorial Porrúa, México, Mexico, 2 edition, 1991.

# 8

# Prediction of Soil Organic Carbon for Ethiopian Highlands Using Soil Spectroscopy

**Tadele Amare,[1,2] Christian Hergarten,[1] Hans Hurni,[1] Bettina Wolfgramm,[1] Birru Yitaferu,[2] and Yihenew G. Selassie[3]**

[1] *Centre for Development and Environment, University of Bern, Hallerstrasse 10, CH-3012 Bern, Switzerland*
[2] *Amhara Regional Agricultural Research Institute, P.O. Box 527, Bahir Dar, Ethiopia*
[3] *Bahir Dar University, P.O. Box 79, Bahir Dar, Ethiopia*

Correspondence should be addressed to Tadele Amare; tadele17b@yahoo.com

Academic Editors: M. Bernoux and G. Broll

Soil spectroscopy was applied for predicting soil organic carbon (SOC) in the highlands of Ethiopia. Soil samples were acquired from Ethiopia's National Soil Testing Centre and direct field sampling. The reflectance of samples was measured using a FieldSpec 3 diffuse reflectance spectrometer. Outliers and sample relation were evaluated using principal component analysis (PCA) and models were developed through partial least square regression (PLSR). For nine watersheds sampled, 20% of the samples were set aside to test prediction and 80% were used to develop calibration models. Depending on the number of samples per watershed, cross validation or independent validation were used. The stability of models was evaluated using coefficient of determination ($R^2$), root mean square error (RMSE), and the ratio performance deviation (RPD). The $R^2$ (%), RMSE (%), and RPD, respectively, for validation were Anjeni (88, 0.44, 3.05), Bale (86, 0.52, 2.7), Basketo (89, 0.57, 3.0), Benishangul (91, 0.30, 3.4), Kersa (82, 0.44, 2.4), Kola tembien (75, 0.44, 1.9), Maybar (84. 0.57, 2.5), Megech (85, 0.15, 2.6), and Wondo Genet (86, 0.52, 2.7) indicating that the models were stable. Models performed better for areas with high SOC values than areas with lower SOC values. Overall, soil spectroscopy performance ranged from very good to good.

## 1. Introduction

Ethiopia is one of the largest countries in Africa. The vast majority of the society (~80%) depends on agriculture which contributes around 42% of the growth domestic product (GDP) [1]. Despite its strategic importance for the country's economic development, the agricultural sector suffers from low efficiency, population pressure, ineffective land management, and unfavourable land use practices leading to widespread land degradation. Still, the country is strongly committed to environmental protection, rehabilitation, and sustainable land management as evident from Article 92 of its constitution [2], and various internationally supported initiatives have been launched aiming to stop and reverse this trend. Among others, the Growth and Transformation Plan (GTP) is the major one. To address these challenges, efficient and affordable land management practices are required in all

regions of the country and their effectiveness should be analysed and monitored. Alongside, there is pronounced need to improve knowledge on soil resources and to collect reliable data on soil's state and dynamics, allowing for operational assessment and monitoring of this important resource more precisely [3]. Such information is critically important for research and development interventions.

Traditionally, such information was collected through comparatively expensive and slow wet chemistry analysis methods. Every year, many soil samples are collected and analysed for determination of soil properties critical to soil management and crop husbandry using the conventional (wet chemistry) laboratory methods. However, the country is facing a critical challenge to continue with conventional methods as the cost for chemicals is very high and the efficiency of conventional laboratories is low. Soil spectroscopy as an alternative approach has yet to be established in

Ethiopia; therefore, conventional methods of soil analysis currently remain the only available option. The advent of spectroscopy in soil science provided a promising prospect and is nowadays successfully applied in many parts of the world since its early beginnings [3]. Soil spectroscopy is an analytical technique used for soil analysis [4]. It is fast and efficient in producing comprehensive soil information in a short span of time [3–7]. Many soil attributes can be determined with a single scan [4, 6, 8]. Moreover soil spectroscopy requires only a small amount of soil; it is a nondestructive procedure allowing samples to be stored and reanalysed. Finally, it requires less complex infrastructure as compared to the conventional soil analysis. Despite the aforementioned benefits, the technology is not yet widely used in Ethiopia. To our knowledge, the only scientific report on soil spectroscopy in Ethiopia was published by Vågen et al. [9]. In view of the country's urgent need for more efficient technologies, a study on the applicability of soil spectroscopy for the prediction of SOC was conducted in the highlands of Ethiopia. Soil organic carbon is an important indicator for soil health and fertility, serving alongside as a measure for the potential of carbon storage in the soils in relation to climate change mitigation [10]. Therefore, the objective of the study was to evaluate the feasibility of VNIR soil spectroscopy for predicting SOC in the highlands of Ethiopia.

## 2. Materials and Methods

*2.1. Study Area.* The chosen study sites cover most of the agroecological zones, climate conditions, geologies, and soils of the Ethiopian highlands (Figure 1 and Table 1).

The study sites including Bale Mountain, Basketo, Maybar, and Wondogenet are characterised by bimodal type of rainfall, which is manifested in the eastern, northeastern, and south-eastern parts of the country. The other study sites have one prolonged rainy season [11]. The general characteristics of the study sites are given in Table 1. The southern and western parts of the country are mostly under perennial crops; the northern parts of the country are dominated by cereal crops. The farming systems are characterised by small-scale crop-livestock mixed types. In all of the study sites, sustainable land management practices have been introduced in the course of land rehabilitation and reclamation campaigns.

*2.2. Soil Sample Sets.* For this study, a total of 1159 soil samples were chemically and spectrally analysed. The major share of soil samples (713) was acquired from the soil archive of Ethiopia's National Soil Testing Centre (NSTC), representing the soils of the dominant highland areas of the country. We explored the soil archive for the existing soil samples together with the analysed soil parameters, method of analysis, year of sampling, and origin of the samples. Then, sites for this study were selected based on availability of organic carbon data and their representation to the highlands of Ethiopia. Nevertheless, it was not possible to get samples representing the whole range of the highlands collected during the same year and hence we took samples collected from 2006 to 2010 as follows: Benishangul in 2006; Megech in 2008; Bale

and Wondogenet in 2009; and Kola tembien, Kersa, and Basketo in 2010. In addition, 446 soil samples were collected in October-December 2010 from Anjeni and Maybar sites.

Soil samples in the catchments of Anjeni and Maybar (Figure 1) were collected according to a field sampling procedure presented in Amare et al. [12]. The sampling design was optimized to account for the spatial variability of soil properties due to land use and land management activities, as well as for pedogenetic factors varying across spatial and temporal scales. Samples from all other sites analysed and stored at the NSTC were collected following similar sampling designs as outlined previously, addressing soil characterisation and effects of land use and management practices in the watersheds.

*2.3. Chemical Soil Analysis.* The soil organic carbon content of all samples from the NSTC was determined using the wet oxidation Walkley-Black method [13], a standard procedure in Ethiopia. The additionally collected samples from Anjeni and Maybar were dried under shadow conditions, ground, and sieved to pass through 2 mm sieve and then soil organic carbon was determined similarly to the other sample sets. All samples were analysed in the same laboratory following the same procedure to errors due to laboratory procedure and helped to work with a homogeneous sample set.

*2.4. Spectral Measurement.* Pretreatment and spectral measurements were carried out in line with the protocol developed by the global soil spectroscopy group [14, 15]. The reflectances of all soil samples were measured at Ethiopia's National Soil Testing Centre in Addis Abeba, February 2011, using a FieldSpec 3 diffuse reflectance spectrometer (Analytical Spectral Devices, Boulder, CO, USA). The spectra were measured from 350 to 2500 nm at 1 nm interval. A high-intensity mug light source was used to illuminate samples from the bottom through borosilicate Duran glass petri dishes. Roughly, 1 cm layer of soil was poured into the sample holders. A panel consisting of spectralon (Labsphere, North Sutton, USA) was used as a white reference standard between each measurement. The spectrometer was optimised regularly after measuring 10 samples and between shifting sets of different soil samples. Two replicate spectra with 90-degree rotation were collected in order to increase the precision of the measurements. Whenever high variations occurred between the scans of the same sample, the measurement was taken again. After checking the spectral signatures of each repeated measurement, average spectra of the two readings were computed. Spectral inconsistencies (splices) were observed for all samples at 1000 and 1830 nm and were corrected by applying an offset. In order to reduce data and the processing time, only every 10th reading starting from 380 nm was kept for further analysis [14]. Spectral regions below 380 nm or above 2450 nm were removed because they were affected by noise [3].

*2.5. Data Analysis and Model Development.* The geological background, soil type, and agro-ecology of the study areas are diverse (Table 1 and Figure 1). Not only between, but

TABLE 1: Overview description of the study sites.

| Sampling site | Agroecological zone[*] | Rainfall (mm) | Temperature range (°C)[**] | Dominant soil types[***] | Geological background |
|---|---|---|---|---|---|
| Anjeni | Wet Weyna Dega | 1650 | 9–23 (16) | Alisols, Nitisols, Cambisols, Regosols, Acrisols, Luvisols, and Leptosols | Tertiary Basalts |
| Bale mountains | Dega-High Dega | 870–1064 | 8.6–15.2 (13.1) | Andosols | Volcanic rocks with volcanic ash material |
| Basketo | Weyna Dega | ~1300 | 12.8–27.7 (18) | Nitisols and Andosols | Ignimbrites and Rhyolites |
| Benishangul | Weyna Dega/ Kolla | 1116 | 9.5–34.4 (25) | Nitisols | Tertiary Gneisses and Basalts |
| Kersa | Dega/Weyna Dega | ~1400 | ~18–25 (19) | Vertisols and Nitisols | Mostly tertiary basaltic rocks |
| Kolatembien | Weyna Dega | 600 | ~16–20 (18) | Regosols, Cambisols, and Leptosols | Antalo Limestones |
| Maybar | Moist Weyna Dega/moist Dega | 1150 | 12.3–21.4 (16.4) | Phaeozems, Gleysols, Fluvisols, and Leptosols | Alkali-olive Basalts and Tuffs |
| Megech | Weyna Dega | 700–1160 | 14.2–26.9 (20) | Vertisols, Fluvisols, and Cambisol | Pleistocene to recent alluvial deposits |
| Wondogenet | Weyna Dega | 948 | 12.6–27.3 (19) | Luvisols, Andosols, and Nitisols | Ignimbrites, Basalts, and Tuffs |

[*]Hurni, 1998 [11], [**]the mean annual values are in the parentheses, [***]based on FAO-Unesco soil classification.

also within a given watershed, environmental conditions are highly variable (Table 2). Due to this fact, a separate modelling strategy was followed in this study, developing one model for each watershed. According to the main objectives of this study, which was to test the applicability of soil spectroscopy for various environmental settings in Ethiopian highlands, this choice was further motivated by findings and discussions given by [16, 17]. For exploratory data analysis, the R statistics language [18] was used while modelling was done in Unscrambler 10.1 (CAMO, Oslo, Norway).

The dataset was screened for outliers using PCA as well as model residuals. One sample from Bale, one sample from Megech, and three samples from Wondogenet study sites were flagged for removal due to high residual variance as well as high leverage and score distance; we speculate that sample handling errors or fundamentally different soil characteristics (out-of-population sample) could be possible reasons for these variations.

Various spectral preprocessing options were tested, and finally preprocessing with first derivative was found to improve the performance of the models best in terms of stability and interpretability, except for the Bale case study site, where second-order derivative was scoring higher. Derivative is a widely used preprocessing technique [16], which has a baseline correction effect and is able to enhance weak signals [16, 19]. Before modelling, all spectra further underwent autoscaling (mean-centering and variance-scaling), which is a recommended standard procedure for most cases [20].

The raw spectra displayed in Figure 2 reflect the variability of the VNIR spectra in the sample set exemplarily for the Bale case study site. Basically, the shape of all VNIR spectra was similar, showing a steep ascent from 400 to 750 nm, which is characteristic for iron oxides [16]. The

dominant absorption regions (reflectance minima) around 1450 and 1900 nm are usually attributable to $OH^-$ and $H_2O$, masking most of other signals [16]. The higher similarity of the spectra of the derivative spectra given in the lower part of Figure 2 is resulting from the removal of additive and multiplicative scatter effects achieved by the derivative preprocessing techniques.

*2.5.1. Calibration Models.* Partial least square regression (PLSR) was used to develop models for the relationship between soil spectral reflectance and soil organic carbon determined by wet chemistry. The number of samples selected for calibration, validation, and prediction varied from watershed to watershed based on the number of samples. For all watersheds except for Anjeni, models were developed with 80% of the samples with cross validation, while 20% of the samples were randomly selected and set aside for model testing. The distributions of SOC for these samples selected randomly for independent model testing were compared and scrutinized with SOC values samples used for calibration and validation using descriptive statistics (Table 2) and statistical tests discussed later.

*2.5.2. Splitting Samples into Calibration, Validation, and Test Sets.* For Anjeni watershed, the number of samples (302) was sufficient to split into calibration and validation sets. Before splitting the sample sets into calibration and validation sets, 20% of the samples (61 samples) were selected randomly for checking purpose. The 61 samples constitute the test set. After selection and separation of samples for testing set, the remaining 80% of the samples (241 samples) were split again randomly into calibration set (145 samples) and validation set (96 samples). This approach separating

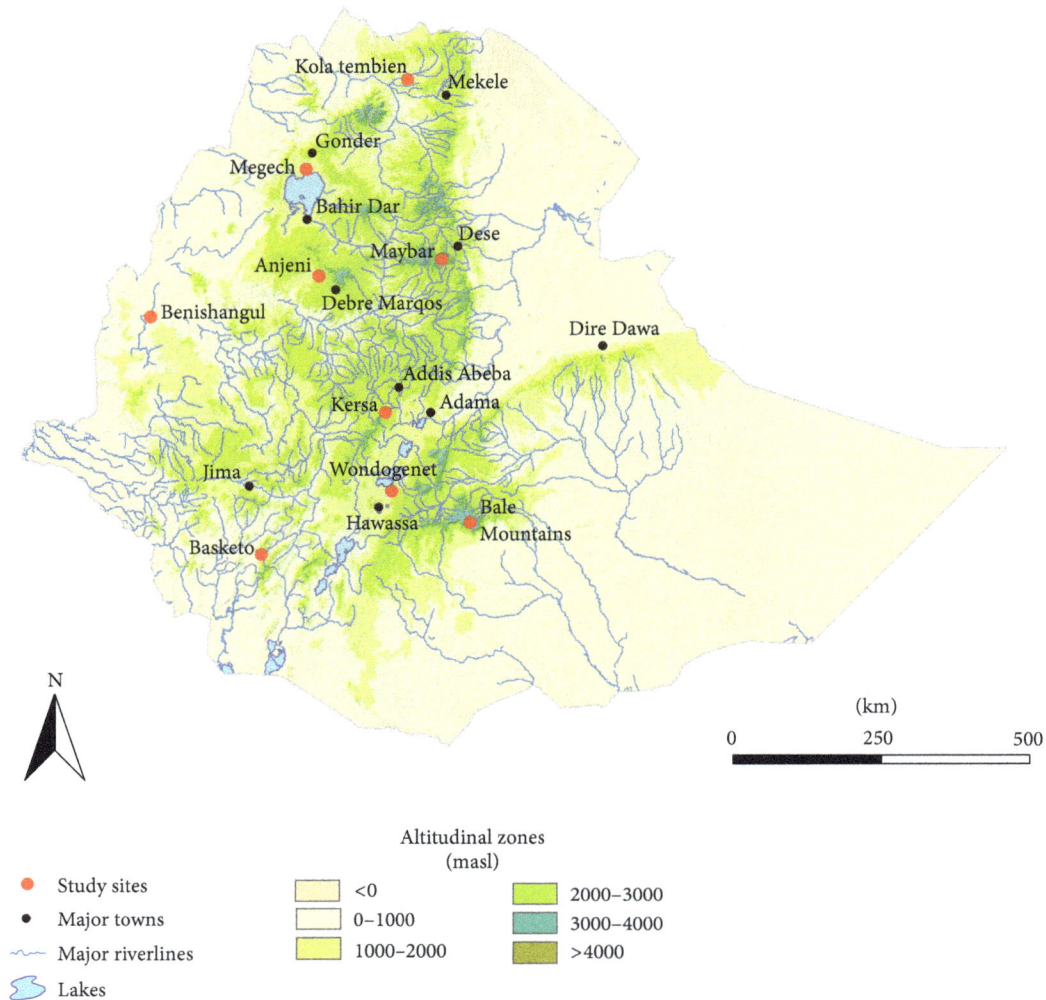

FIGURE 1: Map of Ethiopia indicating the location of the study sites.

calibration, validation, and testing sets is recommended as a robust procedure by Varmuza and Filzmoser [21].

Therefore, the separation of all testing sets was done following the same procedure as given previously. The sample selection of calibration set, validation set, and test set for a given site was based on the size of the samples sets available per study area. The statistical appropriateness of the sample split was scrutinized using the two samples $t$-test for mean homogeneity, Wilcox-sum rank test for distribution homogeneity, and the Levene's test for variance homogeneity. Furthermore, descriptive statistics were calculated for total samples, calibration, validation, and test sets as presented in Table 2.

The statistical parameters ($R^2$, RMSE, and RPD) were calculated as indicated below.

The coefficient of determination ($R^2$) is as follows:

$$R^2 = \frac{\sum_{i=1}^{N}\left(X - \overline{X}\right)\left(Y - \overline{Y}\right)^2}{\sqrt{\sum\left(X - \overline{X}\right)^2 \sum\left(Y - \overline{Y}\right)^2}}, \quad (1)$$

where $X$ is the laboratory measured soil organic carbon (%) for each observation and $\overline{X}$ is the mean value of the laboratory measured while $Y$ is the predicted soil organic carbon (%) and $\overline{Y}$ its mean value.

The root mean square error (RMSE) is as follows:

$$\text{RMSE} = \sqrt{\sum_{i=1}^{N}\left(\frac{Y - X}{N}\right)^2}, \quad (2)$$

where $Y$ is the predicted (fitted) value, $X$ the measured value, and $N$ is the number of observation.

The ratio of performance deviation (RPD) [22] is as follows:

$$\text{RPD} = \frac{\text{SD}}{\text{RMSE}}, \quad (3)$$

where SD is the standard deviation of the measured samples.

Finally, calibration models were evaluated based on coefficient of determination ($R^2$), root mean square error (RMSE), and ratio of performance deviation (RPD) as discussed later in the results part of this paper.

TABLE 2: Statistical summary of the samples analyzed for SOC content (%) for calibration, validation, and test sample sets.

| Study sites | Number of samples | Min. | Max. | Med. | Mean | SDV | 1st quartile | 3rd quartile |
|---|---|---|---|---|---|---|---|---|
| Anjeni | | | | | | | | |
|   Total | 302 | 0.20 | 13.68 | 1.53 | 1.67 | 1.34 | 0.90 | 2.15 |
|   Calibration | 145 | 0.20 | 13.68 | 1.39 | 1.54 | 1.36 | 0.84 | 1.94 |
|   Validation | 96 | 0.21 | 9.90 | 1.55 | 1.70 | 1.27 | 0.92 | 2.22 |
|   Test | 61 | 0.27 | 9.74 | 1.76 | 1.93 | 1.40 | 1.08 | 2.39 |
| Bale | | | | | | | | |
|   Total | 74 | 1.23 | 7.50 | 3.14 | 3.35 | 1.49 | 2.26 | 4.22 |
|   Calibration | 59 | 1.23 | 7.20 | 3.08 | 3.25 | 1.41 | 2.20 | 4.22 |
|   Test | 15 | 1.84 | 7.5 | 3.36 | 3.78 | 1.81 | 2.56 | 4.47 |
| Basketo | | | | | | | | |
|   Total | 102 | 0.28 | 8.05 | 2.18 | 2.30 | 1.60 | 0.94 | 2.77 |
|   Calibration | 81 | 0.28 | 8.05 | 2.23 | 2.36 | 1.69 | 0.89 | 2.81 |
|   Test | 21 | 0.39 | 5.6 | 2.18 | 2.13 | 1.23 | 1.52 | 2.4 |
| Benishangul | | | | | | | | |
|   Total | 78 | 0.10 | 5.46 | 0.70 | 1.19 | 0.98 | 0.49 | 1.80 |
|   Calibration | 64 | 0.10 | 5.46 | 0.72 | 1.19 | 1.01 | 0.44 | 1.73 |
|   Test | 14 | 0.20 | 2.87 | 0.67 | 1.18 | 0.90 | 0.53 | 1.92 |
| Kersa | | | | | | | | |
|   Total | 171 | 0.11 | 6.07 | 1.69 | 1.86 | 1.02 | 1.23 | 2.22 |
|   Calibration | 135 | 0.11 | 6.07 | 1.77 | 1.91 | 1.05 | 1.24 | 2.36 |
|   Test | 36 | 0.20 | 4.42 | 1.63 | 1.70 | 0.89 | 1.20 | 2.05 |
| Kolatembien | | | | | | | | |
|   Total | 65 | 0.28 | 4.59 | 0.83 | 1.10 | 0.86 | 0.53 | 1.29 |
|   Calibration | 51 | 0.28 | 4.59 | 0.81 | 1.05 | 0.85 | 0.48 | 1.21 |
|   Test | 14 | 0.46 | 3.90 | 1.09 | 1.29 | 0.93 | 0.61 | 1.51 |
| Maybar | | | | | | | | |
|   Total | 144 | 0.19 | 7.58 | 1.56 | 1.97 | 1.33 | 1.14 | 2.40 |
|   Calibration | 115 | 0.19 | 7.58 | 1.54 | 1.99 | 1.42 | 1.09 | 2.34 |
|   Test | 29 | 0.68 | 4.27 | 1.71 | 1.89 | 0.93 | 1.35 | 2.40 |
| Megech | | | | | | | | |
|   Total | 147 | 0.16 | 2.54 | 0.88 | 0.92 | 0.38 | 0.74 | 1.11 |
|   Calibration | 118 | 0.16 | 2.54 | 0.89 | 0.93 | 0.39 | 0.73 | 1.11 |
|   Test | 29 | 0.32 | 1.44 | 0.84 | 0.87 | 0.29 | 0.78 | 1.02 |
| Wondogenet | | | | | | | | |
|   Total | 71 | 1.23 | 8.29 | 3.1 | 3.37 | 1.38 | 2.34 | 4.21 |
|   Calibration | 57 | 1.23 | 8.29 | 3.28 | 3.36 | 1.39 | 2.44 | 4.02 |
|   Test | 14 | 1.56 | 6.16 | 2.96 | 3.36 | 1.43 | 2.34 | 4.30 |

SDV: the standard deviation, Min: minimum, Max: maximum, Med: median. All splits of the sample sets given in Table 2 have been tested for statistical conformity. For all splits, the test results confirmed the null hypothesis which is assuming homogeneity of the parameters tested.

## 3. Results and Discussion

*3.1. Soil Organic Carbon Distributions in the Study Sites.* As shown in Table 2, soil organic carbon (from wet chemistry analysis) was highly variable between locations and within each location. The highest variability of soil organic carbon contents was observed at Anjeni with values ranging from 0.2% (soil samples from agricultural fields) to 13.68% (samples collected from the top soil of an old forest belonging to Orthodox Church). The latter value is also the maximum SOC value recorded for the entire sample set used for this study. Such high SOC content results from a combination of high rainfall, favourable temperature, and regular turnover of biomass without forest clearing or biomass removal, based on the Ethiopian Orthodox Church's strict rules and regulations for its own forest management. The church's contribution

to and role in forest protection and reforestation have been studied in detail by Wassie Eshete [23]. The highest mean value for soil organic carbon was observed in Wondogenet (3.37%), followed by Bale (3.35%). Both sites are located in high rainfall areas of southern Ethiopia, where agroforestry-based farming systems with generally higher SOC values is dominating. The Bale site is located at the Bale Mountain National Park. The lowest mean values for SOC were found in Megech (0.92%) and Kola tembien (1.10%). Kola tembien is located in the northern and dryer part of the country, where many areas are affected by land degradation and low SOC values. As such, our SOC content analysis findings listed in Table 2 are consistent with those from earlier soil-related studies conducted in the respective parts of Ethiopia. In northern Ethiopia, Mekuria et al. [24], Van de Wauw et al. [25], Abegaz et al. [26], Abegaz and van Keulen [27], and

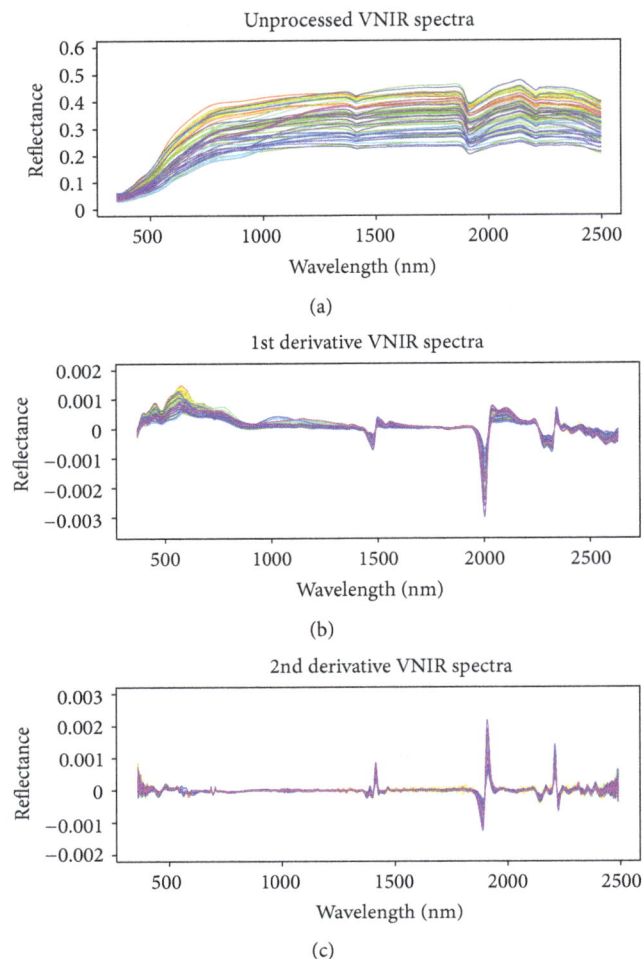

Unprocessed VNIR spectra

(a)

1st derivative VNIR spectra

(b)

2nd derivative VNIR spectra

(c)

FIGURE 2: Spectral preprocessing from raw spectral data to first derivative and second derivatives for Bale samples.

Tesfahunegn et al. [28] reported SOC contents within the range of the results given in Table 2. For the southern parts of the country, Solomon et al. [29], Lemenih et al. [30], Teklay et al. [31], and Ashagrie et al. [32] found similar carbon contents. Yimer et al. [33] and Yimer et al. [34] in the Bale Mountains, while Spaccini et al. [35] and Tulema et al. [36] reported similar results for the central and northeastern highlands. Based on these findings, we conclude that the results presented in Table 2 are adequately well representing the respective areas in terms of the SOC range covered.

*3.2. Spectral Data Analysis Using PCA.* As described previously, the spectral readings from all study sites were analysed using PCA in order to detect outliers and patterns in the data; however, this was also done to detect clusters or subgroups for aggregated model building, comparing the sites to each other. Given the high variability listed in Table 1, PCA seems to be a promising and powerful tool for reducing the amount of redundant data and dimension reduction. Therefore, PC1 and PC2 were calculated from the raw spectra of selected study sites and the scores were displayed in a scatterplot (Figure 3). The cumulative variance explained by the first two

PCs was higher than 90% in all cases. Figure 3 displays the score plots for the study sites of Kersa, Kola tembien, and Maybar, indicating that these watersheds are very different; only little similarity exists between the scores, even if the 95% data coverage ellipse might suggest a small overlap. Similarly, Bale, Kola tembien, and Megech study sites were projected into the PC space and show well-defined regions with few overlapping samples. Samples from Maybar were also separated from samples of Wondogenet and Bale Mountain with only few overlapping. Samples from Megech were also separated from samples of Wondogenet and Bale. While the Megech samples cover rather compact region in the plot, the samples from Kola tembien spread less uniformly over much larger area, which might indicate various levels of complexity and variability within the watershed. In Figure 3 samples Maybar covers compact region with five distinct samples far away from the centre. Still Maybar samples were also separated from Wondogenet and Bale samples with only few overlap. Figure 3 also displays samples from Megech together with Bale and Kersa. This figure clearly shows a certain degree of similarity between Bale and Kersa watersheds, indicating the potential for developing a combined model. The discussed results are not very surprising when one considers the highly diverse environmental conditions present in the different catchments, and the uneven spread of samples the in PCA space of Kola tembien and Bale even suggest to further subset models. Similar results and conclusions have been reported by McDowell et al. [17] and Stenberg et al. [16], stating that subset models score higher in terms of precision and accuracy. The downside of course of subset model is their lack of stability and robustness.

Thus, using PCA, it was possible to reveal the data structure in the samples relating to their sources of origin, with varying levels of overlap. The method is fast due to complete absence of soil analysis by wet chemistry procedure. Odlare et al. [37] have suggested that field characteristics must be significantly different to be distinguished from each other. Similarly, PCA has been used effectively to separate organic and nonorganic wine products [38], soil colors [18], soluble and less soluble elements [39], and sediment sources [40]. The use of PCA also reported to separate soil samples according to management practices such as tillage [41] and composting [42].

*3.3. Model Calibration, Validation, and Testing.* Using PLSR, the results show in general that visible/near-infrared spectroscopy performed well for all different agroecological conditions, soil types, and land management practices of Ethiopia. As shown in Table 3, all values of the coefficient of determination ($R^2$) were in a range of very good to good with references to Stenberg and Rossel [43] and Chang et al. [44].

For all sites except Kola tembien and Kersa, the ratio of performance deviation (RPD) was above 2.5 for validation— a value that is considered very adequate for developing stable prediction models for soil organic carbon [45]. The value of RPD for the present study ranged between 2.79 (Maybar) and 6.04 (Benishangul) for calibration and between 1.92 (Kola tembien) and 3.39 (Benishangul) for the validation set. The

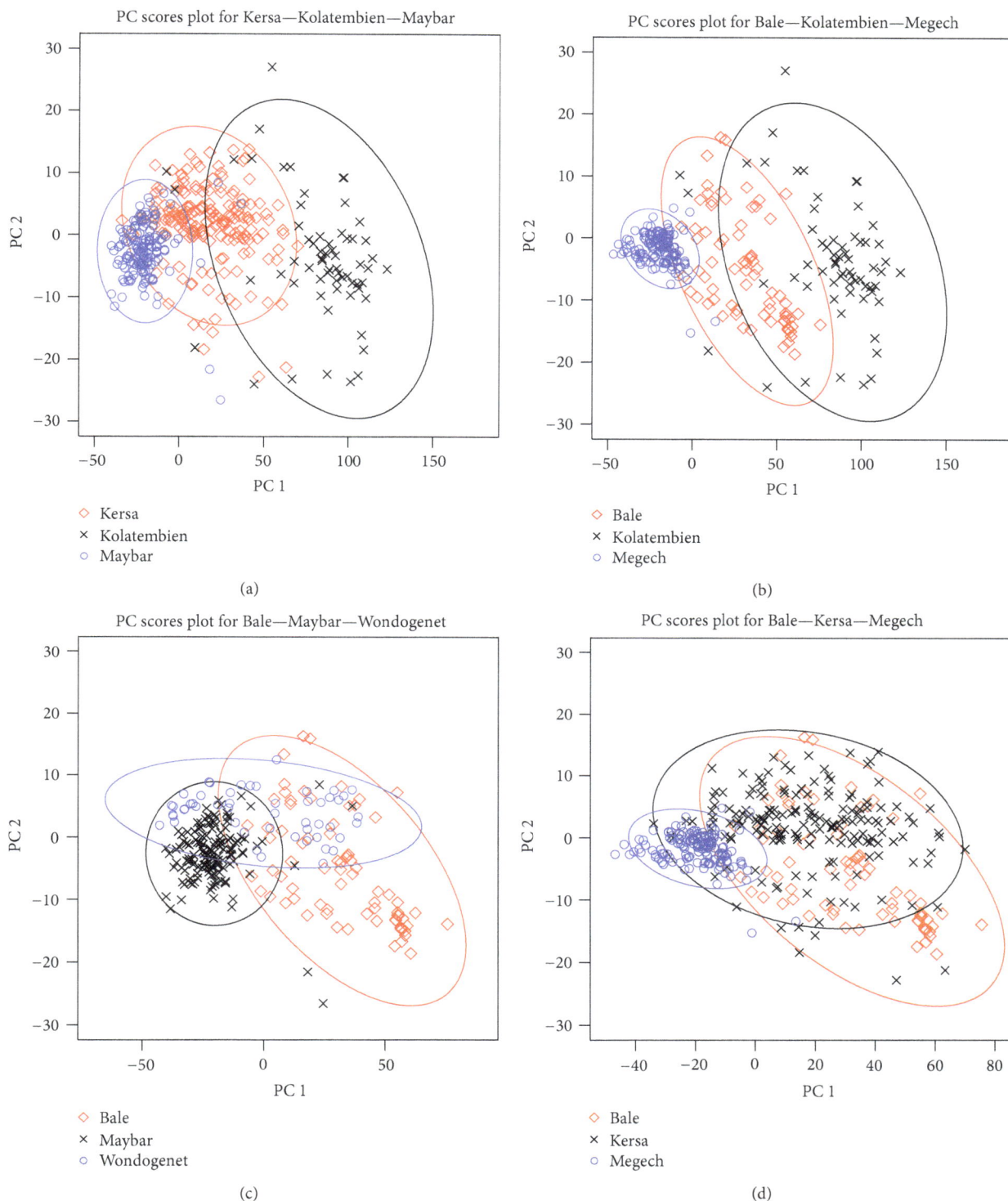

FIGURE 3: PCA scores for the study sites.

comparatively poor performance of the calibration model for Kola tembien may be related to the low levels of SOC, which is related to the high level of soil degradation [46] and high carbonate. This reflects the findings by other authors who reported unstable calibration models for areas coined by low

SOC and high carbonate levels [16]. Benishangul is covered by a single soil type (Nitisols), which developed on Gneisses and Basalts. The area is further dominated by uniform land use (cropland). This low variation of environmental conditions is well expressed by the good model fit. On the other

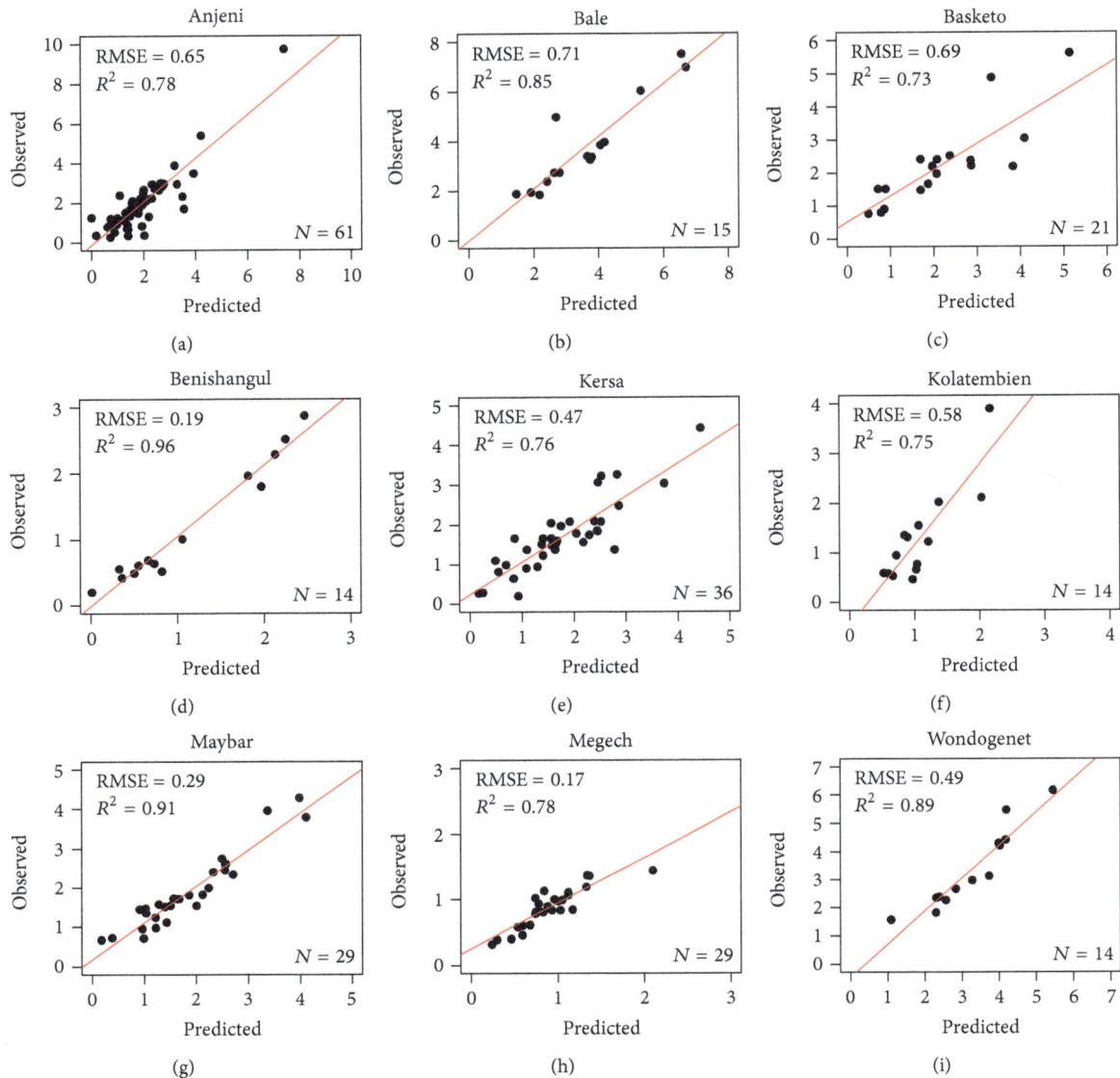

FIGURE 4: Comparison between SOC (%) determined by Walkley-Black method (observed) and spectral model (predicted) for the test sample sets.

hand, Anjeni, watershed in Gojam, provided very acceptable model results despite the high range of SOC (Table 2), high variability of soil types (Table 1), and land uses [47].

The range of the root mean square error (%) varied between 0.11 (Megech) and 0.53 (Maybar) for the calibration and between 0.15 (Megech) and 0.57 (Maybar) for validation. Rossel et al. [45] reported a range of root mean square error (RMSE) comparable to the present results. Rossel and Behrens [48] achieved maximum precision with 0.75% RMSE and 89% $R^2$ for soil organic carbon, using artificial neural network (ANN) with wavelet coefficients—this is comparable to the present results (Table 3). They reported higher RMSE and lower $R^2$ with multiple linear regressions (MLR), PLSR, multivariate adaptive regression splines (MARS), support vector machines (SVM), random forests (RF), and boosted trees (BT)—when compared to ANN—indicating that the

results are fairly reliable. After calibration and validation, the developed models were tested with ~20% of the samples from each study area, which is presented in Figure 4.

## 4. Conclusions and Recommendations

From the results of the study, it is possible to conclude that soil spectroscopy is an effective method for predicting soil organic carbon in the highlands of Ethiopia. There are considerable variations in the performance of the local models, which are mostly linked to the variability of the environmental settings prevailing within the watersheds. Models for the study sites in the moist southern and western parts of the country seem to perform slightly better. In general terms, the relationship between predicted and measured values of soil organic carbon was high for all sites, regardless of the

TABLE 3: Summary of model results.

| | Model | Number of samples | $R^2$ | RMSE (%) | RPD |
|---|---|---|---|---|---|
| Anjeni | Calibration | 145 | 90 | 0.46 | 3.1 |
| | Validation | 96 | 88 | 0.44 | 3.1 |
| Bale | Calibration | 59 | 93 | 0.42 | 3.7 |
| | Validation | 59 | 86 | 0.52 | 2.7 |
| Basketo | Calibration | 81 | 93 | 0.46 | 3.9 |
| | Validation | 81 | 89 | 0.57 | 3.0 |
| Benishangul | Calibration | 64 | 97 | 0.18 | 6.0 |
| | Validation | 64 | 91 | 0.30 | 3.4 |
| Kersa | Calibration | 135 | 90 | 0.35 | 3.2 |
| | Validation | 135 | 82 | 0.44 | 2.4 |
| Kolatembien | Calibration | 51 | 88 | 0.33 | 2.3 |
| | Validation | 51 | 75 | 0.44 | 1.9 |
| Maybar | Calibration | 115 | 87 | 0.53 | 2.8 |
| | Validation | 115 | 84 | 0.57 | 2.5 |
| Megech | Calibration | 118 | 94 | 0.11 | 4.0 |
| | Validation | 118 | 85 | 0.15 | 2.6 |
| Wondogenet | Calibration | 57 | 90 | 0.46 | 3.2 |
| | Validation | 57 | 86 | 0.52 | 2.7 |

$R^2$: the coefficient of determination, RMSE: root mean square error, RPD: ratio of performance deviation.

high variability of soil organic carbon within watersheds like Anjeni. The RPD values for validation ranged from 1.9 (Kola tembien) to 3.4 (Benishangul). This indicates fairly stable models and thus the models developed from these data could be used to build local spectral libraries with sufficient predictive power for local land use and management-related applications. Nevertheless, there is a potential for further improvement of the presented local models, which could be achieved by enlarging the range of the sample set and adding more samples from the respective sites, which would also allow for developing robust calibration models.

Considering the results from the PCA, there is a potential for research in investigating the performance of aggregated models and other modelling strategies, alongside with the use of nonparametric calibration methods as suggested by many authors. While the present study only focussed on soil organic carbon, soil spectroscopy's ability to predict other chemical, biological, and physical soil parameters in Ethiopia remains to be explored.

## Acknowledgments

This paper is based on the work conducted within the framework of Swiss National Centre of Competence in Research (NCCR) North South: Research Partnerships for Mitigating Syndromes of Global Change. The authors wish to thank the Centre for Development and Environment (CDE), University of Bern, for providing the spectrometer. They would also like to express thier gratitude to the Ethiopian National Soil Testing Centre for providing access to the soil archive. Thanks are due to the Amhara Regional Agricultural Research Institute (ARARI) for its support and facilitation. They further extend their gratitude to Berhanu Debele and Deresse Gebrewold for their facilitation and administrative support. Finally, thanks are due to Lorenz Ruth for his keen assistance with spectral measurement.

## References

[1] L. Barrow, P. Mwanakatwe, H. Hashi, H. Zerihun, and V. Ndisale, "Federal Democratic Republic of Ethiopia Country Strategy Paper," 2011.

[2] Ethiopia, *The Constitution of the Federal Democratic Republic of Ethiopia*, Addis Abeba, Ethiopia, 1995.

[3] K. D. Shepherd and M. G. Walsh, "Development of reflectance spectral libraries for characterization of soil properties," *Soil Science Society of America Journal*, vol. 66, no. 3, pp. 988–998, 2002.

[4] K. D. Shepherd and M. G. Walsh, "Infrared spectroscopy—enabling an evidence-based diagnostic surveillance approach to agricultural and environmental management in developing countries," *Journal of Near Infrared Spectroscopy*, vol. 15, no. 1, pp. 1–19, 2007.

[5] D. J. Brown, K. D. Shepherd, M. G. Walsh, M. Dewaynemays, and T. G. Reinsch, "Global soil characterization with VNIR diffuse reflectance spectroscopy," *Geoderma*, vol. 132, no. 3-4, pp. 273–290, 2006.

[6] R. A. V. Rossel, D. J. J. Walvoort, A. B. McBratney, L. J. Janik, and J. O. Skjemstad, "Visible, near infrared, mid infrared or combined diffuse reflectance spectroscopy for simultaneous assessment of various soil properties," *Geoderma*, vol. 131, no. 1-2, pp. 59–75, 2006.

[7] D. Brunet, B. G. Barthès, J.-L. Chotte, and C. Feller, "Determination of carbon and nitrogen contents in Alfisols, Oxisols and Ultisols from Africa and Brazil using NIRS analysis: effects of sample grinding and set heterogeneity," *Geoderma*, vol. 139, no. 1-2, pp. 106–117, 2007.

[8] E. Ben-Dor and A. Banin, "Near-infrared analysis as a rapid method to simultaneously evaluate several soil properties," *Soil Science Society of America Journal*, vol. 59, no. 2, pp. 364–372, 1995.

[9] T. V. Vågen, L. A. Winowiecki, A. Abegaz, and K. M. Hadgu, "Landsat-based approaches for mapping of land degradation prevalence and soil functional properties in Ethiopia," *Remote Sensing of Environment*, vol. 134, pp. 266–275, 2013.

[10] R. T. Conant, S. M. Ogle, E. A. Paul, and K. Paustian, "Measuring and monitoring soil organic carbon stocks in agricultural lands for climate mitigation," *Frontiers in Ecology and the Environment*, vol. 9, no. 3, pp. 169–173, 2011.

[11] H. Hurni, *Agroecological Belts of Ethiopia*, University of Berne, Bern, Switzerland, 1998.

[12] T. Amare, A. Terefe, Y. G. Selassie, B. Yitaferu, B. Wolfgramm, and H. Hurni, "Soil properties and crop yields along the terraces and toposequece of Anjeni Watershed, Central Highlands of Ethiopia," *Journal of Agricultural Science*, vol. 5, no. 2, 2013.

[13] A. Walkley and A. Black, "An examination of degtjareff method for determining soil organic matter and a proposed modification of the chromic acid titration method," *Soil Science*, vol. 37, pp. 29–37, 1934.

[14] R. V. Rossel, "The soil spectroscopy group and the development of a global soil spectral library," in *Proceedings of the European*

*Geosciences Union (EGU) General Assembly Conference*, vol. 11, abstracts 14021, April 2009.

[15] J. Wetterlind, B. Stenberg, and R. A. V. Rossel, "Soil analysis using visible and near infrared spectroscopy," in *Plant Mineral Nutrients: Methods and Protocols, Methods in Molecular Biology*, F. J. M. Maathuis, Ed., 1st edition, 2013.

[16] B. Stenberg, R. A. V. Rossel, A. M. Mouazen, and J. Wetterlind, "Visible and near infrared spectroscopy in soil science," in *Advance in Agronomy*, vol. 107, Chapter 5, pp. 163–215, Academic Press, New York, NY, USA, 2010.

[17] M. L. McDowell, G. L. Bruland, J. L. Deenik, and S. Grunwald, "Effects of subsetting by carbon content, soil order, and spectral classification on prediction of soil total carbon with diffuse reflectance spectroscopy," *Applied and Environmental Soil Science*, vol. 2012, Article ID 294121, 14 pages, 2012.

[18] A. M. Mouazen, R. Karoui, J. Deckers, J. De Baerdemaeker, and H. Ramon, "Potential of visible and near-infrared spectroscopy to derive colour groups utilising the Munsell soil colour charts," *Biosystems Engineering*, vol. 97, no. 2, pp. 131–143, 2007.

[19] R. A. V. Rossel, "ParLeS: software for chemometric analysis of spectroscopic data," *Chemometrics and Intelligent Laboratory Systems*, vol. 90, no. 1, pp. 72–83, 2008.

[20] M. K. Boysworth and K. S. Booksh, "Aspects of Multivariate Calibration Applied to Near-Infrared Spectroscopy," in *Handbook of Near-Infrared Analysis*, D. A. Burns and E. W. Ciurczak, Eds., pp. 207–2228, CRC Press, New York, NY, USA, 3rd edition, 2008.

[21] K. Varmuza and P. Filzmoser, *Introduction To Multivariate Statistical Analysis in Chemometrics*, vol. 64, no. 4, CRC Press, New York, NY, USA, 2009.

[22] P. C. Williams, *Near Infrared Technology in the Agriculture and Food Industries*, P. Williams and K. Norris, Eds., American Association of Cereal Chemists, St. Paul, Minn, USA, 1st edition, 1987.

[23] A. Wassie Eshete, *Ethiopian Church Forests? Opportunities and Challenges for Restoration*, Wageningen University, The Netherlands, 2007.

[24] W. Mekuria, E. Veldkamp, M. Haile, J. Nyssen, B. Muys, and K. Gebrehiwot, "Effectiveness of exclosures to restore degraded soils as a result of overgrazing in Tigray, Ethiopia," *Journal of Arid Environments*, vol. 69, no. 2, pp. 270–284, 2007.

[25] J. Van de Wauw, G. Baert, J. Moeyersons et al., "Soil-landscape relationships in the basalt-dominated highlands of Tigray, Ethiopia," *Catena*, vol. 75, no. 1, pp. 117–127, 2008.

[26] A. Abegaz, H. van Keulen, and S. J. Oosting, "Feed resources, livestock production and soil carbon dynamics in Teghane, Northern Highlands of Ethiopia," *Agricultural Systems*, vol. 94, no. 2, pp. 391–404, 2007.

[27] A. Abegaz and H. van Keulen, "Modelling soil nutrient dynamics under alternative farm management practices in the Northern Highlands of Ethiopia," *Soil and Tillage Research*, vol. 103, no. 2, pp. 203–215, 2009.

[28] G. B. Tesfahunegn, L. Tamene, and P. L. G. Vlek, "Catchment-scale spatial variability of soil properties and implications on site-specific soil management in northern Ethiopia," *Soil and Tillage Research*, vol. 117, pp. 124–139, 2011.

[29] D. Solomon, J. Lehmann, T. Mamo, F. Fritzsche, and W. Zech, "Phosphorus forms and dynamics as influenced by land use changes in the sub-humid Ethiopian highlands," *Geoderma*, vol. 105, no. 1-2, pp. 21–48, 2002.

[30] M. Lemenih, E. Karltun, and M. Olsson, "Soil organic matter dynamics after deforestation along a farm field chronosequence in southern highlands of Ethiopia," *Agriculture, Ecosystems and Environment*, vol. 109, no. 1-2, pp. 9–19, 2005.

[31] T. Teklay, A. Nordgren, and A. Malmer, "Soil respiration characteristics of tropical soils from agricultural and forestry land-uses at Wondo Genet (Ethiopia) in response to C, N and P amendments," *Soil Biology and Biochemistry*, vol. 38, no. 1, pp. 125–133, 2006.

[32] Y. Ashagrie, W. Zech, G. Guggenberger, and T. Mamo, "Soil aggregation, and total and particulate organic matter following conversion of native forests to continuous cultivation in Ethiopia," *Soil and Tillage Research*, vol. 94, no. 1, pp. 101–108, 2007.

[33] F. Yimer, S. Ledin, and A. Abdelkadir, "Concentrations of exchangeable bases and cation exchange capacity in soils of cropland, grazing and forest in the Bale Mountains, Ethiopia," *Forest Ecology and Management*, vol. 256, no. 6, pp. 1298–1302, 2008.

[34] F. Yimer, S. Ledin, and A. Abdelkadir, "Changes in soil organic carbon and total nitrogen contents in three adjacent land use types in the Bale Mountains, south-eastern highlands of Ethiopia," *Forest Ecology and Management*, vol. 242, no. 2-3, pp. 337–342, 2007.

[35] R. Spaccini, J. S. C. Mbagwu, P. Conte, and A. Piccolo, "Changes of humic substances characteristics from forested to cultivated soils in Ethiopia," *Geoderma*, vol. 132, no. 1-2, pp. 9–19, 2006.

[36] B. Tulema, J. B. Aune, F. H. Johnsen, and B. Vanlauwe, "The prospects of reduced tillage in tef (Eragrostis tef Zucca) in Gare Arera, West Shawa Zone of Oromiya, Ethiopia," *Soil and Tillage Research*, vol. 99, no. 1, pp. 58–65, 2008.

[37] M. Odlare, K. Svensson, and M. Pell, "Near infrared reflectance spectroscopy for assessment of spatial soil variation in an agricultural field," *Geoderma*, vol. 126, no. 3-4, pp. 193–202, 2005.

[38] D. Cozzolino, M. Holdstock, R. G. Dambergs, W. U. Cynkar, and P. A. Smith, "Mid infrared spectroscopy and multivariate analysis: a tool to discriminate between organic and non-organic wines grown in Australia," *Food Chemistry*, vol. 116, no. 3, pp. 761–765, 2009.

[39] S. Xiong, Y. Zhang, Y. Zhuo, T. Lestander, and P. Geladi, "Variations in fuel characteristics of corn (Zea mays) stovers: general spatial patterns and relationships to soil properties," *Renewable Energy*, vol. 35, no. 6, pp. 1185–1191, 2010.

[40] G. Alaoui, M. N. Léger, J.-P. Gagné, and L. Tremblay, "Assessment of estuarine sediment and sedimentary organic matter properties by infrared reflectance spectroscopy," *Chemical Geology*, vol. 286, no. 3-4, pp. 290–300, 2011.

[41] B. Govaerts, K. D. Sayre, and J. Deckers, "A minimum data set for soil quality assessment of wheat and maize cropping in the highlands of Mexico," *Soil and Tillage Research*, vol. 87, no. 2, pp. 163–174, 2006.

[42] M. V. Gil, L. F. Calvo, D. Blanco, and M. E. Sánchez, "Assessing the agronomic and environmental effects of the application of cattle manure compost on soil by multivariate methods," *Bioresource Technology*, vol. 99, no. 13, pp. 5763–5772, 2008.

[43] B. Stenberg and R. A. V. Rossel, "Diffuse reflectance spectroscopy for high-resolution soil sensing," in *Proximal Soil Sensing*, R. A. V. Rossel, A. B. McBratney, and B. Minasny, Eds., vol. 1 of *Progress in Soil Science*, pp. 29–47, Springer, The Netherlands, 2010.

[44] C.-W. Chang, D. A. Laird, M. J. Mausbach, and J. Hurburgh C.R., "Near-infrared reflectance spectroscopy—principal components regression analyses of soil properties," *Soil Science Society of America Journal*, vol. 65, no. 2, pp. 480–490, 2001.

[45] R. A. V. Rossel, R. N. McGlynn, and A. B. McBratney, "Determining the composition of mineral-organic mixes using UV-vis-NIR diffuse reflectance spectroscopy," *Geoderma*, vol. 137, no. 1-2, pp. 70–82, 2006.

[46] K. Vancampenhout, J. Nyssen, D. Gebremichael et al., "Stone bunds for soil conservation in the northern Ethiopian highlands: impacts on soil fertility and crop yield," *Soil and Tillage Research*, vol. 90, no. 1-2, pp. 1–15, 2006.

[47] W. Alemu, T. Amare, B. Yitaferu, Y. G. Selassie, B. Wolfgramm, and H. Hurni, "Impacts of soil and water conservation on land suitability to crops: the case of Anjeni Watershed, Northwest Ethiopia," *Journal of Agricultural Science*, vol. 5, no. 2, 2013.

[48] R. A. V. Rossel and T. Behrens, "Using data mining to model and interpret soil diffuse reflectance spectra," *Geoderma*, vol. 158, no. 1-2, pp. 46–54, 2010.

# 9

# Assessing Sediment-Nutrient Export Rate and Soil Degradation in Mai-Negus Catchment, Northern Ethiopia

**Gebreyesus Brhane Tesfahunegn[1,2] and Paul L. G. Vlek[2]**

[1] College of Agriculture, Aksum University, Shire Campus, P.O. Box 314, Shire, Ethiopia
[2] Center for Development Research, University of Bonn, Walter-Flex-Straße 3, 53113 Bonn, Germany

Correspondence should be addressed to Gebreyesus Brhane Tesfahunegn; gebre33@gmail.com

Academic Editors: T. J. Cutright, J. A. Entry, M. Goss, C. Martius, and W. R. Roy

Even though soil degradation challenges sustainable development, the use of degradation indicators such as nutrient export (NE) and nutrient replacement cost is not well documented at landform level. This study is aimed to investigate the extent of soil degradation, NE rates, and their replacement cost across landforms in the Mai-Negus catchment, northern Ethiopia. Different erosion-status sites (*aggrading*, *stable*, and *eroded*) in the landforms were identified, and soil samples were randomly collected and analysed. Nutrient export, replacement cost, and soil degradation were calculated following standard procedures. This study showed that soil degradation in the *eroded* sites ranged from 30 to 80% compared to the corresponding *stable* site soils, but the highest was recorded in the mountainous and central ridge landforms. Average NE of 95, 68, 9.1, 3.2, 2.5, and 0.07 $kg\,ha^{-1}\,y^{-1}$ for soil calcium, carbon, nitrogen, potassium, magnesium, and phosphorus, respectively, was found from the landforms. Significantly strong relationships between NE and sediment yield in the landforms were observed. Annual nutrient replacement costs varied among the landforms though the highest was in the reservoir (€9204 in May 2010). This study thus suggests that while introducing antierosion measures, priority should be given to erosion sources to the reservoir such as mountainous and central ridge landforms.

## 1. Introduction

Soil erosion is a challenge for sustainable agricultural development in many developing countries [1, 2]. The problem is more serious in the Ethiopian highlands such as the Tigray region [3–5]. Inappropriate agricultural practices, high population pressure from human and livestock, higher rainfall intensity, and rugged topography have been reported as the main facilitators for having severe erosion [6, 7].

In the Tigray region, average soil loss by erosion on cultivated land is more than $49\,t\,ha^{-1}\,y^{-1}$ [8], which exceeds the average soil loss of $42\,t\,ha^{-1}\,y^{-1}$ for Ethiopia as a whole [3]. Such soil loss through water erosion is almost always accompanied by losses of essential soil nutrients. Erosion is selective for fine soil particles, which are relatively richer in soil nutrients [9, 10]. In line with this, Stoorvogel and Smaling [11] and UNDP [12] reports showed that compared to rates in sub-Saharan Africa, Ethiopia has the highest soil

nutrient outflow rates of $60\,kg\,ha^{-1}$ ($30\,kg\,ha^{-1}$ nitrogen and $15$–$20\,kg\,ha^{-1}$ phosphorous), while inflows from fertilizers are very low ($<10\,kg\,ha^{-1}$). In the long term, such soil nutrient losses by erosion adversely affect soil productivity of the source areas and are expected to lead to an increase in nutrients in the deposition areas [9, 10].

Nutrient export rates can well describe the level of soil nutrient degradation in the source soils and the enrichments in the deposition sites [13]. Despite such utility, the rate of soil nutrient export as a means of assessing degradation from different landforms in a catchment considering different erosion-status sites has not yet received research attention. However, such research is pertinent to the development of site-specific strategies targeting the source sites.

Other studies have shown that erosion and deposition processes can significantly contribute to the variability of fine soil particles and the associated nutrient exports from catchment topography [14, 15]. Soil erosion and sediment

TABLE 1: Biophysical description of the landforms in the Mai-Negus catchment, northern Ethiopia.

| Landform | Area[e] (%) | Land-use cover (%)[a] | | | | Lithology %[b] | | | Slope, deg[c] | Elevation, m[d] |
|---|---|---|---|---|---|---|---|---|---|---|
| | | Arable | Grazing | Bush and Wood[f] | Others[g] | BM | LP | ST | | |
| Rolling hills | 10.0 | 80 | 10 | 4 | 6 | 100 | n.a. | n.a. | 3–16 | 2150–2240 |
| Mountainous | 14.5 | 36 | 34 | 26 | 4 | 35 | 65 | n.a. | 4–79 | 2350–2650 |
| Central ridge | 25.5 | 70 | 13 | 5 | 12 | 5 | 95 | n.a. | 3–25 | 2230–2450 |
| Valley | 19.9 | 65 | 31 | 2 | 2 | 77 | 23 | n.a. | 0–6 | 2070–2100 |
| Plateau | 9.8 | 47 | 16 | 30 | 7 | 58 | 42 | 9 | 3–10 | 2500–2550 |
| Escarpment | 19.1 | 47 | 31 | 14 | 5 | 84 | 7 | n.a. | 3–30 | 2270–2540 |
| Reservoir | 1.2 | n.a. | n.a. | n.a. | n.a. | 100 | n.a. | n.a. | 0 | 2060–2080 |

BM: basic metavolcanics; LP: lava pyroclastic; ST: sandstone; n.a.: not applicable.
[a]The proportion of the land covers was derived from a Landsat image of November 2007 overlaid by the landform map.
[b]This was derived by overlaying the landform map with the geology map of Ethiopia.
[c]Developed from digital elevation model (DEM).
[d]Derived from DEM.
[e]Total catchment area is 12.40 km$^2$, and reservoir area is 0.15 km$^2$.
[f]Includes closed area, plantations, and natural vegetation.
[g]Includes settlements, rock-out crop, and marginalized areas.

delivery processes, which are responsible for high sediment transport and the associated export of sediment-bound nutrients to deposition areas in a catchment, are influenced by landscape characteristics [13]. In line with this, studies on nutrient balances in Ethiopia indicate that the balance at farm level is more positive than plot level [11, 16, 17], while losses from some fields may be of benefit to other fields. This may be attributed to the effect of nutrient redistribution by erosiondeposition processes or to active nutrient transfers by farmers.

Studies of erosion-induced changes in soil properties at the field scale have constantly shown that soil texture, surface soil organic carbon, nitrogen, and phosphorus concentrations are higher in areas of soil deposition compared to areas of soil removal by erosion [18]. Assessing the effect of erosion on soil properties variability as a result of nutrient export from catchment landforms can provide more meaningful results. This study aims to (1) assess soil nutrient depletion and soil physical degradation comparing the eroded and stable-sites soil, (2) examine the rate of soil nutrient export considering the *aggrading* and *eroded* erosion-status sites in the landforms of the Mai-Negus catchment, northern Ethiopia, and (3) quantify the costs of soil degradation associated with nutrient losses through water erosion at landform and catchment scale and extrapolate the findings to the Tigray region and highlands of the country.

## 2. Materials and Methods

*2.1. Study Site.* This study was conducted in the Mai-Negus catchment in the Tigray region of northern Ethiopia (Figure 1). The catchment has an area of 1240 ha and altitudes ranging from 2060 to 2650 m a.s.l. It has a mean annual temperature of 22°C and precipitation of 700 mm. Most of the rainfall (>70%) occurs in July and August. Land use is predominantly arable, with teff (*Eragrostis tef*) being the major crop along with different-sized areas of pasture land and scattered patches of trees, bushes, and shrubs. The major

rock types are lava pyroclastic and metavolcanic. Soils are mainly leptosols on very steep positions, cambisols on middle to steep slopes, and vertisols on flat areas.

*2.2. Terrain Assessment.* In this study, field reconnaissance surveys were carried out to gain an overall image of the catchment characteristics (Table 1). Data were collected from June to December 2009. The landforms in the catchment (Figure 1) were developed in ArcGIS software using field survey-based data in combination with the information from the topographic map. Considering elevation, slope, geology, and geomorphologic characters (surface flows, alluvial, and colluvial deposition), the catchment topography is classified into six main landforms (Figure 1), namely, valley (19% of the catchment area), plateau (8%), rolling hills (9%), central ridge (27%), escarpments (29%), and mountainous (6%).

The reservoir was considered a separate landform in the catchment and is located at the toeslope of the catchment. Sediment deposition in the reservoir was used to estimate the level of soil degradation, nutrient export rate, and the associated replacement costs from the entire catchment. Deposition sites in the other landforms of the catchment were used to assess soils deposited on the way to the reservoir after being transported from the original place towards the outlets of the landforms. Generally, during the field survey, three sites, namely, *deposition/aggrading, stable*, and *eroded* erosion-status sites in the six landforms of the study catchment were identified using standard geomorphological procedures of erosion-deposition indicators.

*2.3. Erosion-Status Site Selection and Soil Sampling Points.* Representative erosion-status sites (*aggrading, stable*, and *eroded*) and the corresponding soil sampling points (Figure 1) were selected in four steps. A reconnaissance survey on land use, lithology, soil types, slopes, and elevation ranges of the landforms (Table 1), followed by informal discussions with farmers and development agents to gain insight on land use history as well as land- and crop-management

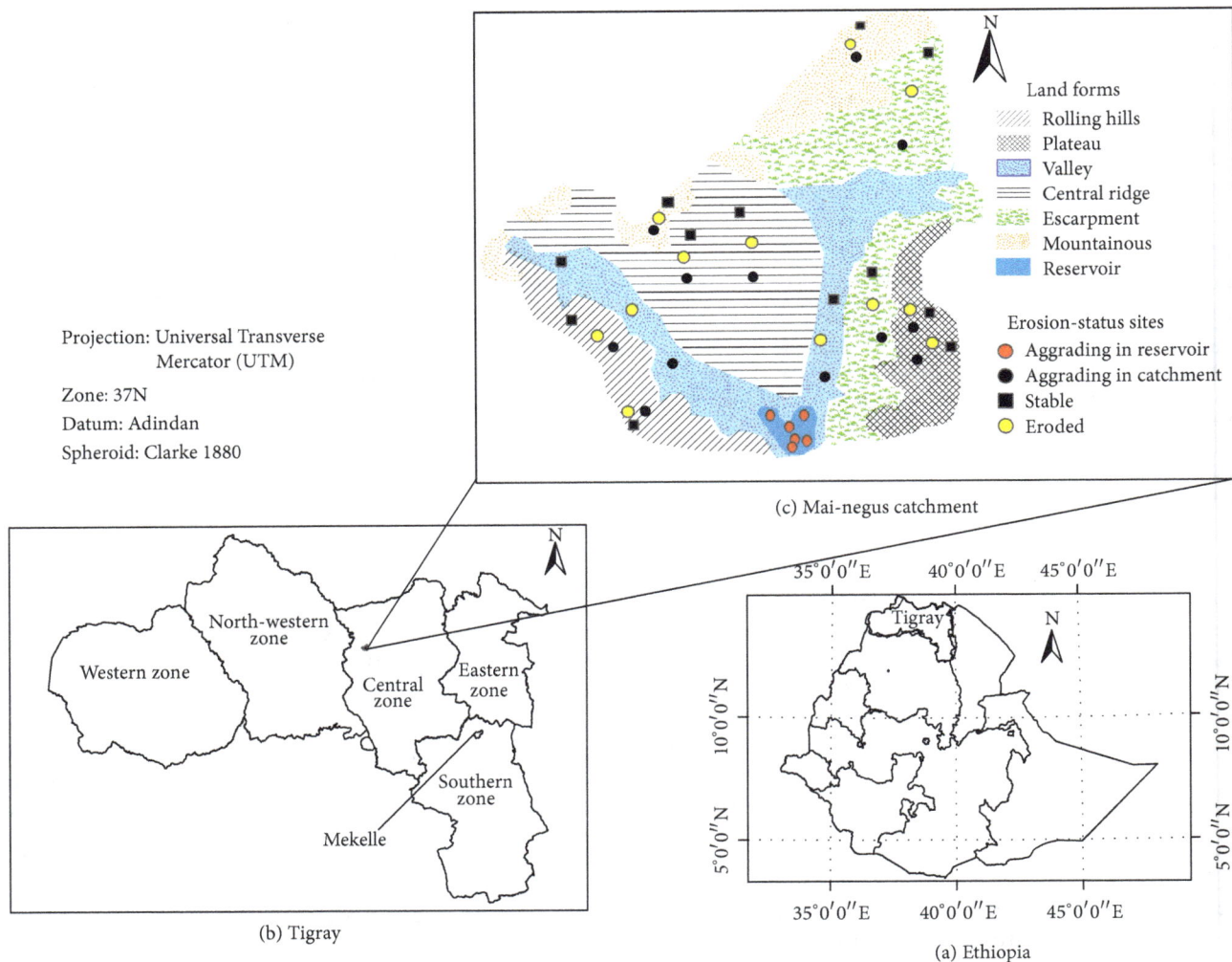

FIGURE 1: Study areas: Ethiopia (a), Tigray (b), and Mai-Negus catchment, (c) (landforms) with representative soil sampling points for *aggrading (deposition), stable,* and *eroded* erosion-status sites.

practices, was carried out. Subsequently, erosion-status sites were selected using soil morphology such as thickness of alluvial/colluvial deposits and degree of truncation of the top soil horizon (soil profile) and erosion indicators (rills, gullies, surface sheet wash, exposure of roots and stones, and depositions). In the *stable* sites, slopes are flat to gentle, and there is little evidence of soil truncation or deposition, indicating that soil loss and gain is more or less balanced. The *eroded* sites were identified based on a combination of erosion indicators and features associated with soil profile truncation. When *aggrading* sites were selected, depression areas that received sediment from upper slopes and erosion channels were considered. Finally, representative soil/sediment sampling points of the erosion-status sites were located and georeferenced. Composite samples were randomly collected from sampling grid areas ranging from 150 to 300 m$^2$ for each soil sampling point in the erosion-status site. Each site had two soil sampling points, and 5–8 composite samples were collected at 0–20 cm soil depth. The composite samples

(weighted 500 gm each) were pooled and mixed thoroughly in a basket manually. A subsample of 500 g was taken for analysis after being air dried and sieved to pass through 2 mm sieve. A total of 36 soil samples (6 landforms × 3 erosion-status sites in each landform × 2 sampling points), and 6 sediment samples from the reservoir resulted in a grand total of 42 samples for analysis.

The sampling point selection in the reservoir took into account the sediment depth variability, flow source, outflow, and length of time where water was in the reservoir. There was no water at the selected sampling points in the reservoir on the second week of June 2009. For the sampling points identified in the reservoir, 6 representative pits with depths ranging from 1.0 to 2.5 m were dug down to reach the interface with the original soil. Then three composite samples were collected from the entire depth of each pit. The samples from each pit were pooled and mixed thoroughly in a basket, and a subsample of 500 g was taken for analysis.

### 2.4. Soil/Sediment Analysis.

Soil/sediment samples were determined for texture using the Bouyoucos hydrometer method [19], bulk density (BD) by core method [20], organic carbon by Walkley-Black method [21], available phosphorus (Pav) by Olsen method [22], and total nitrogen (TN) and total phosphorus (TP) by Kjeldahl Digestion method [23]. Cation exchange capacity (CEC) is determined by ammonium acetate extraction buffered at pH 7 [24]. Exchangeable bases (calcium, Ca; magnesium, Mg; potassium, K) were analyzed after extraction using 1 M ammonium acetate at pH 7.0. Iron (Fe) and zinc (Zn) were determined using 0.005 M diethylene triamine pentaacetic acid extraction by the method described in Baruah and Barthakur [25].

### 2.5. Estimation of the Extent of Soil Degradation.

Soil nutrient and soil physical degradation were assessed based on the differences in the content of the soil nutrients and fine soil particles in the *stable* and *eroded* sites. The *stable* site soils were used as a reference. The magnitude of degradation was similarly calculated as the ratio of the difference in the content of the soil properties between the *stable* and *eroded* sites to the reference soils.

### 2.6. Quantification of Nutrient Export Rate Associated with Water Erosion.

The nutrient export (NE) to the reservoir associated with runoff and sediment and to the hypothetical deposition areas on the way to the reservoir at the outlets of the landforms was calculated using equations defined in Verstraeten and Poesen [26] as

$$NE_r = \frac{SM * NC}{A * Y * NTE * 10},$$

$$SM = SV * dBD,$$

$$NE_l = SSY\left(\frac{NC}{NTE * 10}\right), \qquad (1)$$

$$SSY = \frac{SM}{A * Y},$$

where $NE_r$ is the nutrient export ($kg\,ha^{-1}\,y^{-1}$) from the entire catchment to the reservoir, $NE_l$ is the nutrient export ($kg\,ha^{-1}\,y^{-1}$) to the outlet of each landform in the catchment, SM is total sediment mass (ton), NC is average nutrient content (ton of nutrient per kg of sediment), $A$ is drainage area (ha), $Y$ is duration (age) of sediment accumulation which was 16 years for the reservoir, SV is total sediment volume ($m^3$), dBD is dry bulk density (ton $m^{-3}$), and SSY is area-specific sediment yield ($t\,km^{-2}\,y^{-1}$). The nutrient trap efficiency (NTE) of the reservoir or the depositional areas in the landforms is assumed to be equal to sediment trap efficiency (STE in %). The parameters used for the calculation of nutrient export rates are presented in Table 2.

An empirical model was used for the verification of the information gathered through discussions with the farmers on the condition of outflow from the reservoir for determining STE. The STE was calculated on the basis of the formula given in Verstraeten and Poesen [26] as

$$STE = 100\left(1 - \frac{1}{1 + 0.0021D\,(C/A)}\right), \qquad (2)$$

where $C$ is reservoir storage capacity ($m^3$) and $A$ is catchment area ($km^2$). $D$ has constant values ranging from 0.046 to 1 with a mean value of 0.1. In this study, considering the scarcity of local information for defining a value of $D$, a $D$ value of 0.1 was used.

The nutrient trap efficiency was considered to be equal to the sediment trap efficiency (STE) as (1) no spillage occurred except in one season in 2007, (2) the stored water is used for irrigation during the dry period once the sediments and nutrients have settled in the reservoir, and (3) soil nutrients are mainly transported bound to sediment, as soil nutrient solubility is low in the soils of the region [13]. Such assumptions imply that both nutrients and sediments are assumed to have the same chance of being trapped in the reservoir. Similar assumptions were also used in past studies while estimating the STE of different reservoirs in the Tigray region catchments [13].

The area-specific sediment yield (SSY) of the reservoir was determined based on the field survey of sediment depth measured. Spatially distributed sediment depths (bold dots) were located within the reservoir (Figure 2) based on visual observation of sediment size, pattern, and depth. Augering was also conducted to assess the sediment thickness (pit depth). Differentiation between bottom *in situ* material and newly deposited sediment was easy due to sediment stratification. The sampling pits were georeferenced and then interpolated using *Thiessen* polygonsin ArcGIS to calculate

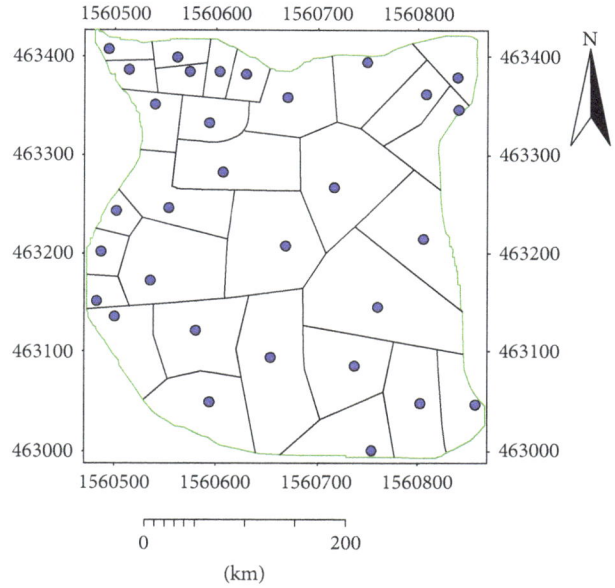

FIGURE 2: Spatial distribution of sediment depths measured (blue dots) and polygons (areas) influenced by the dots in the reservoir of the Mai-Negus catchment, northern Ethiopia.

TABLE 2: Sediment yield and other related parameters of the different landforms in the Mai-Negus catchment, northern Ethiopia.

| Landform | SV (m$^3$) | SM (t) | A (Km$^2$) | dBD (t m$^{-3}$) | STE (%) | SSY (t Km$^{-2}$ y$^{-1}$) |
|---|---|---|---|---|---|---|
| Rolling hills | 1,024 | 1,566 | 1.25 | 1.53 | 63 | 1,254 |
| Mountainous | 3,579 | 5,762 | 1.82 | 1.61 | 55 | 3,165 |
| Central ridge | 4,761 | 7,568 | 3.20 | 1.59 | 57 | 2,366 |
| Valley | 4,327 | 5,884 | 2.52 | 1.36 | 66 | 2,335 |
| Plateau | 965 | 1,408 | 1.21 | 1.46 | 62 | 1,164 |
| Escarpment | 3,212 | 4,786 | 2.40 | 1.49 | 69 | 1,994 |
| Reservoir[a] | 343,776 | 388,467 | 12.40 | 1.13 | 99 | 1,958 |
| Average | 51,663 | 59,349 | 3.54 | 1.45 | 67 | 2,034 |
| Standard deviation | 128,818 | 145,143 | 3.07 | 0.17 | 14.80 | 690 |

SV: annual sediment volume; SM: annual sediment mass; A: drainage area; dBD: dry bulk density; STE: sediment trap efficiency; SSY: area specific sediment yield.
[a]Sediment in volume and mass is total amount of sediment accumulated in the reservoir over 16 years (1994–2009). SSY coming from the whole catchment area (12.40 km$^2$) was calculated after the depth of the spatially distributed pits opened in the reservoir sediment was measured, georeferenced, and then interpolated through Thiessen polygons in a GIS environment.

SSY for each polygon using the equation defined in Tamene [8]. However, the SSY of the landforms in the study catchment was estimated from the Soil and Water Analysis Tool (SWAT) model simulation result of the subbasins. The model takes into account transfers of SSY using routing procedure between landforms, for example, from mountain to plain land. The SWAT model was calibrated, validated, and assessed for uncertainty using Nash-Sutcliffe coefficient (>0.5) and coefficient of determination (>0.6). Such model efficiency values indicated that the model is adequate for prediction in the study catchment conditions. The details of the model evaluation result can be found in Tesfahunegn [27].

Distributed sediment delivery ratio (SDR) estimation is necessary for different landscape positions (landforms). A simple approach to estimate SDR is defined by Tamene [8] as

$$SDR_i = \left( \frac{HD_i}{SL_i} \right) 10^\wedge, \qquad (3)$$

where $SDR_i$ = SDR of a land cell (landform); $HD_i$ = elevation difference between a land cell at a given point and the associated main stream outlet cell in the landform; $SL_i$ = length of the flow path between the inlet and the main channel outlet of a landform. To estimate the distributed SDR values for each landform using (3), the study catchment was first subdivided into different landforms. The STE of the landforms was found after the SDR was estimated and assumed to be equivalent to the sediments not delivered into the outlets of the landforms.

Annual sediment mass deposition (SM, ton) was calculated by multiplying SSY (t km$^{-2}$ y$^{-1}$) by the area of each landform. There was no sediment trapping structure in the six landforms during the study period. In the presence of impoundments, the trap efficiency can be higher than the assumption used in this study. Generally, the procedure for estimating sediment trap efficiency at the outlets of the landforms was considered to be similar to the sediment not delivered, and comparison of SM among the landforms for priority setting is thus scientifically unbiased. In addition,

in the absence of other reliable alternative procedures for determining STE in landforms within catchments and sub-catchments without reservoirs and ungauged measurements, the use of the present approach and assumption can support for site-specific decision-making processes.

The measured sediment volumes were converted to sediment masses using the mean value of the bulk density determined for each landform. To determine the mean bulk density, duplicate undisturbed sediment core samples were collected from the six pits at the dry-bed reservoir. For the other landforms, mean bulk density was determined using core samples collected from the deposition sites at 0–20 cm depth.

*2.7. Costs of Nutrient Loss through Erosion.* To assess the costs of sediment-associated nutrient losses, a "replacement cost" approach has already been applied to compare among different catchments [13, 28]. However, there is still an information gap with regard to the application of such approach to landforms within a catchment with and without reservoir. In this study, the loss of nutrients associated with water erosion was calculated using the equivalent price of commercial fertilizers, for example, urea and di-ammonium phosphate (DAP). First, the nitrogen (N) and phosphorous (P) export rate from each landform and the entire catchment to the reservoir was assessed. Second, the P value was converted to an equivalent $P_2O_5$ value and then to DAP fertilizer. Third, the amount of N for DAP was estimated from the N exported. Fourth, the remaining N exported was converted to urea fertilizer. Finally, the cost of nutrient export was estimated in terms of the money required for replacing the nutrient losses as urea and DAP fertilizer. The local price in May 2010 for urea and DAP was used to estimate the replacement cost of the exported nutrients. The price of 100 kg of DAP was 360 Ethiopian Birr currency (ETB) (€21.18), and that of 100 kg urea was 320 ETB (€18.82). The replacement cost in the catchment was projected to the scale of the Tigray region and Ethiopian highlands.

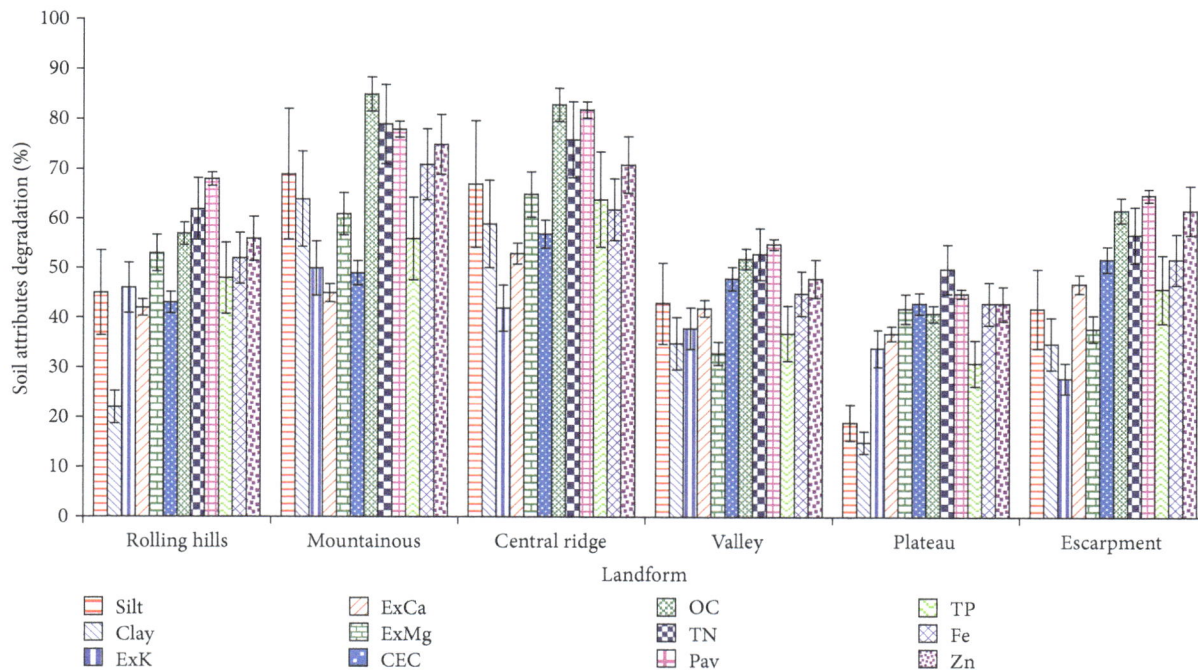

FIGURE 3: Extent of soil degradation using selected soil properties (%) in the landforms of Mai-Negus catchment, northern Ethiopia. Y Error bars are standard deviation of samples. Note: ExK, exchangeable potassium; ExCa, exchangeable calcium; ExMg, exchangeable magnesium; CEC, cation exchange capacity; OC, organic carbon; TN, total nitrogen; Pav, available phosphorus; TP, total phosphorus; Fe, iron; Zn, zinc.

*2.8. Data Analysis.* Data were subjected to one-way analysis of variance (ANOVA) using SPSS 18.0 release software [29]. The landform was considered as a group variable. Normality and homogeneity assumptions of ANOVA were checked. Means were separated using the least significant difference (LSD) and tested by all-pairwise comparisons at probability level, $P \leq 0.05$. Both SPSS and Excel sheets were also used for descriptive statistics and regression analysis.

## 3. Results and Discussion

*3.1. Extent of Soil Degradation across the Landforms.* To assess the extent of soil degradation in the landforms, soil parameters in the *eroded* sites were compared with that of the *stable* site soils (reference soils). Soil parameters values below the level of the *stable* site soils indicated that the extent of soil degradation is higher than aggradation. Accordingly, a high level of soil degradation was observed across the landforms for all soil parameters, but the highest was observed in the mountainous and central ridge landforms (Figure 3). The level of soil degradation for most soil indicators in the *eroded* sites of the central ridge and mountainous landforms ranged from 50 to 80% compared to the corresponding *stable* site soils. The soil nutrients of OC, TN, and Pav were among the most degraded parameters across the landforms. In agreement to this study, the finding in Haileslassie et al. [30] indicated that soil erosion accounted for losses of TN by 70% and Pav by 80% while assessing nutrient balance at a regional scale in Ethiopia.

The landforms with less degraded soils were the plateau followed by the valley and escarpment, where soil nutrients degraded in the range of 30–40% as compared to the *stable* site soils (Figure 3). The magnitude of physical degradation as silt was higher than that of clay and ranged from 19% in the plateau to 69% in the mountainous and central ridge landforms. The reason could be attributed to higher susceptibility of silt to erosion processes. However, this study showed that soil nutrient degradation is higher than soil physical degradation in the study catchment landforms. Thus, appropriate interventions that improve the soil nutrients should be introduced in the landforms.

*3.2. Evaluation of Nutrient Export Rate Associated with Sediment Yield.* This study showed that soil nutrient export rates were significantly ($P \leq 0.05$) varied among most landforms (Table 3). On average, nutrient export rates (kg ha$^{-1}$ y$^{-1}$) of 95 for Ca, 68 for OC, 9.1 for TN, 3.2 for K, 2.5 for Mg, and 0.07 for Pav were observed in the landforms. But the export rate from the entire catchment to the reservoir was significantly higher than the average of all landforms and even the values of the individual landforms (Table 3). This could be attributed to the fact that the reservoir is the largest sink for sediment-attached soil nutrients exported from the hotspot erosion sources in the entire catchment. This was followed by the valley, which received sediment coming from the upstream landforms. Sediment from all other landforms is routed through the valley, and then finally to the reservoir. Generally, this study indicated that from the erosion source areas, large

TABLE 3: Nutrient export rate associated with sediment yield across the landforms in Mai-Negus catchment, northern Ethiopia.

| Landform | Rate of nutrient export (kg ha$^{-1}$ y$^{-1}$) | | | | | |
|---|---|---|---|---|---|---|
| | K | Ca | Mg | OC | TN | Pav |
| Rolling hills | 1.36e | 35.43e | 1.21d | 23.76e | 4.23d | 0.032de |
| Mountainous area | 3.81b | 126.8c | 2.91b | 86.49b | 10.36b | 0.087c |
| Central ridge | 3.73b | 115.7c | 2.82b | 57.73c | 8.66bc | 0.096b |
| Valley | 3.91b | 135.9b | 2.98b | 98.11b | 11.56b | 0.102b |
| Plateau | 1.52d | 25.87e | 0.79e | 29.39e | 3.92d | 0.022e |
| Escarpment | 1.97c | 57.27d | 2.2c | 42.37d | 7.11c | 0.041d |
| Reservoir | 6.14a | 166.8a | 4.79a | 135.7a | 17.74a | 0.135a |
| Average | 3.21 | 94.8 | 2.53 | 67.6 | 9.08 | 0.074 |
| Standard deviation | 1.71 | 54.8 | 1.32 | 40.9 | 4.78 | 0.042 |

K: potassium; Ca: calcium; Mg: magnesium; OC: organic carbon; TN: total nitrogen; Pav: available phosphorus.

amounts of soil nutrients are exported to the deposition sites in the same or other landforms. Such erosion processes are aggravated by terrain characteristics (e.g., slope), high gully networks, poor soil cover, and conservation measures, which have been reported as the main controlling factors for the rate of sediment yield variability in the Tigray catchments [8, 31].

The nutrient export rate from the valley was higher than the remaining landforms except the reservoir even though the valley is located at a place where losses can be compensated by imports from the upstream landforms (e.g., mountainous and central ridge). The higher nutrient export from the valley seems to be inconsistent with the point that this landform showed the largest sediment deposits (due to flat slope) next to the reservoir. But the fact for being higher nutrient export by erosion is that the valley received sediment coming from all the other landforms by sediment routing procedure, and such amount of sediment is beyond the capacity of this landform to deposit. Thus, introducing appropriate interventions to the upstream landforms (sources of erosion) that reduce sediment due to water erosion can be the remedy for decreasing the higher nutrient export from the valley landform.

The nutrient export rates from the other landforms showed a well-defined pattern; for example, the mountainous followed by the central ridge landform exported large amounts of nutrients to the valley. Most rates of soil nutrients exported to the aggrading sites in the mountainous area, central ridge, and valley landforms were not significantly ($P > 0.05$) different (Table 3). This indicated that the mountainous and central ridge could be the major sources of the sediment-attached nutrients exported to the valley landform and thereby to the reservoir. The nutrient export rate to the reservoir associated with sediment is higher in the present study catchment when compared to the mean of 13 other catchments in the Tigray region reported by Haregeweyn et al. [13]. However, the nutrient export rate in this study is lower than the findings reported in other sub-Saharan Africa such as Ghana by Amegashie [32].

A significant ($P < 0.05$) and high coefficient of determination ($R^2$), that is, 0.87, 0.85, 0.83, 0.78, 0.73, and 0.68 for Ca, Pav, K, TN, OC, and Mg, respectively, was observed between the nutrients exported and area-specific

sediment yield (SSY). This indicated that the SSY accounted for between 68 and 87% of the variability in the nutrients exported across the landforms. The significant and strong relationships between nutrient export and SSY illustrated that erosion could be the main responsible factor for sediment-attached nutrient export variability among the landforms in the catchment, regardless of the inherent variability of the original soils. Generally, hydrological processes such as erosion can be affected by variability in soil properties, topography, land use/cover, and human-induced changes, and these in turn affect the soil nutrient distribution in landforms [8, 27].

The $R^2$ (0.68) for the relationship between Mg and SSY in this study is higher than the $R^2$ (0.29) reported in the same study region but in different catchments in Haregeweyn et al. [13]. Some, among the 13 catchments, showed lower nutrient export rates than those estimated for the landforms of this study, and others showed higher nutrient export rates. Despite such differences, the nutrient export rates calculated in the previous and present studies deviated by <15%, which indicates that the present results hardly under- or overestimate the nutrient export associated with sediment yield. Hence, scaling-up of the present approach to catchments or subcatchments without reservoirs to assess sediment-attached nutrient export is applicable for management planning.

### 3.3. Cost of Soil Nutrient Degradation Caused by Erosion.
This study presented the replacement costs related to N and P exported by water erosion from the different landforms of the catchment (Table 4). Taking the average value of all landforms may underestimate the replacement cost for the entire catchment as the nutrient transported to the aggrading sites in the landforms was calculated based on the assumption of minimum sediment delivery. The sum of all landforms may also overestimate the replacement cost, because part of the exported nutrients from the upper landforms can be routed to the reservoir and the remaining deposited on the way to the reservoir. We therefore restricted the calculation to nutrients exported to the reservoir from the entire catchment while extrapolating the result of this study.

TABLE 4: Replacement costs for erosion-associated nutrients exported as N and P from the landforms and the entire catchment.

| Landform | N exported[a] (kg ha$^{-1}$ y$^{-1}$) | Converted to urea (kg ha$^{-1}$ y$^{-1}$) | Replacement cost as urea (ETB ha$^{-1}$ y$^{-1}$) | P exported (kg ha$^{-1}$ y$^{-1}$) | Converted to DAP (kg ha$^{-1}$ y$^{-1}$) | Replacement cost as DAP (ETB ha$^{-1}$ y$^{-1}$) | Total RC[b,c] (ETB ha$^{-1}$ y$^{-1}$) | Total RC per landform[d] (ETB y$^{-1}$) |
|---|---|---|---|---|---|---|---|---|
| Rolling hills | 4.23 | 9.20 | 29.52 | 0.03 | 0.16 | 0.57 | 30.09 | 3,761 |
| Mountainous | 10.36 | 22.52 | 72.32 | 0.08 | 0.43 | 1.55 | 73.86 | 13,443 |
| Central ridge | 8.66 | 18.83 | 60.52 | 0.09 | 0.47 | 1.71 | 62.22 | 19,910 |
| Valley | 11.56 | 25.13 | 80.71 | 0.10 | 0.50 | 1.81 | 82.52 | 20,795 |
| Plateau | 3.92 | 8.52 | 27.33 | 0.02 | 0.11 | 0.39 | 27.72 | 3,354 |
| Escarpment | 7.11 | 15.46 | 49.58 | 0.04 | 0.20 | 0.73 | 50.31 | 12,074 |
| Reservoir | 17.74 | 38.57 | 123.79 | 0.14 | 0.67 | 2.40 | 126.19 | 156,476 |
| Average | 9.08 | 19.75 | 63.39 | 0.07 | 0.36 | 1.31 | 64.70 | 32,830 |

DAP: diammonium phosphate; ETB: currency, Ethiopian Birr; 17 ETB ~ €1 in May 2010 on average.
[a] Part of the N exported is converted to the N in DAP.
[b] RC: replacement cost for both N and P exported as urea and DAP.
[c] The price of each 100 kg of urea and DAP in 2010 crop season was 320 and 360 ETB, respectively.
[d] Column 8 multiplied by the area of each landform which is found in Table 2.

The total replacement cost as urea and DAP for the soil nutrients exported (N and P) from the erosion source areas of the entire study catchment to the reservoir was 126 ETB ha$^{-1}$ y$^{-1}$ (€7.40 in May 2010). This value was followed by the valley landform, where the cost was 82.50 ETB ha$^{-1}$ y$^{-1}$ (€5.00). The lowest total replacement cost for the same nutrients exported to the outlet of the plateau landform was 27.70 ETB ha$^{-1}$ y$^{-1}$ (€1.60), followed by the rolling hills of 30.00 ETB ha$^{-1}$ y$^{-1}$ (€1.80) (Table 4). The average replacement cost for all landforms including the reservoir was 64.70 ETB ha$^{-1}$ y$^{-1}$ (€3.80), which was lower than the replacement cost for the reservoir.

The replacement costs estimated for rolling hills, plateau, and escarpment landforms in this study are consistent with the costs estimated by FAO [33] for cropped areas, as they are within the range of 29 to 44 ETB ha$^{-1}$ y$^{-1}$ (€1.70–2.60 in May 2010). But when compared to the replacement costs quantified for the other landforms, the report by FAO [33] underestimated the cost that replaces the nutrients exported (N and P) as a result of erosion. Alternatively, the replacement cost in the remaining landforms including the reservoir was within the range of the replacement costs estimated for the 13 reservoir catchments in the Tigray region by Haregeweyn et al. [13], regardless of the differences in the foreign currency exchange rate across years.

The total nutrient replacement cost for the entire study catchment (Mai-Negus) was calculated using the cost required to replace the nutrient exported to the reservoir multiplied by the catchment area of 1240 ha, that is per annum 156,476 of ETB (€9204 in May 2010). Extrapolation the nutrient replacement cost to the Tigray regional state with a total area of 53,000 km$^2$ ranged from 632 million ETB (€37.2 million) to 700 million ETB (€41.2 million) and for the Ethiopian highlands that covers 490,000 km$^2$ is estimated from 6.2 billion ETB (€364 million) to 7.0 billion ETB (€412 million), which are quite impressive figures for researchers, planners, and decision and policy makers to come up with

suitable erosion controlling measures. The replacement cost for the Ethiopian highlands accounts for about 8% of the Ethiopian national annual budget in 2010/11, which is almost twice the Tigray region (northern Ethiopia) annual budget. The above-mentioned replacement costs will become higher if other nutrient losses (e.g., OC, cations) and the losses in physical soil attributes are included. In general, the results of this study can be considered as an indication of the need to implement well-defined and appropriate land use redesign, and soil and water management practices in order to reduce costs related to nutrient replacement, while maintaining sustainable soil productivity in the landforms. The use of inorganic fertilizer as part of the soil management practices is indispensable to replace such nutrients losses. However, this has to be complemented with organic amendments as organic matter sources are locally available. It is also now widely recognized among experts and policy makers that the increasing application of fertilizer at the current prices is not affordable by most farmers and possibly by the government and requires foreign exchange to import fertilizers.

## 4. Conclusions

This study showed that the level of soil degradation ranged from 30 to 80% in the source soils (*eroded* site) when compared to the *stable* sites by the effect of soil erosion. But the severity of degradation was higher in the central ridge and mountainous landforms, which are the main sources of sediment to downstream landforms. The nutrient export rates of OC, TN, Pav, K, Ca, and Mg showed variability among the landforms, even though significantly higher values were exported into the reservoir. A strong and positive relationship between nutrient export rates and sediment yields also indicated that erosion plays the prime role for nutrient export into the *aggrading* sites in the landforms. In addition, this study indicated that an annual nutrient replacement cost varied across the landforms, with the highest in the reservoir followed by the valley, mountainous, and central ridge landforms. This study thus supports in identifying and prioritizing landforms which are sources of higher nutrient export rates and thereby higher soil degradation, while introducing remedial technologies. The technologies should be targeted to improving soil resistance to erosion and retaining the soil in place just after detachment by raindrops and before transport by runoff from the erosion source areas in the landforms of the study catchment.

## Acknowledgments

The authors gratefully acknowledge the financial support by DAAD/GTZ (Germany) through the Center for Development Research (ZEF), University of Bonn (Germany), and the support of Aksum University (Ethiopia) for the first author's field work. The authors also greatly appreciate the assistance offered by the local farmers and extension agents during the field study.

## References

[1] UNEP (United Nations Environment Program) & UNESCO (United Nations Educational, Scientific and Cultural Organization), *Provision Map of Present Land Degradation Rate and Present State of Soil*, FAO, Rome, Italy, 1980.

[2] H. Eswaran, R. Lal, and P. F. Reich, "Land degradation: an overview," in *Response to Land Degradation*, E. M. Bridges, I. D. Hannam, L. R. Oldeman, F. W. T. Penning de Vries, J. S. Scherr, and S. Sombatpanit, Eds., pp. 20–35, Science Publisher, Enfield, NH, USA, 2001.

[3] H. Hurni, "Land degradation, famine, and land resource scenarios in Ethiopia," in *World Soil Erosion and Conservation*, D. Pimentel, Ed., pp. 27–62, Cambridge University Press, 1993.

[4] W. Mekuria, E. Veldkamp, M. Haile, J. Nyssen, B. Muys, and K. Gebrehiwot, "Effectiveness of exclosures to restore degraded soils as a result of overgrazing in Tigray, Ethiopia," *Journal of Arid Environments*, vol. 69, no. 2, pp. 270–284, 2007.

[5] P. L. G. Vlek, Q. B. Le, and L. Tamene, "Assessment of land degradation, its possible causes and threat to food security in Sub-Saharan Africa," in *Advances in Soil Sciences Food Security and Soil Quality*, R. Lal and B. A. Stewart, Eds., pp. 57–86, CRC Press, Boca Raton, Fla, USA, 2010.

[6] P. Dubale, "Soil and water resources and degradation factors affecting productivity in Ethiopian highland agro-ecosystems," *Northeast African Studies*, no. 8, pp. 27–52, 2001.

[7] S. Damene, L. Tamene, and P. L. G. Vlek, "Performance of exclosure in restoring soil fertility: a case of Gubalafto district in North Wello Zone, northern highlands of Ethiopia," *Catena*, vol. 101, pp. 136–142, 2013.

[8] L. Tamene, *Reservoir siltation in the drylands of northern Ethiopia: causes, source areas and management options [Ph.D. thesis]*, Center for Development Research, University Bonn, Bonn, Germany, 2005.

[9] W. D. Ellison, "Fertility Erosion," *The Land*, vol. 9, p. 487, 1950.

[10] G. M. Hashim, K. J. Coughlan, and J. K. Syers, "On site nutrient depletion: an effect and a cause of soil erosion," in *Soil Erosion at Multiple Scales: Principles and Methods For Assessing Causes and Impacts*, F. W. T. Penning de Vries, F. Agus, and J. Kerr, Eds., pp. 207–221, CAB International, 1998.

[11] J. J. Stoorvogel and E. M. A. Smaling, "Assessment of soil nutrient depletion in Sub-Saharan Africa, 1983–2000," Report 28, DLO Winand Staring Center for Integrated Land, Soil and Water Research, Wageningen, The Netherlands, 1990.

[12] UNDP (United Nations Development Programme), *Technical Note on the Environment*, DCGE, Addis Ababa, Ethiopia, 2002.

[13] N. Haregeweyn, J. Poesen, J. Deckers et al., "Sediment-bound nutrient export from micro-dam catchments in Northern Ethiopia," *Land Degradation and Development*, vol. 19, no. 2, pp. 136–152, 2008.

[14] J. R. Stone, J. W. Gilliam, D. K. Cassel, R. B. Daniels, L. A. Nelson, and H. J. Kleiss, "Effect of erosion and landscape position on the productivity of Piedmont soils," *Soil Science Society of America Journal*, vol. 49, no. 4, pp. 987–991, 1985.

[15] W. R. Kreznor, K. R. Olson, W. L. Banwart, and D. L. Johnson, "Soil, landscape, and erosion relationships in a northwest Illinois watershed," *Soil Science Society of America Journal*, vol. 53, no. 6, pp. 1763–1771, 1989.

[16] E. Elias, S. Morse, and D. G. R. Belshaw, "Nitrogen and phosphorus balances of Kindo Koisha farms in southern Ethiopia," *Agriculture, Ecosystems and Environment*, vol. 71, no. 1–3, pp. 93–113, 1998.

[17] I. Scoones, Ed., *Dynamics and Diversity: Soil Fertility and Farming Livelihoods in Africa: Case Studies From Ethiopia, Mali and Zimbabwe*, Earthscan Publications Ltd, London, UK, 2001.

[18] S. K. Papiernik, T. E. Schumacher, D. A. Lobb et al., "Soil properties and productivity as affected by topsoil movement within an eroded landform," *Soil and Tillage Research*, vol. 102, no. 1, pp. 67–77, 2009.

[19] G. W. Gee and J. W. Bauder, "Particle-size analysis," in *Methods of Soil Analysis. Part 1*, A. Klute, Ed., pp. 383–411, American Society of Agronomy, Soil Science Society of America, Madison, Wis, USA, 2nd edition, 1986.

[20] G. R. Blake and K. H. Hartge, "Bulk Density," in *Methods of Soil Analysis. Part 1*, A. Klute, Ed., vol. 9 of *Agronomy Monograph*, pp. 363–375, American Society of Agronomy, Madison, Wis, USA, 2nd edition, 1986.

[21] J. M. Bremmer and C. S. Mulvaney, "Nitrogen total," in *Method of Soil Analysis. Part 2. Chemical and Microbiological Properties*, A. L. Page, Ed., vol. 9 of *Agronomy Monograph*, pp. 595–624, American Society of Agronomy, Madison, Wis, USA, 1982.

[22] S. R. Olsen and L. E. Sommers, "Phosphorus," in *Method of Soil Analysis. Part 2. Chemical and Microbiological Properties*, A. L. Page, Ed., vol. 9 of *Agronomy Monograph*, pp. 403–430, American Society of Agronomy, Madison, Wis, USA, 1982.

[23] J. M. Anderson and J. S. I. Ingram, *Tropical Soil Biology and Fertility. A Handbook of Methods*, CAB International, Wallingford, UK, 1993.

[24] J. D. Rhoades, "Cation exchange capacity," in *Methods of Soil Analysis. Part 2*, A. L. Page, R. H. Miller, and D. R. Keeney, Eds., vol. 9 of *Agronomy Monograph*, pp. 149–157, American Society of Agronomy, Madison, Wis, USA, 1982.

[25] T. C. Baruah and H. P. Barthakur, *A Text Book of Soil Analysis*, Vikas Publishing House, New Delhi, India, 1999.

[26] G. Verstraeten and J. Poesen, "Regional scale variability in sediment and nutrient delivery from small agricultural watersheds," *Journal of Environmental Quality*, vol. 31, no. 3, pp. 870–879, 2002.

[27] G. B. Tesfahunegn, "Soil erosion modeling and soil quality evaluation for catchment management strategies in northern Ethiopia," Ecology and Development Series No. 83, Center for

Development Research, University of Bonn, Bonn, Germany, 2011.

[28] J. Bojö, "The costs of land degradation in Sub-Saharan Africa," *Ecological Economics*, vol. 16, no. 2, pp. 161–173, 1996.

[29] SPSS (Statistical Package for Social Sciences), *Statistical Package For Social Sciences. Release 18.0*, SPSS Inc., 2011.

[30] A. Haileslassie, J. Priess, E. Veldkamp, D. Teketay, and J. P. Lesschen, "Assessment of soil nutrient depletion and its spatial variability on smallholders' mixed farming systems in Ethiopia using partial versus full nutrient balances," *Agriculture, Ecosystems and Environment*, vol. 108, no. 1, pp. 1–16, 2005.

[31] J. Nyssen, J. Poesen, J. Moeyersons, M. Haile, and J. Deckers, "Dynamics of soil erosion rates and controlling factors in the Northern Ethiopian highlands—towards a sediment budget," *Earth Surface Processes and Landforms*, vol. 33, no. 5, pp. 695–711, 2008.

[32] B. K. Amegashie, *Assessment of catchment erosion, sedimentation and nutrient export into small reservoirs from their catchments in the upper east region of Ghana [M.S. thesis]*, Kwame Nkrumah University of Science and Technology, Kumasi, Ghana, 2009.

[33] FAO (Food and Agriculture Organization of the United Nations), "Ethiopian highland reclamation study," Final Report, FAO, Rome, Italy, 1986.

# 10

# The Variations in the Soil Enzyme Activity, Protein Expression, Microbial Biomass, and Community Structure of Soil Contaminated by Heavy Metals

**Xi Zhang,**[1] **Feng Li,**[1,2] **Tingting Liu,**[1] **Chen Xu,**[1] **Dechao Duan,**[1] **Cheng Peng,**[1] **Shenhai Zhu,**[1] **and Jiyan Shi**[1]

[1] *Institute of Environmental Science and Technology, Zhejiang University, Hangzhou, Zhejiang 310058, China*
[2] *College of Materials and Environmental Engineering, Hangzhou Dianzi University, Hangzhou, Zhejiang 310018, China*

Correspondence should be addressed to Jiyan Shi; shijiyan@zju.edu.cn

Academic Editors: W. Peijnenburg, D. G. Strawn, and J. Thioulouse

Heavy metals have adverse effects on soil ecology. Given the toxicity of heavy metals, there is an urgent need to select an appropriate indicator that will aid in monitoring their biological effects on soil ecosystems. By combining different monitoring techniques for various aspects of microbiology, the effects of heavy metals on soil microorganisms near a smelter were studied. Our goal was to determine whether proteins could be a proper indicator for soil pollution. This study demonstrated that the activities of acid phosphatase and dehydrogenase, as well as the levels of microbial biomass carbon and proteins, were negatively affected by heavy metals. In addition, significantly negative correlations were observed between these microbial indicators and heavy metals. Denaturing gradient gel electrophoresis analysis was used in this study to demonstrate that heavy metals also have a significantly negative effect on soil microbial diversity and community structure. The soil protein expression was similar across different soils, but a large quantity of presumably low molecular weight protein was observed only in contaminated soil. Based on this research, we determined that the soil protein concentration was more sensitive to heavy metals than acid phosphatase, dehydrogenase, or microbial biomass carbon because it was more dramatically decreased in the contaminated soils. Therefore, we concluded that the soil protein level has great potential to be a sensitive indicator of soil contamination. Further research is essential, particularly to identify the low molecular weight protein that only appears in contaminated soil, so that further insight can be gained into the responses of microbes to heavy metals.

## 1. Introduction

Heavy metals (e.g., Cd, Cr, Cu, Pb, and Zn) can be introduced into soils from several sources, such as waste from mines and smelters, atmospheric deposition, animal manures, sewage sludge, and, in some circumstances, inorganic fertilizer. Once these elements enter the soil, they can remain for extremely long periods and are difficult to remove [1]. Heavy metals are known to be toxic to most organisms when present in excessive concentrations [2]. Given the toxicity of these elements, there is an urgent need to monitor their biological effects on soil ecosystems.

Soil microbes play vital roles in the recycling of plant nutrients, the maintenance of soil structure, and the detoxification of poisonous chemicals [3]. Therefore, the diversity of microbial communities and the activity of microorganisms are important indicators of soil quality. Alterations in the composition of microbial communities have often been considered effective indicators for soil contamination [4]. Many reports have shown that either long-term or short-term exposure of soil to heavy metals results in a reduction of microbial diversity and microorganism activities [3, 5].

Although soil enzyme activity and microbial biomass are the most tested parameters in soil quality monitoring, these parameters still have many limitations [2, 6]. They need to be supplemented by other valid microbial indicators, working together to fully assess soil contamination. The recent

The Variations in the Soil Enzyme Activity, Protein Expression, Microbial Biomass, and Community Structure of Soil Contaminated by Heavy Metals

117

emergence of soil proteomics could provide a new perspective on soil microbial activities [7]. Proteins are the chief actors in organism metabolism processes, and they fall into a variety of categories; for example, enzymes are proteins with a catalytic function. Proteins work together to complete various biological functions. Soil proteins primarily come from microorganisms as well as from flora and fauna tissues. The structure of the microbial community, biomass, and microbial status often determine the synthesis levels of soil proteins. Therefore, any change to the microbial composition, which can be caused by environmental changes such as contamination, could reflect in the composition and expressions of soil proteins [8]. Soil proteomics studies aim to investigate the spatial and temporal changes of proteins extracted from soils because proteins are the functional components of the microbial genomic expression products. Such studies more conclusively determine the ecological functions of soil microbes and their roles in soil pollutant transportation and transformation [9–11]. Therefore, proteomics analyses have great potential in soil pollution assessment.

Currently, soil proteomics are predominantly used in studies of the biogeochemical cycle [12–14] and the rhizosphere soil microecosystem [15, 16]. However, there is little literature on the monitoring of contaminated soil, and the existing studies were all carried out under controlled conditions [8, 17]. Unlike a controlled laboratory, natural soil is a complicated ecosystem, and various factors combine to influence the transportation and transformation of pollutants as well as the structure of the microorganism community. Consequently, we were interested in examining the modification of soil proteins under heavy metals stress and determining whether proteins can be a useful indicator of natural soil pollution.

In this study, a typical heavy metal smelter was selected. We intended to study the effects of heavy metals on soil microbes by combining different biochemical analysis techniques. Three compounds were used for heavy metal extraction, $CaCl_2$, EDTA, and DTPA, to assess the bioavailability of trace metals in soil [18, 19]. Soil microbiological and biochemical properties (including enzyme activities, protein contents, and microbial biomass carbon levels) were measured to evaluate the effects of heavy metals. A soil microbial community analysis was carried out using molecular biology techniques. Partial 16S rDNA genes were amplified from soil microorganism community DNA by the polymerase chain reaction (PCR), using primers that bind to evolutionarily conserved regions within the bacteria and actinomycete genes. The diverse PCR amplified products were transformed to genetic fingerprints using denaturing gradient gel electrophoresis (DGGE) [3, 20]. The soil protein composition was analyzed by SDS-polyacrylamide gel electrophoresis (SDS-PAGE) [7].

## 2. Materials and Methods

*2.1. Soil Research Region and Soil Sampling.* The soil area selected for study is downwind of a copper-zinc smelter in Fuyang County, Zhejiang Province, China. The soil in the area has been contaminated by heavy metals, primarily Cu and Zn. This area was formerly used to cultivate crops. Soil samples were taken in April 2012 at nine points along a pollution gradient of Cu and Zn (Table 1). At each point, three replicated samples were taken, and each sample was composed of three soil cores 5 cm in diameter and 20 cm in depth (approximately 500 g). Samples were taken randomly from different areas at each site. Each replicate soil sample was homogenized thoroughly. The site 2,000 m away from the smelter was chosen as the reference site, and the soil collected from it was marked "CK."

*2.2. Sample Treatment and Analysis.* Fresh soil samples were sieved ($\Phi \leq 2$ mm) by a nylon sieve to remove stones, large pieces of plant material, and soil animals. Portions of the samples were kept moist in the dark at 4°C to determine the microbial biomass and then stored at −70°C to extract soil DNA. Other portions were freeze-dried and passed through a 0.25 mm nylon sieve prior to soil protein extraction. The remaining soil was air-dried under cool conditions (approximately 20°C), sieved through a 0.25 mm nylon sieve, and stored at 4°C to analyze the enzyme activity, soil pH, organic matter, total heavy metals, and bioavailable heavy metals.

The soil pH was determined with distilled water in a ratio (soil : water) of 1 : 2.5 (w/v) using a pH meter (Orion 5-Star, Thermo). The soil was heated at 105°C for 12 h and then combusted at 550°C overnight to measure the organic matter content [21]. The samples were digested using the method described by Tang et al. [22]. The dry sample (0.2 g) was weighed and digested with a mixture of nitric acid ($HNO_3$), hydrofluoric acid (HF), and perchloric acid ($HClO_4$). The total concentrations of Cd, Cr, Cu, Pb, and Zn were measured by flame atomic absorption spectrometry (MKIL-M6, Thermo).

The soil bioavailable heavy metal fractions were estimated by extraction with 0.01 M $CaCl_2$ (soil : solution, 1 : 5), 0.05 M EDTA (soil : solution, 1 : 5), and DTPA (0.005 M DTPA, 0.01 M $CaCl_2$, and 0.1 M TEA at pH 7.3) (soil : solution, 1 : 2.5). The soil suspensions were shaken at 200 rpm for 2 h, centrifuged at 4,000 ×g for 10 min, and filtered [23–25]. The Cd, Cr, Cu, Pb, and Zn contents of the bioavailable fraction in each filtrate were determined by flame atomic absorption spectrometry.

*2.3. Enzyme Activity Determination.* We chose acid phosphatase and dehydrogenase as the subjects for our investigation of enzyme changes under heavy metal stress because they are closely related to the soil phosphorous cycle and to microbial metabolic activities [26, 27]. To determine acid phosphatase activity, dry soil was incubated in disodium phenyl phosphate for 2 h, and the results are expressed as the micrograms of $P_2O_5$ released per 100 g of dry soil using 4-aminoantipyrine colorimetry at 510 nm. To measure dehydrogenase activity, dry soil was incubated in triphenyltetrazolium chloride (TTC) for 24 h, and the results are expressed as the micrograms of triphenyl formazan (TPF) released per 1 g of dry soil. The concentration of TPF in the extract was measured using a colorimeter at 485 nm [27, 28].

TABLE 1: Selected soil physicochemical properties and the soil concentrations of heavy metals.

| Soil number | CK | | | Distance from the smelter | | | | | | | | |
| | | | | 800 m | | | 400 m | | | 200 m | | |
| | 1 | 2 | 3 | 1 | 2 | 3 | 1 | 2 | 3 | 1 | 2 | 3 |
|---|---|---|---|---|---|---|---|---|---|---|---|---|
| pH | 5.94 | 5.76 | 5.72 | 7.57 | 7.34 | 6.23 | 6.61 | 8.00 | 6.56 | 6.52 | 6.95 | 7.10 |
| Organic matter (%) | 4.76 | 5.36 | 4.69 | 4.06 | 4.64 | 3.23 | 6.84 | 5.42 | 6.21 | 7.65 | 6.78 | 6.13 |
| Cu (mg·kg$^{-1}$) | | | | | | | | | | | | |
| CaCl$_2$ | 0.10 | 0.05 | 0.05 | 0.17 | 0.28 | 0.17 | 0.68 | 0.68 | 0.33 | 1.54 | 0.81 | 0.65 |
| DTPA | 4.45 | 4.78 | 5.16 | 33.41 | 48.67 | 21.59 | 124.29 | 85.39 | 102.08 | 138.29 | 129.54 | 110.12 |
| EDTA | 8.13 | 8.05 | 8.72 | 83.51 | 107.06 | 36.40 | 200.03 | 143.65 | 154.78 | 293.92 | 238.30 | 200.66 |
| Total | 52 | 46 | 46 | 314 | 414 | 175 | 711 | 746 | 511 | 1147 | 794 | 624 |
| Zn (mg·kg$^{-1}$) | | | | | | | | | | | | |
| CaCl$_2$ | 0.96 | 0.93 | 1.14 | 0.46 | 0.44 | 5.93 | 12.32 | 8.44 | 8.74 | 46.02 | 20.69 | 13.48 |
| DTPA | 3.44 | 3.75 | 4.46 | 18.36 | 25.56 | 17.83 | 55.86 | 62.40 | 41.71 | 179.49 | 148.50 | 139.94 |
| EDTA | 5.33 | 5.69 | 6.77 | 43.45 | 63.33 | 30.84 | 75.43 | 121.80 | 63.59 | 680.38 | 394.31 | 468.81 |
| Total | 116 | 113 | 111 | 260 | 364 | 187 | 599 | 596 | 545 | 2472 | 1456 | 1421 |
| Pb (mg·kg$^{-1}$) | | | | | | | | | | | | |
| CaCl$_2$ | 0.60 | 0.65 | 0.57 | 0.59 | 0.59 | 0.57 | 0.58 | 0.59 | 0.63 | 0.70 | 0.61 | 0.60 |
| DTPA | 3.38 | 4.21 | 4.25 | 6.46 | 13.06 | 5.58 | 25.79 | 89.79 | 23.99 | 71.09 | 101.70 | 78.98 |
| EDTA | 8.60 | 9.73 | 9.56 | 16.79 | 29.97 | 10.68 | 71.45 | 169.41 | 56.51 | 360.61 | 233.78 | 174.38 |
| Total | 83 | 76 | 75 | 94 | 176 | 77 | 281 | 616 | 230 | 1199 | 770 | 557 |
| Cd (mg·kg$^{-1}$) | | | | | | | | | | | | |
| CaCl$_2$ | 0.04 | 0.05 | 0.05 | 0.03 | 0.04 | 0.28 | 0.31 | 0.24 | 0.25 | 0.55 | 0.30 | 0.19 |
| DTPA | 0.22 | 0.21 | 0.21 | 0.35 | 0.69 | 0.57 | 1.63 | 1.13 | 1.41 | 1.73 | 1.15 | 0.99 |
| EDTA | 0.23 | 0.24 | 0.24 | 0.50 | 0.99 | 0.78 | 1.92 | 1.79 | 1.65 | 2.32 | 1.45 | 1.25 |
| Total | 1.29 | 1.52 | 1.14 | 2.54 | 4.17 | 3.16 | 5.82 | 7.26 | 5.29 | 7.15 | 4.79 | 4.11 |
| Cr (mg·kg$^{-1}$) | | | | | | | | | | | | |
| CaCl$_2$ | N.D. | N.D. | N.D. | N.D. | N.D. | N.D. | N.D. | N.D. | N.D. | N.D. | N.D. | N.D. |
| DTPA | N.D. | N.D. | N.D. | N.D. | N.D. | N.D. | N.D. | N.D. | N.D. | N.D. | N.D. | N.D. |
| EDTA | N.D. | N.D. | N.D. | N.D. | N.D. | N.D. | N.D. | N.D. | N.D. | N.D. | N.D. | N.D. |
| Total | 96 | 115 | 115 | 108 | 154 | 116 | 67 | 63 | 75 | 44 | 73 | 87 |

Note: values presented in the table are the arithmetic mean of three replicates. N.D.: not detected.

*2.4. Microbial Biomass Carbon.* The amount of microbial biomass carbon (biomass C) in the soil was measured by a modified fumigation extraction (FE) method [29]. Soil samples were divided into two parts: one part was fumigated with ethanol-free chloroform for 24 h at 25°C in the dark, and the second part was stored in the dark for 24 h at 25°C but not fumigated. The organic carbon was extracted from both soils using 0.5 M K$_2$SO$_4$ (soil : solution, 1 : 4). Soil extractions were shaken at 150 rpm for 30 min and filtered, and then the total organic carbon was determined using a TOC analyzer (TOC-102A, Analytik Jena). The microbial biomass carbon (MBC) was calculated using the following equation:

$$MBC = [TOC \text{ (fumigated soil)} - TOC \text{ (nonfumigated soil)}] \times 2.22. \quad (1)$$

The microbial biomass was expressed as milligrams of biomass C per kilogram of soil.

*2.5. Soil Protein Extraction.* The citrate-SDS sequential extraction method [15, 16] was used to extract the soil proteins. Specifically, 3.0 g of dried soil was mixed with 15 mL of 0.25 M citrate buffer (pH 8.0), and the homogenate was shaken at 1,200 rpm in room temperature for 4 h. Then, the suspension was centrifuged for 15 min at 15,000 ×g and 4°C and filtered through filter paper (approximately 30 to 50 μm). Next, the soil was extracted using 15 mL of SDS buffer, which contained 1.25% (w/v) SDS, 0.1 M Tris-HCl (pH 6.8), and 20 mM dithiothreitol (DTT). This SDS soil mixture was shaken for 1 h at 1,200 rpm and room temperature, and then it was centrifuged for 15 min at 15,000 ×g and 4°C. For protein recovery, both the citrate extract and the SDS extract were mixed with buffered phenol (pH 8.0) at a volume ratio (extract : phenol) of 3 : 1 and centrifuged at 15,000 ×g and 4°C for 30 min. After centrifugation, the phenol phase was precipitated at −20°C overnight with five volumes of cold 0.1 M ammonium acetate dissolved in methanol. The proteins were recovered by centrifugation at 20,000 ×g and 4°C for 20 min. The pellets were washed once with cold methanol and twice with cold acetone, and then they were air-dried for further use.

*2.6. Protein Content Determination and Protein Separation.* Protein pellets were solubilized in 500 μL of lysis buffer, which contained 9 M urea, 4% w/v CHAPS, 1% w/v DTT, 0.5% ampholyte, and 1 mM PMSF. The concentration of protein in the supernatant was determined by the Bradford method [30], and the protein was stored at −70°C.

The Variations in the Soil Enzyme Activity, Protein Expression, Microbial Biomass, and Community Structure of Soil
Contaminated by Heavy Metals

119

Samples of the extracted proteins were added to an equal volume of loading buffer, which contained 100 mM Tris-HCl (pH 6.8), 4% (w/v) SDS, 20% glycerol, 0.5% (w/v) bromophenol blue, and 100 mM DTT, and then they were heated in water at 95°C for 5 min prior to SDS-PAGE [8]. Subsequently, discontinuous SDS-PAGE was performed using the Mini-PROTEAN 3 Electrophoresis Cell (Bio-Rad) with a 4% stacking gel and a 12% separating gel. The process was run at a constant 75 V/gel through the stacking gel and a constant 150 V/gel through the separating gel. A prestained protein ladder (approximately 10 to 170 kDa, Fermentas) was loaded as a molecular weight marker, and each lane was loaded with the same quantity of protein. After separation, the gels were stained by silver staining [31] for further comparisons.

*2.7. DNA Extraction.* The extraction of the total soil DNA was accomplished using a Sangon DNA isolation kit, according to the manufacturer's protocol (SK8233).

*2.8. PCR-DGGE Microbial Community Analysis.* The primers F357 and R518, 5'-CCT ACG GGA GGC AGC AGC-3' and 5'-ATT ACC GCG GCT GCT GG-3' [20], respectively, were used in this study for the amplification of bacterial 16S rDNA genes. The primers Com2xf and Ac1186r-(pH), 5'-AAA CTC AAA GGA ATT GAC GG-3' and 5'-CTT CCT CCG AGT TGA CCC-3' [32], respectively, were used for the amplification of actinomycete 16S rDNA genes. A GC clamp (CGC CCG CCG CGC CCC GCG CCC GGC CCG CCG CCC CCG CCC C) was added to the forward primers to facilitate the DGGE [20]. The PCR reaction mixture (Takara) contained 5 $\mu$L of 10x reaction buffer (0.1 M Tris-HCl at pH 8.3 and 0.5 M KCl), 6 $\mu$L of MgCl$_2$ (25 $\mu$M), 1 $\mu$L of each primer (20 $\mu$M), 4 $\mu$L of each dNTP (2.5 mM), 0.5 $\mu$L of Taq DNA polymerase (5 U·$\mu$L$^{-1}$), and 2 $\mu$L of DNA; milli-Q water was added to reach a total reaction volume of 50 $\mu$L. The PCR protocol used to amplify the soil bacterial and actinomycete 16S rDNA gene fragments was as follows: a 5 min initial denaturation step at 94°C, followed by 20 cycles of 94°C for 1 min, 65°C for 1 min, and 72°C for 1 min, then, 10 cycles of 94°C for 1 min, 55°C for 1 min, and 72°C for 1 min, with a final extension at 72°C for 7 min (S1000 Thermal Cycler, Bio-Rad). Prior to DGGE, the PCR products were checked by electrophoresis in 1.0% (w/v) agarose gels stained with ethidium bromide.

For the DGGE analysis, the PCR products generated from each sample were separated on an 8% acrylamide gel using the Bio-Rad Dcode System with a linear denaturant gradient range from 30% to 60%. DGGE was performed in 1x TAE buffer at 60°C and 150 V for 8 h. Gels were stained with SYBR-Green I, and the gels were scanned (Bio-Rad).

*2.9. Analysis of DGGE Profiles.* The digitized DGGE images were analyzed by Quantity One image analysis software (Version 4.62, Bio-Rad). The Shannon index ($H$) was used to estimate the soil microbial diversity based on the intensity and number of bands using the following equation:

$$\text{Shannon index } (H) = -\sum \frac{n_i}{N} \ln \frac{n_i}{N}, \quad (2)$$

where $i$ is the number of bands in each lane of the DGGE gel, $n_i$ is the peak height of band $i$, and $N$ is the sum of the peak heights in a given lane of the DGGE gel [33].

Principal component analysis (PCA) was performed by SPSS 16.0 [34], and WPGAMA cluster analysis was performed by Quantity One 4.62 to calculate the similarities of the gel patterns.

*2.10. Data Analysis.* All measurements of soil pH, enzyme activity, and levels of organic matter, heavy metals, microbial biomass carbon, and proteins were performed in triplicate. All of the values reported are the average of three determinations. All of the statistical analyses were performed using SPSS software version 16.0. One-way analysis of variance (ANOVA) was used for statistical comparisons, and the Pearson coefficient was used for correlation analysis. A value of $P < 0.05$ was considered statistically significant.

# 3. Results

*3.1. Soil Properties.* Selected soil physicochemical properties, the total heavy metal contents, and the bioavailable fraction of heavy metals are shown in Table 1. Compared with the contaminated area, soil samples from the reference site were mildly more acidic, with a pH range from 5.72 to 5.94, and had a lower organic matter content between 4.69% and 5.36%. In the field contaminated by heavy metals, the soil pH varied from 6.23 to 8.00, and the range of soil organic matter was from 3.23% to 7.65%. The amount of organic matter in each sample was lower the further the sample was from the smelter.

In all of the soil samples, the Cr concentration was either below or only slightly above the natural background concentration of 90 mg·kg$^{-1}$, as defined by the Chinese Environmental Quality Standards for Soil. In addition, the bioavailable Cr fraction was not detected by the apparatus. Therefore, the effects of Cr stress on soil microorganisms were not considered.

The soils varied greatly in their concentrations of total Cu (approximately 175 to 1,147 mg·kg$^{-1}$), Zn (approximately 187 to 2,472 mg·kg$^{-1}$), Pb (approximately 77 to 1,199 mg·kg$^{-1}$), and Cd (approximately 2.54 to 7.26 mg·kg$^{-1}$) with their distances from the smelter. The highest concentrations of Cu, Zn, and Pb were found in the samples nearest to smelter (200 m), but the highest Cd level was found in the sample 400 m away (Table 1).

*3.2. Concentration of Bioavailable Heavy Metals.* The fraction of metal, that is, in its bioavailable form, is crucial to understanding metal ecological toxicity. The bioavailability of heavy metals is related to their chemical forms in soil. Several fractions or compartments of the soil act as reservoirs for bioavailable metals [3]. Extractions by CaCl$_2$, DTPA, and EDTA are widely used methods of soil analysis, providing an operationally defined soil compartment that is characterized by its mobility and bioavailability. CaCl$_2$ is used to estimate the soluble and exchangeable metals. DTPA and EDTA are predominantly used to study the metals that have bonded

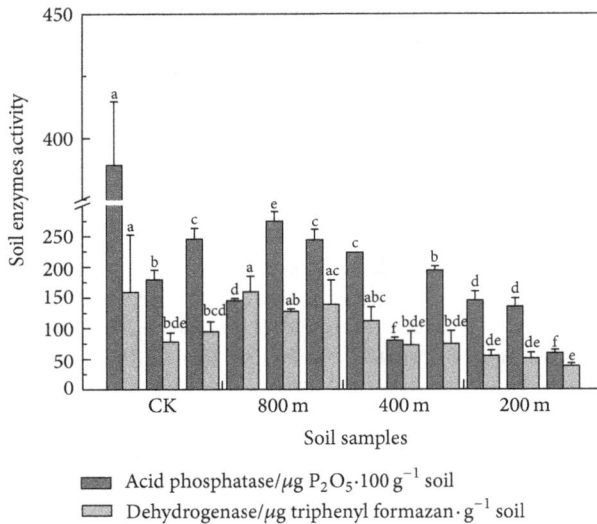

FIGURE 1: The effect of heavy metals on the activities of soil enzymes. Note: different letters in same enzyme indicate significant differences ($P < 0.05$).

TABLE 2: The Pearson correlation coefficients between the microbial parameters and the concentrations of heavy metals.

| | Acid phosphatase | Dehydrogenase | Microbial biomass C | Protein |
|---|---|---|---|---|
| Cu | | | | |
| CaCl$_2$ | −0.500 | −0.597* | −0.632* | −0.705* |
| EDTA | −0.576* | −0.628* | −0.834** | −0.884** |
| DTPA | −0.568 | −0.649* | −0.839** | −0.874** |
| Total | **−0.586*** | **−0.594*** | **−0.729*** | **−0.866*** |
| Zn | | | | |
| CaCl$_2$ | −0.400 | −0.584* | −0.604* | −0.543 |
| EDTA | −0.554 | −0.685* | −0.683* | −0.596* |
| DTPA | −0.613* | −0.734** | −0.748** | −0.698* |
| Total | **−0.531** | **−0.680*** | **−0.716*** | **−0.653*** |
| Pb | | | | |
| CaCl$_2$ | −0.239 | −0.485 | −0.303 | −0.219 |
| EDTA | −0.586* | −0.709** | −0.624* | −0.643* |
| DTPA | −0.719** | −0.760** | −0.583* | −0.669* |
| Total | **−0.569** | **−0.693*** | **−0.604*** | **−0.645*** |
| Cd | | | | |
| CaCl$_2$ | −0.379 | −0.515 | −0.551 | −0.542 |
| EDTA | −0.486 | −0.544 | −0.679* | −0.814** |
| DTPA | −0.425 | −0.539 | −0.726** | −0.794** |
| Total | **−0.528** | **−0.494** | **−0.576*** | **−0.817*** |

Note: ** correlation is significant at the 0.01 level (2-tailed).
* Correlation is significant at the 0.05 level (2-tailed).

with organic matter and the overall phytoavailability of heavy metals [18, 19].

The concentrations of bioavailable Cu, Zn, Pb, and Cd in the soil samples varied significantly (Table 1). The amount of bioavailable heavy metals in the soil decreased with increasing distance from the smelter, and the bioavailable fractions of heavy metals were significantly positively correlated with the total heavy metals ($P < 0.05$). In all of the soil samples, the concentration of CaCl$_2$-extractable metals < DTPA-extractable metals < EDTA-extractable metals. These results correlated with the extraction abilities of the reagents [35]. The concentration of CaCl$_2$-extractable heavy metals was far less than those of the DTPA- and EDTA-extractable heavy metals. This result indicates that heavy metals bonded with organic matter and were predominantly in their bioavailable form, agreeing with the conclusion that heavy metals in soil are primarily associated with soil organic matter [36, 37]. It is this bioavailable fraction of heavy metals that greatly affects soil microorganisms [1].

3.3. Soil Enzyme Activities. The variations in the activities of acid phosphatase and dehydrogenase in the soil samples are shown in Figure 1. Good correlations were observed between enzyme activity and distance from the smelter. Compared with the reference area (CK), acid phosphatase activity was inhibited in the sites contaminated by heavy metals. Soil acid phosphatase activity tended to decrease with increasing heavy metals. A significant decrease was observed in the samples only 200 m away from the smelter. Conversely, dehydrogenase activity was slightly increased in the soil samples 800 m away from the smelter and decreased as the distance from the smelter decreased. This result indicates that a low level of heavy metals will stimulate soil microbes to synthesize dehydrogenase, promoting microbial metabolic activity.

Soil acid phosphatase activity was significantly negatively correlated with EDTA-extractable Cu and Pb; DTPA-extractable Zn and Pb; and total Cu. Similarly, dehydrogenase activity was significantly negatively correlated with CaCl$_2$-extractable Cu and Zn; EDTA-extractable Cu, Zn, and Pb; DTPA-extractable Cu, Zn, and Pb; and total Cu, Zn, and Pb (Table 2).

3.4. Microbial Biomass Carbon in Soil. The amount of microbial biomass carbon (MBC) in the contaminated soil samples ranged from 44.6 to 105.0 mg·kg$^{-1}$, as measured by the FE method, which are lower values than those of the reference site (Figure 2). A significant decrease was observed in the soil samples only 200 m away from the smelter, with the MBC increasing from the plot nearest to the plot farthest away (800 m). The amount of soil MBC was significantly negatively correlated with CaCl$_2$-extractable Cu and Zn; EDTA-extractable Cu, Zn, Pb, and Cd; DTPA-extractable Cu, Zn, Pb, and Cd; and total Cu, Zn, Pb, and Cd (Table 2). These results indicate that the concentrations of DTPA- and EDTA-extractable heavy metals are better predictors of the effects that heavy metals have on MBC compared with the concentrations of CaCl$_2$-extractable heavy metals.

3.5. Soil Microorganism Community Composition. The idea of using a DGGE analysis of soil microbial communities to

The Variations in the Soil Enzyme Activity, Protein Expression, Microbial Biomass, and Community Structure of Soil
Contaminated by Heavy Metals

121

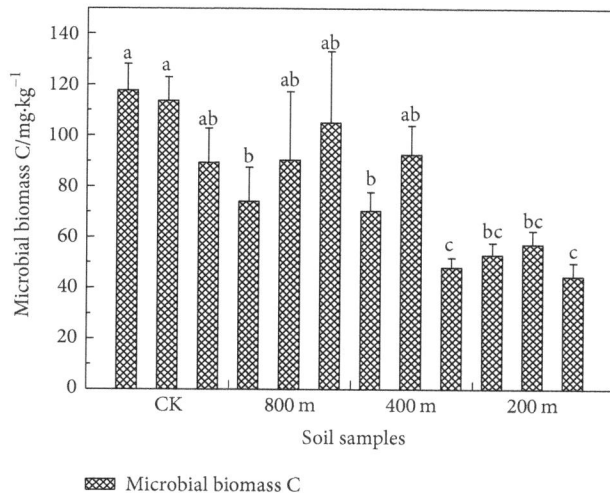

FIGURE 2: The effect of heavy metals on the microbial biomass carbon in soil. Note: the different letters indicate significant differences ($P < 0.05$).

investigate soil microbial diversity was first introduced by Muyzer et al. (1993). DGGE proved to be capable of accurately measuring soil microbial diversity, so it has become the predominant technique in soil microbial ecology for profiling soil microbial communities and monitoring shifts in microbial community composition due to environmental changes [38]. The DGGE profiles of the soil bacteria and actinomycetes are shown in Figures 3(a) and 4(a). Although many bands were found in all lanes, obvious differences in the intensities of the soil microbial communities were clearly observed (Figures 3(a) and 4(a)). As shown in Figure 3(a), the number of bands in the soil was significantly increased with an increasing distance from the smelter; in other words, the number of bands in the DGGE profile increased as the heavy metals presence decreased. A similar result was also found in the soil actinomycete community composition. Compared with the soils from the reference site and from 800 m away, the number of bands in the plots situated at 400 m and 200 m were significantly decreased (Figure 4(a)).

Subsequently, the DGGE gels were interpreted by the Shannon index, principal component analysis (PCA), and cluster analysis to examine the number, presence, and relative intensity of the bands. The Shannon index indicated that the diversity of the soil microorganisms (bacteria and actinomycetes) was decreased significantly with an increase in the heavy metals concentration (Table 3). PCA and the cluster analysis are also useful methods for analyzing DGGE profiles. The PCA plots show a clear separation due to the different distances from the smelter, demonstrating the altered structure and diversity of the microbial community. Using the presence of the bands as input data, the first two principal components (PC1 and PC2) were sufficient to explain 79.5% and 78.0% of the variance in the soil bacteria and the actinomycetes, respectively (Figures 3(b) and 4(b)). The similarity between the soil bacterial and actinomycete communities of the reference site and those of the soil taken

around the smelter was obtained from a cluster analysis of the DGGE profile and was determined to be less than 50% (Figures 3(c) and 4(c)). The cluster analysis indicated that the soil microorganism community structure was greatly affected by the presence of heavy metals.

*3.6. Soil Protein Expression.* The amount of protein in the contaminated soil samples ranged from 15.6 to 104 $\mu g \cdot g^{-1}$, far less than the amount found at the reference site (Figure 5(a)). The concentration of protein decreased drastically when the soil was slightly contaminated by heavy metals at a distance of 800 m from the smelter. The quantity of soil proteins was significantly negatively correlated with soil CaCl$_2$-extractable Cu; EDTA-extractable Cu, Zn, Pb, and Cd; DTPA-extractable Cu, Zn, Pb, and Cd; and total Cu, Zn, Pb, and Cd (Table 2).

The protein profiles are shown in Figure 5(b). The results indicate that no discernible differences exist between the composition and location of the citrate-extracted protein band of the reference site and those of the soil samples from 800 m and 400 m. The bands from these three sample sets were almost entirely located between approximately 55 and 70 kD. However, some type of protein, that is, 40 kD in size, was only found in the soil 200 m from the smelter. Additionally, although the SDS buffer extracts showed similar protein patterns between different soil samples, the expression of some proteins was significantly different. Compared with the contaminated soil samples, the uncontaminated samples had more abundant proteins overall. In the contaminated soil, with its increased concentration of heavy metals, there were fewer large molecular weight (>35 kD) proteins and more low molecular weight proteins (approximately 15 kD). At the base of each contaminated profile was a large agglomeration of presumably low molecular weight proteinaceous material that produced dark areas on the gels. Therefore, these low molecular weight proteins were largely aggregated.

## 4. Discussion

The activity and community composition of soil microbes are closely related to soil fertility and environmental quality. Additionally, soil biological parameters may have potential for use as early and sensitive indicators of soil ecological stress and restoration. Currently, the amount of microbial biomass carbon and enzyme activity are the metrics predominantly used to provide information about the biochemical processes occurring in the soil [3, 39]. The soil microbial biomass is the total mass of microorganisms living in the soil, which is the living part of the soil organic matter [1]. Soil phosphatase plays an essential role in the mineralization of organic phosphorus, while dehydrogenase is an intracellular enzyme that is involved in microbial oxidoreductase metabolism [3, 39].

In our study, the amount of microbial biomass C in the soil decreased when the concentration of heavy metals increased, although the difference was not significant when the heavy metals were at a low level. This finding is not surprising. Microorganisms differ in their sensitivity to metal toxicity, and the development of metal-tolerant strains, as well as shifts in the community structure, could compensate for

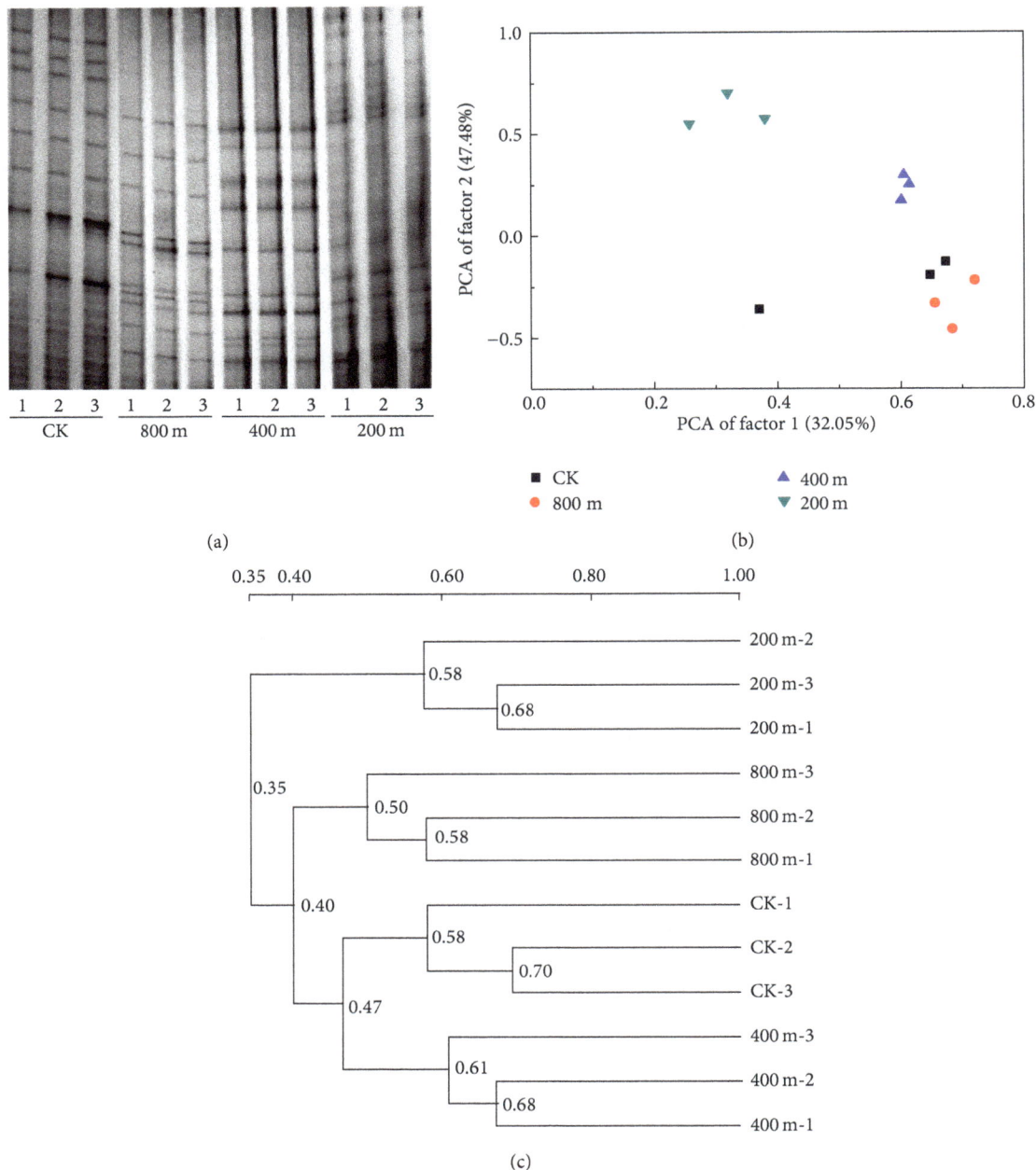

(a)

(b)

(c)

FIGURE 3: The effect of heavy metals on soil bacterial communities. (a) The DGGE profiles of the bacterial communities inhabiting each soil sample. (b) A principle component analysis based on the DGGE profiles of soil bacteria. (c) A cluster analysis based on the DGGE profiles of soil bacteria.

the loss of more sensitive populations [2]. The reduction in soil microbial biomass C might also explain the inhibition of enzyme activity observed in the heavy metal contaminated soil.

It is important to differentiate between extracellular and intracellular enzymes when we study the changes of enzymes under pollutant stress. The dehydrogenase activity, which is only present in viable cells and essentially depends on the metabolic state of the soil biota, may therefore be considered a direct measure of soil microbial activity. In contrast, phosphatase activity can occur extracellularly as well as within a

living cell [1, 3, 6, 26, 39]. For example, Brookes et al. [40] reported less dehydrogenase activity in metal-contaminated soil than in similar uncontaminated soil, while the soil phosphatase was unaffected. This might explain why, in our study, the dehydrogenase activity was also more sensitive to the heavy metal stress than the acid phosphatase activity.

Furthermore, the dehydrogenase activity was significantly correlated with soil microbial biomass C ($r = 0.611$, $P < 0.05$). These results suggest that microbial biomass C and dehydrogenase can be useful measures of the level of heavy metal contamination in a soil sample [26, 39].

The Variations in the Soil Enzyme Activity, Protein Expression, Microbial Biomass, and Community Structure of Soil Contaminated by Heavy Metals

123

(a)

(b)

(c)

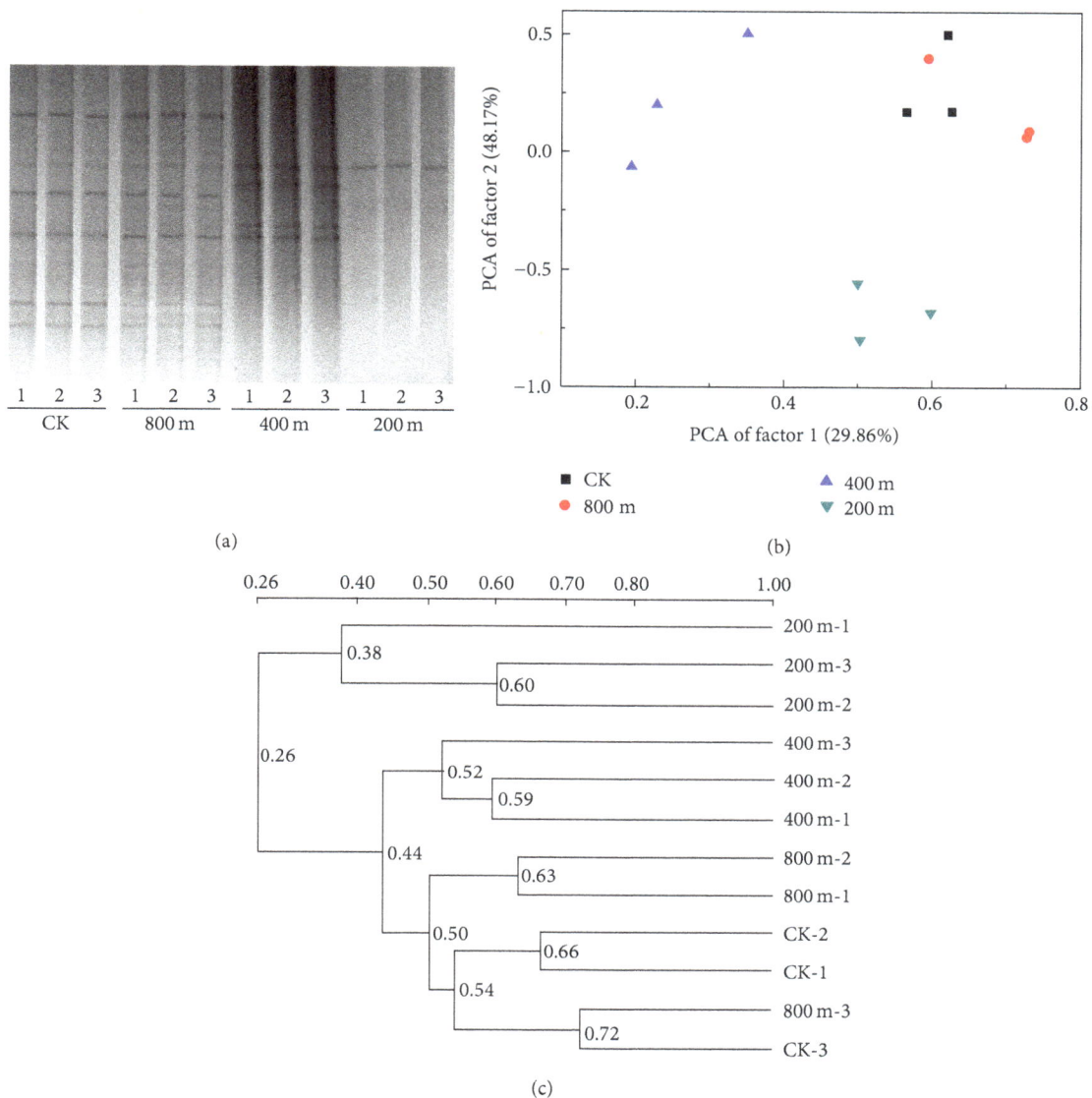

FIGURE 4: The effect of heavy metals on soil actinomycetes communities. (a) The DGGE profiles of the actinomycete communities inhabiting each soil sample. (b) A principle component analysis based on the DGGE profiles of soil actinomycetes. (c) A cluster analysis based on the DGGE profiles of soil actinomycetes.

The microbial biomass and enzyme activity measurements have their limitations in soil pollution studies, however. For example, the changes were not notable when heavy metal concentrations were low and highly soluble Cu generated chromogen interference to dehydrogenase activity measurements [41]. In addition, the changes in the microorganism community structure cannot be evaluated by such techniques. Therefore, the use of soil enzyme and microbial biomass as indicators for soil contaminated by heavy metals still needs further discussions.

The changes in the metabolic profiles indicate the possibility that heavy metal contamination results in a community that is more variable and less stable [3]. In our study, the genetic structure and protein expression of the indigenous soil microbial communities were evaluated by an advanced molecular technique that is based on PCR-DGGE and SDS-PAGE. The PCR-DGGE method has become a predominant molecular technique to study the changes in soil microorganisms due to contamination and agricultural practices [3, 20, 42]. In contrast, the use of SDS-PAGE to monitor contaminated soil is still in its infancy [7, 8, 17]. The results of the DGGE profiles, as analyzed by principal component analysis (PCA) and cluster analysis in this study, make it clear that the number and intensity of DNA bands in the different samples changed significantly. This significant difference between the samples indicates that the soil microbial community structure changed greatly in response to heavy metals contamination (Table 3, Figures 3-4). We also detected an increased relative abundance in the microbial populations with increasing distance from the

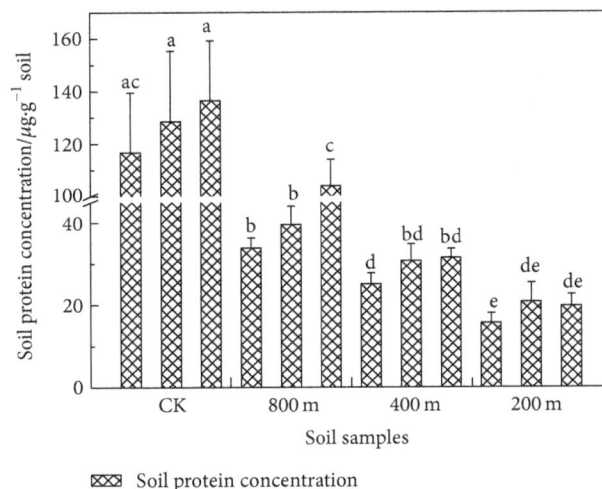

(a)

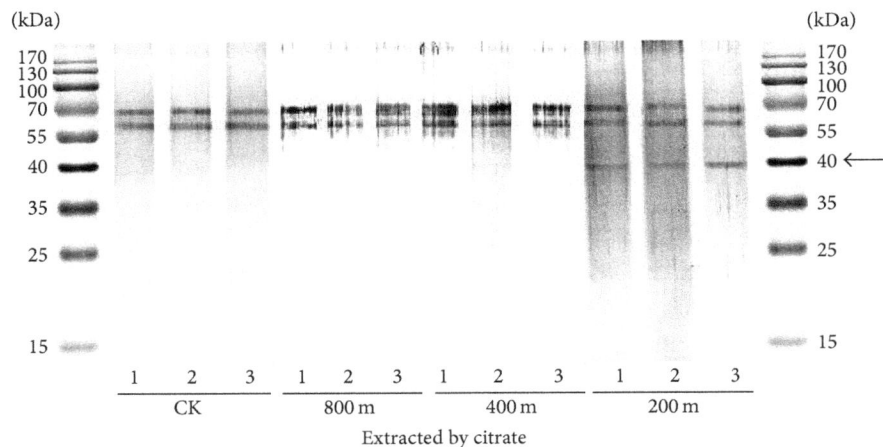

Extracted by citrate

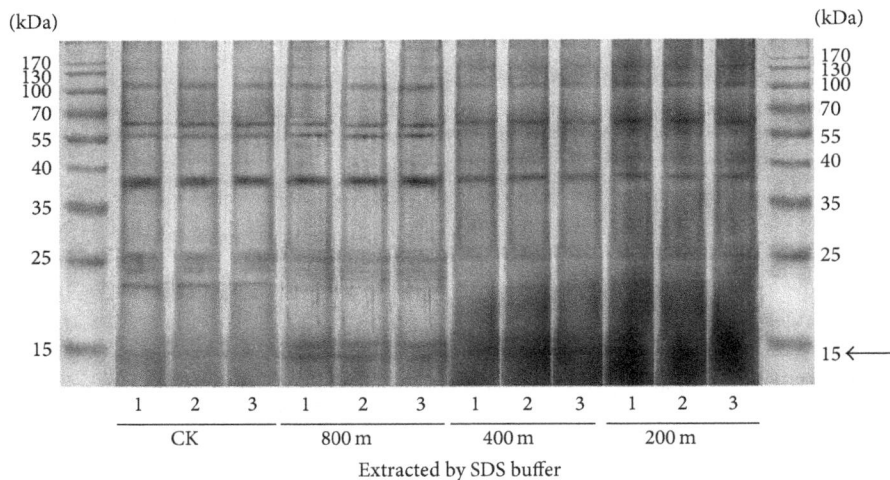

Extracted by SDS buffer

(b)

FIGURE 5: The effect of the concentration of heavy metals on soil protein. (a) The effect of the concentration of heavy metals on the amount of soil proteins. Note: Different letters indicate significant differences ($P < 0.05$). (b) The effect of the heavy metals concentration on soil protein expressions.

The Variations in the Soil Enzyme Activity, Protein Expression, Microbial Biomass, and Community Structure of Soil
Contaminated by Heavy Metals

125

TABLE 3: The Shannon diversity index for the soil samples contaminated by heavy metals.

| Distance | Soil bacteria | Soil actinomycetes |
|---|---|---|
| CK | $3.11 \pm 0.11^a$ | $0.52 \pm 0.02^a$ |
| 800 m | $2.91 \pm 0.12^b$ | $0.40 \pm 0.03^b$ |
| 400 m | $2.90 \pm 0.07^b$ | $0.35 \pm 0.01^c$ |
| 200 m | $2.48 \pm 0.08^c$ | $0.23 \pm 0.02^d$ |

Note: values presented in the table are the mean $\pm$ standard deviation ($n = 3$).
Values in each column followed by different letters are significantly different ($P < 0.05$).

smelter, demonstrating that heavy metals exerted a major effect on bacterial diversity by promoting changes in species composition (represented by the position of the bands) and in species richness (represented by the number of occurring bands). Namely, contamination inhibited certain bacterial groups and stimulated others, and it changed the overall microbial diversity [2, 3, 38, 43, 44].

Subsequently, the content and composition of the soil proteins were studied to investigate their changes under heavy metals stress. Compared with the reference site, the soil protein levels decreased significantly due to heavy metals contamination, and the percentage decrease in total soil proteins was larger than the decrease detected in soil enzyme activity or microbial biomass carbon (Figure 5(a)). The soil protein concentrations were significantly negatively correlated with heavy metals (Table 2). These results indicate that proteins are more sensitive to the changes caused by pollutants, making protein concentration a great potential indicator of soil pollution. The variation in soil proteins may be due to modifications in the microbial community composition [7, 8]. This hypothesis was confirmed by the DGGE profiles in our study (Figures 3(a) and 4(a)). Although protein has potential to be an indictor for soil contamination, its reliability still needs more studies since Bradford method is easily affected by humic substances present in protein solution [45].

The collected soil proteins were separated by SDS-PAGE, and we found that several bands were common to all soils. We also found more presumably low molecular weight proteins in the contaminated soil than in the soil of the reference site (Figure 5(b)). This protein could be related to the presence of heavy metals and could be a microbial response to the metals [8, 46]. For example, many eukaryotic microbes are known to produce metallothioneins (a low molecular weight protein) in response to metal (particularly Cd) exposure [47, 48]. Another explanation may be that sufficient metal exposure will result in the immediate death of cells due to a disruption of their essential functions; thus, the cells are lysed, and their proteins are released into the environment, where they are quickly degraded into low molecular weight materials [2, 8]. Obviously, more work is needed to identify this low molecular weight protein and understand its role in the microbial responses to heavy metals stress. This protein should also be investigated as an indicator for soil pollution.

Although PCR-DGGE and SDS-PAGE are useful for monitoring the changes in microbial community structures, they still have drawbacks. These approaches are easily interfered with by impurities that are coextracted with the soil DNA and proteins, primarily humic substances [7, 38]. The DNA and proteins that are mixed in with the organic matter are not easily separated by DGGE or SDS-PAGE. In addition, relatively small populations of microbes and low abundances of proteins may not be identified by DGGE and SDS-PAGE, respectively [8, 38]. In this study, for example, both the DGGE and SDS-PAGE profiles have a dark background and smeared dark areas at the bottom of SDS-PAGE gel, which make the accuracy and reliability of the analysis questionable. Consequently, the molecular techniques used in this study have much room for improvement, and more work is required to develop effective extraction methods for soil DNA and proteins.

No single technique can provide a comprehensive depiction of the soil microbial situation. Therefore, soil microbial activity and diversity are difficult to elucidate. To obtain a better understanding of soil microbial ecology, we need to integrate different methods to create a comprehensive analysis.

## 5. Conclusions

By combining different monitoring techniques for different aspects of microbiology, we can obtain a better understanding of the soil microorganisms that live in soils contaminated by heavy metals. Our study demonstrated that heavy metals have a significant negative effect on the activities of soil acid phosphatase and dehydrogenase as well as on the levels of proteins and microbial biomass carbon. Compared with the other potential contamination markers tested, the protein level showed the most dramatic decrease in slightly contaminated soil, indicating that it may be more sensitive to heavy metals. Subsequently, the denaturing gradient gel electrophoresis analysis used in this study demonstrated that heavy metals had a significant negative effect on soil microbial diversity and community structure. Furthermore, the soil protein expression was similar in different soils, but a large quantity of presumably low molecular weight protein was observed only in heavy metal contaminated soil. Based on the research described in this paper, we can conclude that the soil protein level has great potential to be a sensitive indicator of soil contamination. Further research is essential to identify the low molecular weight protein that only appears in contaminated soil.

## Acknowledgments

This work was financially supported by the National Natural Science Foundation of China (11179025), National Natural Science Foundation for the Youth of China (21007055), Science Foundation for Postdoctor of China (20100471714, 201104718), and Program for New Century Excellent Talents in University (NCET-11-0455).

## References

[1] P. C. Brookes, "The use of microbial parameters in monitoring soil pollution by heavy metals," *Biology and Fertility of Soils*, vol. 19, no. 4, pp. 269–279, 1995.

[2] K. E. Giller, E. Witter, and S. P. Mcgrath, "Toxicity of heavy metals to microorganisms and microbial processes in agricultural soils: a review," *Soil Biology and Biochemistry*, vol. 30, no. 10-11, pp. 1389–1414, 1998.

[3] Y. Wang, J. Shi, H. Wang, Q. Lin, X. Chen, and Y. Chen, "The influence of soil heavy metals pollution on soil microbial biomass, enzyme activity, and community composition near a copper smelter," *Ecotoxicology and Environmental Safety*, vol. 67, no. 1, pp. 75–81, 2007.

[4] G. Renella, M. Mench, L. Landi, and P. Nannipieri, "Microbial activity and hydrolase synthesis in long-term Cd-contaminated soils," *Soil Biology and Biochemistry*, vol. 37, no. 1, pp. 133–139, 2005.

[5] A. K. Müller, K. Westergaard, S. Christensen, and S. J. Sørensen, "The effect of long-term mercury pollution on the soil microbial community," *FEMS Microbiology Ecology*, vol. 36, no. 1, pp. 11–19, 2001.

[6] C. Trasar-Cepeda, M. C. Leirós, S. Seoane, and F. Gil-Sotres, "Limitations of soil enzymes as indicators of soil pollution," *Soil Biology and Biochemistry*, vol. 32, no. 13, pp. 1867–1875, 2000.

[7] F. Bastida, J. L. Moreno, C. Nicolás, T. Hernández, and C. García, "Soil metaproteomics: a review of an emerging environmental science. Significance, methodology and perspectives," *European Journal of Soil Science*, vol. 60, no. 6, pp. 845–859, 2009.

[8] I. Singleton, G. Merrington, S. Colvan, and J. S. Delahunty, "The potential of soil protein-based methods to indicate metal contamination," *Applied Soil Ecology*, vol. 23, no. 1, pp. 25–32, 2003.

[9] P.-A. Maron, L. Ranjard, C. Mougel, and P. Lemanceau, "Metaproteomics: a new approach for studying functional microbial ecology," *Microbial Ecology*, vol. 53, no. 3, pp. 486–493, 2007.

[10] P. Wilmes and P. L. Bond, "The application of two-dimensional polyacrylamide gel electrophoresis and downstream analyses to a mixed community of prokaryotic microorganisms," *Environmental Microbiology*, vol. 6, no. 9, pp. 911–920, 2004.

[11] P. Wilmes and P. L. Bond, "Metaproteomics: studying functional gene expression in microbial ecosystems," *Trends in Microbiology*, vol. 14, no. 2, pp. 92–97, 2006.

[12] S. Criquet, A. M. Farnet, and E. Ferre, "Protein measurement in forest litter," *Biology and Fertility of Soils*, vol. 35, no. 5, pp. 307–313, 2002.

[13] W. X. Schulze, G. Gleixner, K. Kaiser, G. Guggenberger, M. Mann, and E.-D. Schulze, "A proteomic fingerprint of dissolved organic carbon and of soil particles," *Oecologia*, vol. 142, no. 3, pp. 335–343, 2005.

[14] E. B. Taylor and M. A. Williams, "Microbial protein in soil: influence of extraction method and C amendment on extraction and recovery," *Microbial Ecology*, vol. 59, no. 2, pp. 390–399, 2010.

[15] S. Chen, M. C. Rillig, and W. Wang, "Improving soil protein extraction for metaproteome analysis and glomalin-related soil protein detection," *Proteomics*, vol. 9, no. 21, pp. 4970–4973, 2009.

[16] H.-B. Wang, Z.-X. Zhang, H. Li et al., "Characterization of metaproteomics in crop rhizospheric soil," *Journal of Proteome Research*, vol. 10, no. 3, pp. 932–940, 2011.

[17] D. Benndorf, G. U. Balcke, H. Harms, and M. Von Bergen, "Functional metaproteome analysis of protein extracts from contaminated soil and groundwater," *ISME Journal*, vol. 1, no. 3, pp. 224–234, 2007.

[18] L. A. Brun, J. Maillet, P. Hinsinger, and M. Pépin, "Evaluation of copper availability to plants in copper-contaminated vineyard soils," *Environmental Pollution*, vol. 111, no. 2, pp. 293–302, 2001.

[19] E. Meers, R. Samson, F. M. G. Tack et al., "Phytoavailability assessment of heavy metals in soils by single extractions and accumulation by *Phaseolus vulgaris*," *Environmental and Experimental Botany*, vol. 60, no. 3, pp. 385–396, 2007.

[20] G. Muyzer, E. C. de Waal, and A. G. Uitterlinden, "Profiling of complex microbial populations by denaturing gradient gel electrophoresis analysis of polymerase chain reaction-amplified genes coding for 16S rRNA," *Applied and Environmental Microbiology*, vol. 59, no. 3, pp. 695–700, 1993.

[21] E. Baath, A. Frostegaard, T. Pennanen, and H. Fritze, "Microbial community structure and pH response in relation to soil organic matter quality in wood-ash fertilized, clear-cut or burned coniferous forest soils," *Soil Biology and Biochemistry*, vol. 27, no. 2, pp. 229–240, 1995.

[22] X. Tang, C. Shen, D. Shi et al., "Heavy metal and persistent organic compound contamination in soil from Wenling: an emerging e-waste recycling city in Taizhou area, China," *Journal of Hazardous Materials*, vol. 173, no. 1–3, pp. 653–660, 2010.

[23] E. Lakanen and R. Ervio, "A comparison of eight extractants for determination of plant available micronutrients in soil," *Acta Agraria Fennica*, vol. 123, pp. 223–232, 1971.

[24] W. L. Lindsay and W. A. Norvell, "Development of a DTPA soil test for zinc, iron, manganese, and copper," *Soil Science Society of America Journal*, vol. 42, pp. 421–428, 1978.

[25] P. Quevauviller, "Operationally defined extraction procedures for soil and sediment analysis I. Standardization," *Trends in Analytical Chemistry*, vol. 17, no. 5, pp. 289–298, 1998.

[26] C. Garcia, T. Hernandez, F. Costa, and B. Ceccanti, "Biochemical parameters in soils regenerated by the addition of organic wastes," *Waste Management and Research*, vol. 12, no. 6, pp. 457–466, 1994.

[27] S. Y. Guan, *Soil Enzyme and Its Research Methods*, Agricultural Press, Beijing, China, 1986.

[28] X. Lin, *Principles and Methods of Soil Microbiology Research*, Higher Education Press, Beijing, China, 2010.

[29] E. D. Vance, P. C. Brookes, and D. S. Jenkinson, "An extraction method for measuring soil microbial biomass C," *Soil Biology and Biochemistry*, vol. 19, no. 6, pp. 703–707, 1987.

[30] M. M. Bradford, "A rapid and sensitive method for the quantitation of microgram quantities of protein utilizing the principle of protein dye binding," *Analytical Biochemistry*, vol. 72, no. 1-2, pp. 248–254, 1976.

[31] J. X. Yan, R. Wait, T. Berkelman et al., "A modified silver staining protocol for visualization of proteins compatible with matrix-assisted laser desorption/ionization and electrospray ionization- mass spectrometry," *Electrophoresis*, vol. 21, pp. 3666–3672, 2000.

[32] J. Schäfer, U. Jäckel, and P. Kämpfer, "Development of a new PCR primer system for selective amplification of Actinobacteria," *FEMS Microbiology Letters*, vol. 311, pp. 103–112, 2010.

[33] G. W. Yeates, H. J. Percival, and A. Parshotam, "Soil nematode responses to year-to-year variation of low levels of heavy metals," *Australian Journal of Soil Research*, vol. 41, no. 3, pp. 613–625, 2003.

[34] C. D. Clegg, R. D. L. Lovell, and P. J. Hobbs, "The impact of grassland management regime on the community structure of selected bacterial groups in soils," *FEMS Microbiology Ecology*, vol. 43, no. 2, pp. 263–270, 2003.

[35] C. R. M. Rao, A. Sahuquillo, and J. F. Lopez Sanchez, "A review of the different methods applied in environmental geochemistry for single and sequential extraction of trace elements in soils

The Variations in the Soil Enzyme Activity, Protein Expression, Microbial Biomass, and Community Structure of Soil
Contaminated by Heavy Metals

127

and related materials," *Water, Air, and Soil Pollution*, vol. 189, no. 1–4, pp. 291–333, 2008.

[36] A. Manceau, M.-C. Boisset, G. Sarret et al., "Direct determination of lead speciation in contaminated soils by EXAFS spectroscopy," *Environmental Science and Technology*, vol. 30, no. 5, pp. 1540–1552, 1996.

[37] D. G. Strawn and L. L. Baker, "Molecular characterization of copper in soils using X-ray absorption spectroscopy," *Environmental Pollution*, vol. 157, no. 10, pp. 2813–2821, 2009.

[38] J. Zhou, X. Sun, J. Jiao, M. Liu, F. Hu, and H. Li, "Dynamic changes of bacterial community under the influence of bacterial-feeding nematodes grazing in prometryne contaminated soil," *Applied Soil Ecology*, vol. 64, pp. 70–76, 2013.

[39] J. C. García-Gil, C. Plaza, P. Soler-Rovira, and A. Polo, "Long-term effects of municipal solid waste compost application on soil enzyme activities and microbial biomass," *Soil Biology and Biochemistry*, vol. 32, no. 13, pp. 1907–1913, 2000.

[40] P. Brookes, S. McGrath, D. Klein, and E. Elliott, "Effects of heavy metals on microbial activity and biomass in field soils treated with sewage sludge," in *Environmental Contamination*, pp. 574–583, CEP Consultants, Edinburgh, Scotland, 1984.

[41] K. Chander and P. C. Brookes, "Is the dehydrogenase assay invalid as a method to estimate microbial activity in copper-contaminated soils?" *Soil Biology and Biochemistry*, vol. 23, no. 10, pp. 909–915, 1991.

[42] M. Cea, M. Jorquera, O. Rubilar, H. Langer, G. Tortella, and M. C. Diez, "Bioremediation of soil contaminated with pentachlorophenol by Anthracophyllum discolor and its effect on soil microbial community," *Journal of Hazardous Materials*, vol. 181, no. 1–3, pp. 315–323, 2010.

[43] K. Arnebrant, E. Bååth, and A. Nordgren, "Copper tolerance of microfungi isolated from polluted and unpolluted forest soil," *Mycologia*, vol. 79, no. 6, pp. 890–895, 1987.

[44] E. Baath, "Effects of heavy metals in soil on microbial processes and populations (a review)," *Water, Air, and Soil Pollution*, vol. 47, no. 3-4, pp. 335–379, 1989.

[45] P. Roberts and D. L. Jones, "Critical evaluation of methods for determining total protein in soil solution," *Soil Biology and Biochemistry*, vol. 40, no. 6, pp. 1485–1495, 2008.

[46] M. Hodson, "Effects of heavy metals and metalloids on soil organisms," in *Heavy Metals in Soils*, B. J. Alloway, Ed., pp. 141–160, Springer, Amsterdam, The Netherlands, 2013.

[47] C. A. Blindauer, "Bacterial metallothioneins: past, present, and questions for the future," *Journal of Biological Inorganic Chemistry*, vol. 16, no. 7, pp. 1011–1024, 2011.

[48] M. Mejáre and L. Bülow, "Metal-binding proteins and peptides in bioremediation and phytoremediation of heavy metals," *Trends in Biotechnology*, vol. 19, no. 2, pp. 67–73, 2001.

# 11

# Spatiotemporal Changes of Rainfall Erosivity in Loess Plateau, China

**Mohamed A. M. Abd Elbasit,**[1,2] **Jinbai Huang,**[3] **CSP Ojha,**[4] **Hiroshi Yasuda,**[1] **and Eltayeb O. Adam**[5]

[1] *Arid Land Research Center, Tottori University, 1390 Hamasaka, Tottori 680-0001, Japan*
[2] *Desertification Research Institute, National Center for Research, Khartoum 11111, Sudan*
[3] *College of Hydraulic and Architecture of North East Agricultural University, Harbin 150030, China*
[4] *Department of Civil Engineering, Indian Institute of Technology, Roorkee 247667, India*
[5] *Remote Sensing Authority, National center for Research, Khartoum 11111, Sudan*

Correspondence should be addressed to Mohamed A. M. Abd Elbasit; m abdelbasit@hotmail.com

Academic Editors: J. Artiola, L. A. Dawson, W. Ding, and W. R. Roy

The reason for the severity of soil erosion in Loess Plateau can be attributed to three nonanthropogenic factors: rainfall erosivity, slope gradient, and loess soil. The rainfall erosivity is controlled by the rainfall characteristics. Generally, rainfall characteristics change drastically in space and time. The rainfall erosivity has been investigated using the modified Fournier index (MFI), annual rainfall, and precipitation concentration index (PCI). The study showed a decrease in average MFI by 10%. However, the difference between the MFI in 1960s and 1990s was found to decrease in a large area in Loess Plateau, whereas there was an increase in MFI at the high latitude. The maximum decrease in the rainfall erosivity was higher in the southeast than that in the north and west. The $P_y$ was found to have a trend similar to the MFI, which further indicates that the MFI follows, to a high extent, the annual rainfall trend. The PCI was found to have trend opposite to MFI and $P_y$. The PCI increased in the north and west and decreased toward the southeast. The average temporal difference in the PCI between the 1960s, and 1990s was two percent.

## 1. Introduction

Soil erosion in Loess Plateau, China is a major environmental problem that limiting the area development. The generated sediment is transported by the Yellow River and deposited to Yellow Sea causing various offsite ecological problems. The problem of soil erosion in Loess Plateau received more attention from the government at various levels [1], such as reforestation of hillslopes and check dams construction. The concentration of rainfall in three months, namely, June, July, and August, coupled with sparse land cover, highly erodible soil, and steep slopes in Loess Plateau made the area highly susceptible for soil erosion. The concentration of rainfall in certain period has a large impact on the magnitude of rainfall erosivity. The rainfall erosivity can be defined as the potential of rainfall to cause soil erosion. The rainfall erosivity is a function of rainfall characteristics and rainfall runoff erosivity [2]. The resulted runoff from a rainfall event is highly dependent on rainfall characteristics. Thus, the soil erosion can be linked directly with rainfall characteristics. The rainfall characteristics are highly controlled by the drop size distribution (DSD) [3]; however, data related to DSD are rarely available in arid and semiarid regions. In Loess Plateau, the daily rainfall is proportional to the annual rainfall, which indicates that higher annual rainfall means higher rainfall intensity and rainstorm runoff erosivity [2]. This fact makes the evaluation of rainfall erosivity based on rainfall data with relatively coarse time resolution highly reliable for large-scale erosion evaluation.

The rainfall erosivity has been widely evaluated by the Revised Universal Soil Loss Equation (RUSLE) using rainfall and runoff erosivity factor (R-factor). The R-factor can be

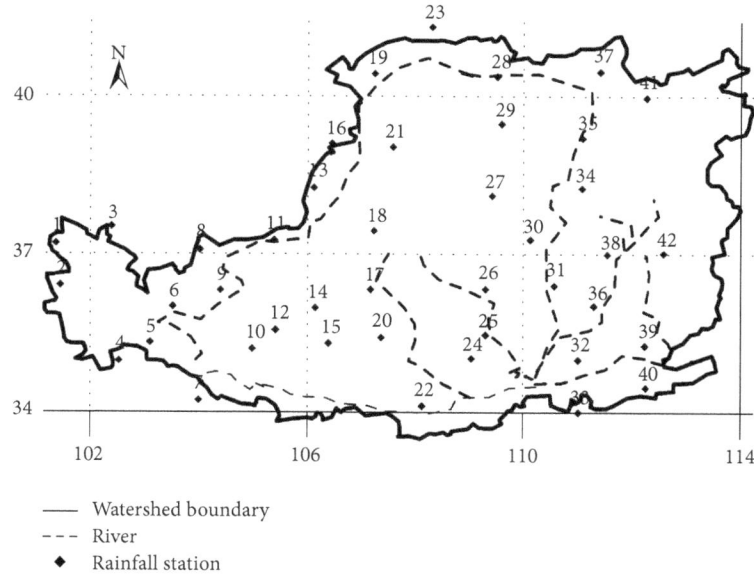

FIGURE 1: Chinese Loess Plateau and location of the rainfall stations considered in this study.

TABLE 1: Forty-year-, and ten-year-based average annual rainfall, the modified Fournier index, and precipitation concentration index for 42 rainfall stations located in Loess Plateau, China.

| Rainfall parameter | 40 years average[a] | 1960s average[b] | 1970s average[b] | 1980s average[b] | 1990s average[b] |
|---|---|---|---|---|---|
| $P_y$ | 417.54 (22.20) | 448.92 (24.11) | 413.98 (22.31) | 420.67 (23.88) | 386.54 (19.47) |
| MFI | 80.07 (3.32) | 85.57 (3.79) | 80.37 (3.54) | 78.92 (3.72) | 75.40 (2.80) |
| PCI | 19.77 (0.45) | 19.75 (0.47) | 19.86 (0.48) | 19.33 (0.47) | 20.16 (0.47) |

The number between parentheses is the standard error; $P_y$ is annual rainfall (mm); MFI is modified Fournier index; PCI is precipitation concentration index.
[a,b]Number of data is 1680, and 420, respectively.

TABLE 2: Correlation coefficient between the rainfall parameters and geographical attributes ($n = 42$).

| Rainfall parameters | Longitude | Latitude | Altitude |
|---|---|---|---|
| $P_y$ | 0.35 (0.02) | −0.68 (<0.001) | −0.07 (ns) |
| MFI | 0.56 (<0.001) | −0.45 (0.002) | −0.20 (ns) |
| PCI | 0.14 (ns) | 0.89 (<0.001) | −0.21 (ns) |

The exponent between parentheses is the $P$ value; $P_y$ is annual rainfall (mm); MFI is modified Fournier index; PCI is precipitation concentration index.

defined as the product of total kinetic energy of storm times its 30-minute maximum rainfall intensity (KE × $I_{30}$), which can be calculated as follows [4]:

$$R\text{-factor} = \frac{1}{n}\sum_{i=1}^{n}\left[\sum_{k=1}^{m}\text{KE}(I_{30})_k\right]_i, \quad (1)$$

where $R$-factor is average rainfall and runoff erosivity (MJ mm ha$^{-1}$ h$^{-1}$ year$^{-1}$); KE is total kinetic energy of single storm (MJ ha$^{-1}$); $I_{30}$ is the maximum 30-minute storm rainfall intensity (mm h$^{-1}$); m is the number of $k$ erosive storms at

each $i$ year; and $n$ is the number of years used to obtain the average $R$-factor [4].

According to (1), rainfall data with half-hour time resolution is required to calculate the $R$-factor which is rarely available in arid and semiarid regions. Moreover, the calculation of $R$-factor required rainfall energy measurement which is usually derived from DSD data. Various relationships were suggested to relate the rainfall intensity with KE under different geographical locations and climate conditions [5–7]. Because of these data requirements, various indices were developed to evaluate the rainfall erosivity using daily, monthly, and annual rainfall data. Daily rainfall data at the rate greater than 9 mm day$^{-1}$ has been used to evaluate the rainfall erosivity in Loess Plateau [8]. On the other hand, researchers used the summation of half-month erosivity index ($M_i$) method derived from daily rainfall rate greater than 12 mm day$^{-1}$ to evaluate the annual rainfall erosivity [9]. The $M_i$ is widely used in China to evaluate the erosivity rainfall [9, 10]. The $M_i$ can be calculated as follows [10]:

$$M_i = \alpha \sum_{j=1}^{k}\left(D_j\right)^{\beta}, \quad (2)$$

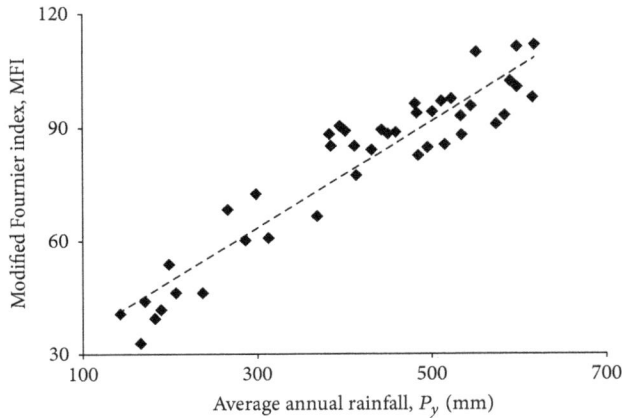

FIGURE 2: Relationship between 40-year average annual rainfall ($P_y$) and average erosivity modified Fournier index (MFI) using 42 rainfall station data located in Loess Plateau China.

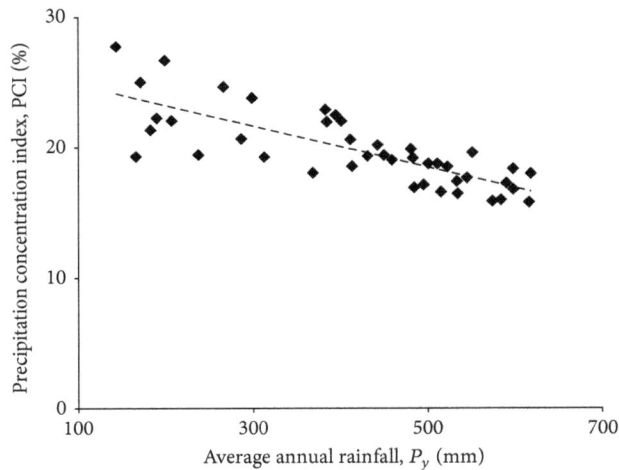

FIGURE 3: Relationship between 40-year-average annual rainfall ($P_y$) and precipitation concentration index (PCI%) using 42 rainfall station data located in Loess Plateau.

where $M_i$ is half-month rainfall erosivity (MJ mm ha$^{-2}$ h$^{-1}$ year$^{-1}$); $D_j$ is effective rainfall for day $j$ in one half-month (mm); $k$ is number of days in half month. The $D_j$ must be greater than or equal to 12 mm day$^{-1}$. The terms $\alpha$ and $\beta$ are empirical parameters calculated as follows:

$$\alpha = 21.586\beta^{-7.1891}$$

$$\beta = 0.8363 + \frac{18.144}{\overline{P}_{d12}} + \frac{24.455}{\overline{P}_{y12}}, \qquad (3)$$

where $\overline{P}_{d12}$ and $\overline{P}_{y12}$ are the average daily rainfall larger than 12 mm and average annual rainfall for days with rainfall higher than 12 mm [10].

The rainfall erosivity can be also assessed using monthly and annual rainfall by employing the Fournier index 11 (Fournier, 1960). The index has been accepted widely in

(a)

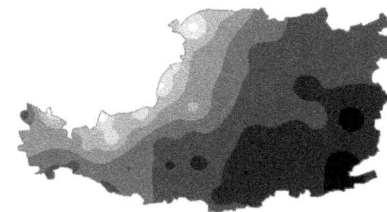

(b)

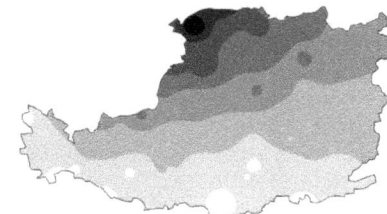

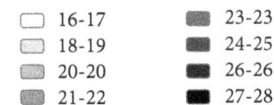

(c)

FIGURE 4: Spatial distribution of average annual rainfall (a), modified Fournier index (b), and precipitation concentration index (c) in the Chinese Loess Plateau determined from 42 stations for the period from 1960–2000.

different locations in the globe as a major erosivity indicator especially under rainfall data scarcity [11–14]. Moreover, recent development in satellite-borne rainfall data, such as Tropical Rainfall Measuring Mission (TRMM) by NASA, provided data with promising temporal and spatial resolution to be used for erosivity analysis in the near future by applying the Fournier index. The FI also has been used by several

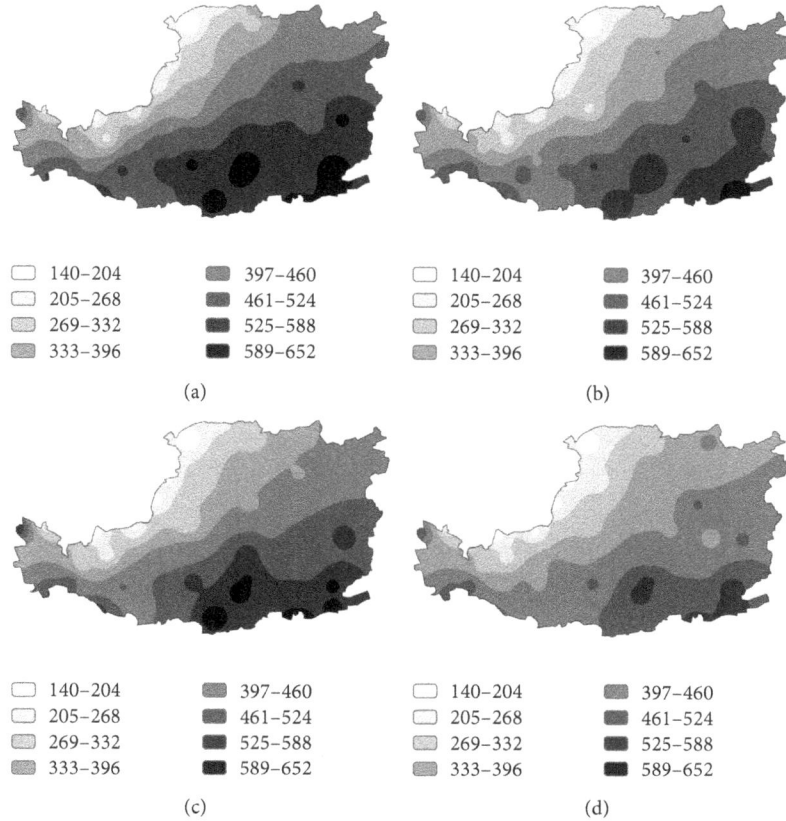

FIGURE 5: Spatiotemporal annual rainfall distribution in the Chinese Loess Plateau. (a) annual rainfall 1960s, (b) annual rainfall 1970s, (c) annual rainfall 1980s, and (d) annual rainfall 1990s.

researchers to determine the impact of different climate change scenarios on rainfall erosivity [15].

The aerial extend and topographic nature of the Chinese Loess Plateau is known by the existence of spatial variability in rainfall characteristics and erosivity. Moreover, Zhang et al. [15] reported that the presence of temporal variability in rainfall erosivity is due to climate change. The $M_i$ index has been used to evaluate the spatial and temporal variability of rainfall erosivity on the Chinese Loess Plateau which employs daily rainfall [10]. In this study, the spatiotemporal variability of rainfall erosivity for the period from 1960–2000 in the Chinese Loess Plateau will be assessed using monthly (FI, $M_i$, $EI_{30}$) and annual rainfall (MFI) indices.

## 2. Materials and Methods

*2.1. Study Area and Rainfall Data Set.* Loess Plateau is located in the northeast of China covering an area equal to $640 \times 10^3 \, \mathrm{km}^2$. The study area covered approximately the whole area of the plateau which lies between latitudes 33.53°N and 41.18°N, and longitudes 101.16°E and 114.23°E (Figure 1). The average elevation of the plateau is 1300 m above the mean sea level. The area is covered by loess soil highly susceptible for erosion by wind and water. The climate can be classified as a continental monsoon climate with annual rainfall range between 200 to 700 mm which increases from

the northwest to the southeast. Rainfall in the Loess Plateau is concentrated in summer, particularly during June, July, August, and September [10]. The concentrated rainfall, low vegetation cover, high erodible soil, and steep slopes made Loess Plateau area with high potential for soil erosion.

Monthly rainfall data set were collected at 84 stations located at Loess Plateau for 40 years starting from 1960 to 2000. Data sets from stations that reported more than 10% missing values have been excluded from this study. Finally, 42 stations out of 84 were selected, and erosivity analysis has been done on these stations (Figure 1). The data were analyzed using Microsoft Spreadsheet Excel application. The data from the 42 stations were interpolated to generate equal size grids (approximately $1.1 \times 1.1$ km) using inverse distance weighted (IDW) method. The generated surfaces were analyzed using ArcGIS 9.3. The temporal variability was assessed using the average 10-year rainfall erosivity indices.

*2.2. Calculations of Erosivity Indices.* The monthly Fournier erosivity index (FI) was calculated using the following [16]:

$$\mathrm{FI} = \frac{p_i^2}{P}, \tag{4}$$

where FI is Fournier Index; $p_i$ monthly rainfall depth in $i$ month; and $P$ is the annual rainfall.

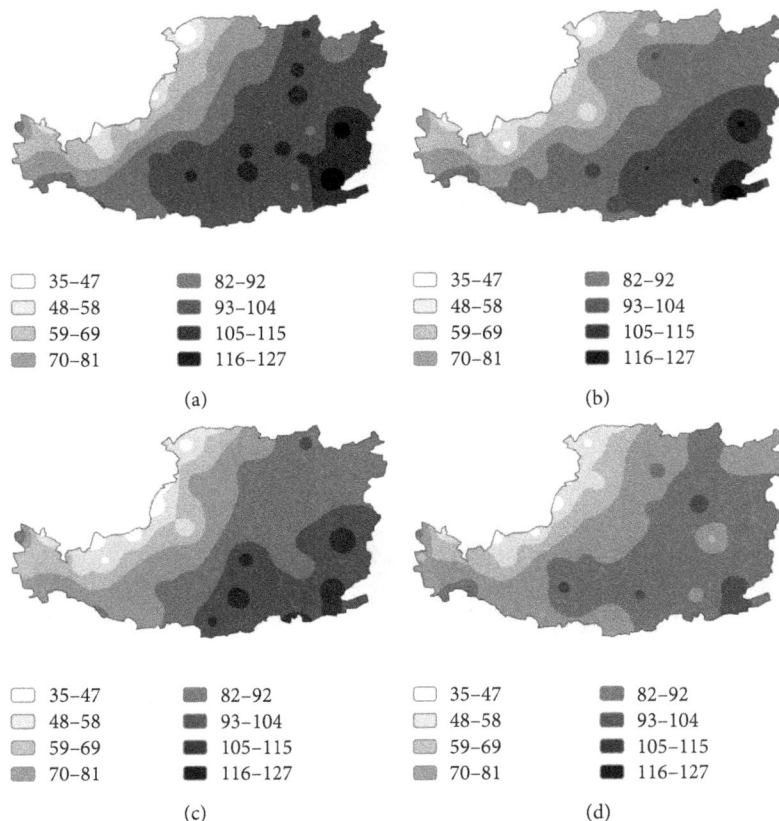

FIGURE 6: Spatiotemporal modified Fournier index distribution in the Chinese Loess Plateau. (a) Modified Fournier index rainfall, 1960s, (b) modified Fournier index rainfall, 1970s, (c) modified Fournier index, 1980s, and (d) modified Fournier index, 1990s.

The modified Fournier index (MFI) was calculated using the following [11]:

$$\text{MFI} = \frac{\sum_{i=1}^{12} p_i^2}{P}, \tag{5}$$

where MFI is the modified Fournier index. The MFI indicates the concentrated impact of rainwater on soil erosion. Higher MFI value means higher rainfall erosivity and vis-a-vis. Generally, the MFI range from 0 to 60 is defined as an indicator of very low erosivity, from 90 to 120 as moderate erosivity, from 120 to 160 as high erosivity, and greater than 160 as very high erosivity.

Another factor that affects the erosivity is the rainfall seasonality, which has been assessed usings precipitation concentration index (PCI), was calculated as follows [15]:

$$\text{PCI} = 100 \sum_{i=1}^{12} \left( \frac{p_i^2}{P^2} \right), \tag{6}$$

where PCI is the precipitation concentration index (%) and the other variables were defined previously. The interpretation on the PCI was carried out as follows: uniform (PCI = 8.3–10%), moderately seasonal (PCI = 10–15%), seasonal (15–20%), highly seasonal (PCI = 20–50%), and irregular (PCI = 50–100%).

## 3. Results and Discussion

*3.1. Temporal Pattern.* Table 1 shows the average $P_y$, MFI, and PCI for 42 selected stations in the Loess Plateau. The average $P_y$ calculated from 40-year data was 417.54 mm. The temporal and spatial average variation in the $P_y$ was less than 6%. The minimum $P_y$ in Loess Plateau was 143.01 mm found on the northwest where the maximum rainfall was 618.16 mm found on the southeast of the Loess Plateau. This result indicates that the maximum variation in the rainfall is more than 113% compared with 6% when compared with the average variation. This result emphasizes that the precipitation of Loess Plateau has a high variability in $P_y$ and it has subsequent impact on rainfall erosivity. The ten-year $P_y$ average showed general decrease in $P_y$. Comparing the $P_y$ in 1970s, 1980s, and 1990s with annual rainfall in 1960s, the $P_y$ showed decrease in 1970s, 1980s, and 1990s by 7.8%, 6.3%, and 13.9%, respectively. Although there was a noticeable general decrease in $P_y$, the rainfall in 1980s showed little increase compared to $P_y$ in 1970s (Table 1). The MFI was found to follow the same decreasing trend compared to 1960s average MFI (Table 1). Similar results on the deceasing of annual rainfall and half-month erosivity index ($M_i$) by 10 and 15%, respectively, have been reported [10]. The MFI was found to decrease by 6.1, 7.8, and 11.9% during 1970s, 1980s, and 1990s, respectively. Referring to these MFI results, the rainfall erosivity in Loess Plateau can be classified in the range

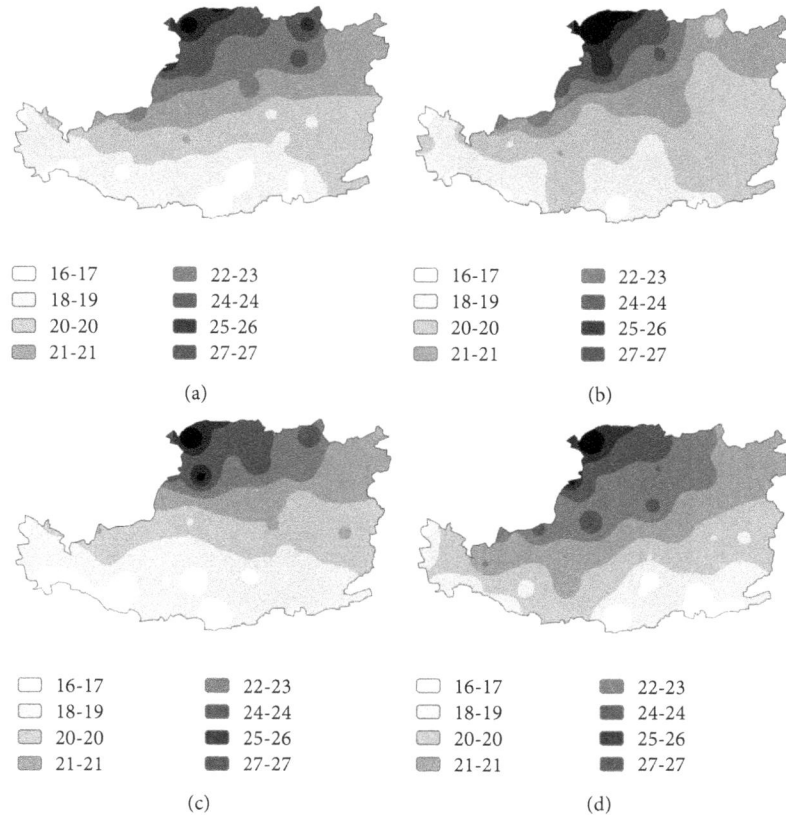

FIGURE 7: Spatiotemporal precipitation concentration index (%) distribution in the chinese loess plateau. (a) Precipitation concentration index, 1960s, (b) precipitation concentration index, 1970s, (c) precipitation concentration index, 1980s, and (d) precipitation concentration index, 1990s.

between low and moderate erosivity [14]. The maximum and minimum MFI was 32.88, and 111.57, respectively. The PCI% was approximately constant (19.8%) which indicated that the rainfall in Loess Plateau follows a seasonal pattern without changes during the study period. The maximum and minimum PCI (%) was 27.74 and 15.77%, respectively, which is between moderate and high seasonality. The $P_y$ had significant positive correlation with the MFI and the correlation coefficient was 0.94 ($P < 0.001$), as shown in Figure 2. On the other hand, the $P_y$ had significant negative correlation with the PCI% and the correlation coefficient was $-0.78$ ($P < 0.001$) (Figure 3). The MFI was found also correlated negatively with the PCI% and the correlation coefficient was $-0.57$ ($P < 0.001$).

### 3.2. Spatial Pattern.
The $P_y$ increases from west to east directions, and decreases from the south to the north (Table 2). The $P_y$ was correlated positively with the longitude (0.35, $P = 0.02$) and negatively with the latitude ($-0.68$, $P < 0.001$). The distribution of rainfall inside the Loess Plateau is controlled by the advance of the southwest equatorial air mass (tropical monsoon) against the polar frontal zone [17]. The MFI was correlated positively with the longitude (0.56, $P < 0.001$) and negatively with the latitude ($-0.45$, $P = 0.002$). On the other hand, the PCI% was correlated negatively with the latitude (0.89, $P < 0.001$). The impact of altitude on

$P_y$, MFI, and PCI% was found insignificant because of the change of altitude negatively with longitude and positively with latitude. This effect works in the opposite direction of the effect of the wet air mass flow (monsoon rainfall). Figure 4 shows the average $P_y$, MFI, and PCI% spatial distribution. The spatial pattern showed high agreement with the statistical analysis discussed above. The rainfall parameters have been found to change with the latitude clearly and slightly with the longitude. The MFI was found in agreement with the $M_i$ distribution reported by [10]. The advantage of the MFI is that it was calculated using monthly rainfall data when compared with $M_i$ which is calculated for daily rainfall.

### 3.3. Erosivity Spatiotemporal Variability.
In the previous two sections, the spatial and temporal changes of rainfall parameters ($P_y$, MFI, and PCI%) were demonstrated. The complexity of rainfall variability in space and time suggested the employment of the mapping techniques in order to develop results readily applicable for soil erosion models. The spatial and temporal changes in $P_y$ are shown in Figure 5. The $P_y$ was found to decrease from northwest to the southeast with decreases in the areas covered by the darker colors and expansion of the area in light color in the north. This trend was also observed in the MFI (Figure 6). This indicated that there was a high association between the $P_y$ and MFI. On the other hand, the PCI% was found to have contradicting trends

(a)

(b)

(c)

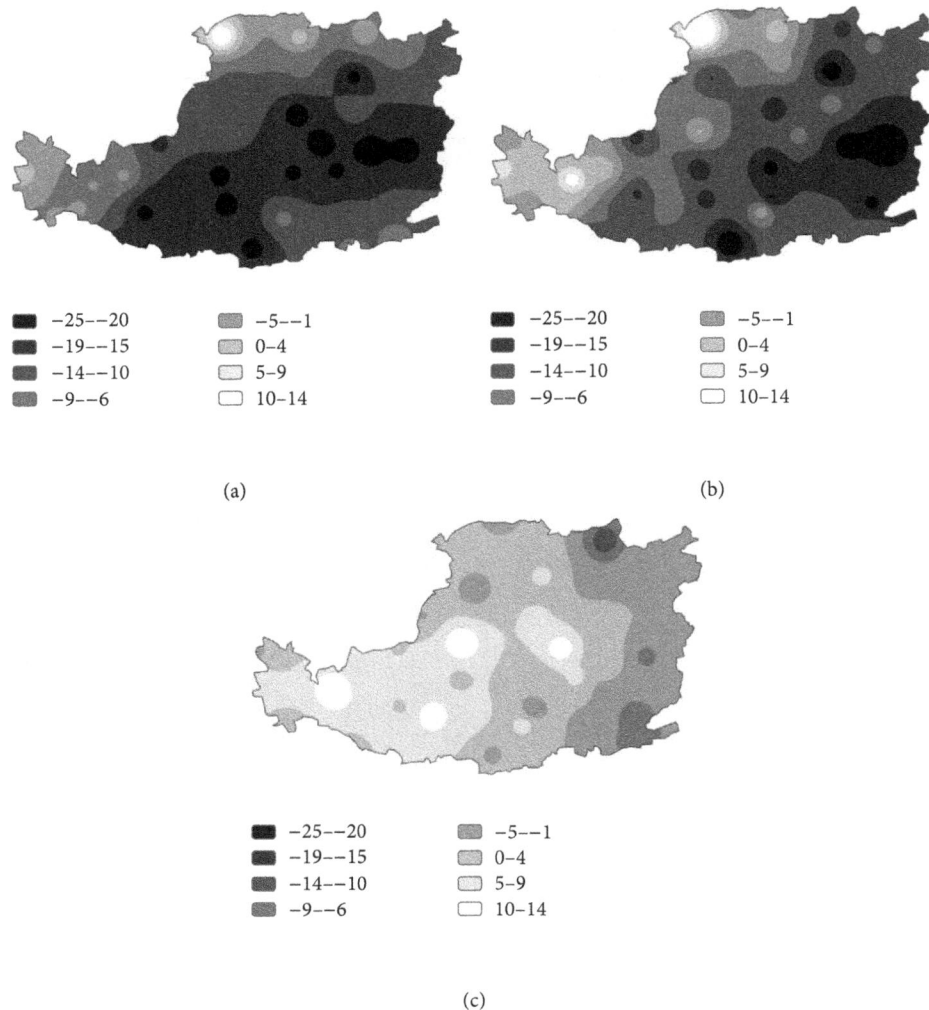

FIGURE 8: Spatial distribution of differences (%) between 1960s and 1990s of the annual rainfall (a), the modified Fournier index (b), and the precipitation concentration index (c) in the Chinese Loess Plateau determined from 42 stations.

to that observed in the $P_y$ and MFI (Figure 7). This relationship can be confirmed by the negative correlation between the PCI from one side and the $P_y$, and MFI on the other side. The differences in the $P_y$ and MFI between 1960s and 1990s showed similar trend which increased in southeast and decreased toward the north and west (Figure 8). The PCI% difference between 1960s and 1990s showed an increase in the southeast and a decrease toward the northeast (Figure 8).

## 4. Conclusions

The rainfall erosivity in the Chinese Loess Plateau showed a decrease of 10% between 1960s and 1990s. This decrease can be attributed to the decrease in rainfall in the region. The decrease in rainfall erosivity covered the majority of the Loess Plateau areas which suggested a subsequent decrease in sediment generation. However, the decrease in rainfall in the Loess Plateau may also negatively affect the vegetation cover development in the area which may expose more land to soil erosion. The general spatial trend of the rainfall erosivity

and annual rainfall was found increasing in the southeast and decreasing towards the north and northwest. This trend did not change significantly during the study period. The rainfall seasonality had a trend contradicting with the annual rainfall and erosivity trend, which increases in the north and northwest and decreases toward the southeast.

## Acknowledgments

This study was financially supported by the Japan Society for the Promotion of Science. The authors gratefully acknowledge the JSPS Core University Program and the Global COE Program of the Ministry of Education, Culture, Sports, Science, and Technology, Japan.

## References

[1] L. D. Chen, J. Wang, B. J. Fu, and Y. Qiu, "Land-use change in a small catchment of northern Loess Plateau, China," *Agriculture, Ecosystems and Environment*, vol. 86, no. 2, pp. 163–172, 2001.

[2] X. Jiongxin, "Precipitation-vegetation coupling and its influence on erosion on the Loess Plateau, China," *Catena*, vol. 64, no. 1, pp. 103–116, 2005.

[3] M. A. M. Abd Elbasit, H. Yasuda, A. Salmi, and H. Anyoji, "Characterization of rainfall generated by dripper-type rainfall simulator using piezoelectric transducers and its impact on splash soil erosion," *Earth Surface Processes and Landforms*, vol. 35, no. 4, pp. 466–475, 2010.

[4] K. G. Renard, G. R. Foster, G. A. Weesies, D. K. McCool, and D. C. Yoder, *Predicting Soil Erosion by Water: A Guide To Conservation Planning with the Revised Universal Soil Loss Equation (RUSLE)*, issue 703 of Agriculture handbook, USDA, Washington, DC, USA, 1997.

[5] W. H. Wischmeier and D. D. Smith, "Rainfall energy its relationship to soil loss," *Transaction of American Geophysics Union*, vol. 39, pp. 285–291, 1958.

[6] N. W. Hudson, *The influence of rainfall on the mechanics of soil erosion with particular reference to northern Rhodesia [M.S. thesis]*, University of Cape Town, 1965.

[7] A. W. Jayawardena and R. B. Rezaur, "Drop size distribution and kinetic energy load of rainstorms in Hong Kong," *Hydrological Processes*, vol. 14, no. 6, pp. 1069–1082, 2000.

[8] B. J. Fu, W. W. Zhao, L. D. Chen et al., "Assessment of soil erosion at large watershed scale using RUSLE and GIS: a case study in the Loess Plateau of China," *Land Degradation and Development*, vol. 16, no. 1, pp. 73–85, 2005.

[9] W. B. Zhang, Y. Xie, and B. Y. Liu, "Rainfall erosivity estimation using daily rainfall amounts," *Scientia Goegraphica Sinica*, vol. 22, pp. 705–711, 2002 (Chinese).

[10] Z. Xin, X. Yu, Q. Li, and X. X. Lu, "Spatiotemporal variation in rainfall erosivity on the Chinese Loess Plateau during the period 1956–2008," *Regional Environmental Change*, vol. 11, no. 1, pp. 149–159, 2011.

[11] H. M. Arnoldus, "An approximation of the rainfall factor in the universal soil loss equation," in *Assessment of Erosion*, M. de Boodt and D. Gabriels, Eds., pp. 127–132, John Wiley and Sons, Chichsester, UK, 1980.

[12] K. Oduro-Afriyie, "Rainfall erosivity map for Ghana," *Geoderma*, vol. 74, no. 1-2, pp. 161–166, 1996.

[13] C. Munka, G. Cruz, and R. M. Caffera, "Long term variation in rainfall erosivity in Uruguay: a preliminary fournier approach," *Geo Journal*, vol. 70, no. 4, pp. 257–262, 2007.

[14] N. A. Elagib, "Changing rainfall, seasonality and erosivity in the hyper-arid zone of Sudan," *Land Degradation Development*, vol. 22, no. 6, pp. 505–512, 2010.

[15] G. H. Zhang, M. A. Nearing, and B. Y. Liu, "Potential effects of climate change on rainfall erosivity in the Yellow River basin of China," *Transactions of the American Society of Agricultural Engineers*, vol. 48, no. 2, pp. 511–517, 2005.

[16] H. Fournier, *Climat et erosion*, Presses Universitaires de France, Paris, France, 1960.

[17] A. Kurashima and Y. Hiranuma, "Synoptic and climatological study on the upper moist tongue extending from southeast Asia to east Asia," in *Water Balance of Monsoon Asia*, M. M. Yoshino, Ed., pp. 152–169, University of Tokyo Press, Tokyo, Japan, 1971.

# 12

# Predicting Saturated Hydraulic Conductivity by Artificial Intelligence and Regression Models

**R. Rezaei Arshad,[1] Gh. Sayyad,[1] M. Mosaddeghi,[2] and B. Gharabaghi[3]**

[1] *Department of Soil Science, Faculty of Agrriculture, Shahid Chamran University of Ahvaz, Ahvaz, Iran*
[2] *Department of Soil Science, College of Agriculture, Isfahan University of Technology, Isfahan, Iran*
[3] *School of Engineering, University of Guelph, Guelph, ON, Canada*

Correspondence should be addressed to R. Rezaei Arshad; rezae805@gmail.com

Academic Editors: G. Benckiser, D. Hui, H. K. Pant, and D. Zhou

Saturated hydraulic conductivity ($K_s$), among other soil hydraulic properties, is important and necessary in water and mass transport models and irrigation and drainage studies. Although this property can be measured directly, its measurement is difficult and very variable in space and time. Thus pedotransfer functions (PTFs) provide an alternative way to predict the $K_s$ from easily available soil data. This study was done to predict the $K_s$ in Khuzestan province, southwest Iran. Three Intelligence models including (radial basis function neural networks (RBFNN), multi layer perceptron neural networks (MLPNN)), adaptive neuro-fuzzy inference system (ANFIS) and multiple-linear regression (MLR) to predict the $K_s$ were used. Input variable included sand, silt, and clay percents and bulk density. The total of 175 soil samples was divided into two groups as 130 for the training and 45 for the testing of PTFs. The results indicated that ANFIS and RBFNN are effective methods for $K_s$ prediction and have better accuracy compared with the MLPNN and MLR models. The correlation between predicted and measured $K_s$ values using ANFIS was better than artificial neural network (ANN). Mean square error values for ANFIS, ANN, and MLR were 0.005, 0.02, and 0.17, respectively, which shows that ANFIS model is a powerful tool and has better performance than ANN and MLR in prediction of $K_s$.

## 1. Introduction

Soil hydraulic properties such as saturated hydraulic conductivity ($K_s$) govern many soil hydrological processes; therefore, they are very important and even necessary in water and mass transport models and irrigation and drainage studies [1]. Direct measurement of soil hydraulic properties including $K_s$ is costly and time-consuming and becomes impractical due to spatial and temporal variabilities when hydrologic predictions are needed for large areas. Also it requires sophisticated measurement devices and skilled operators [2]. In the past few decades, as an alternative, indirect approximation of hydraulic properties from some basic and easily measured soil properties (such as clay, sand, and silt contents, and bulk density) using pedotransfer functions (PTFs) has received considerable acceptance [3–7]. "Pedotransfer function" was first introduced for empirical regression equations relating water and solute transport parameters to the basic soil properties that are available in soil survey [8].

The $K_s$ is an important soil hydraulic property often estimated using PTFs. Different methods such as regression models [3, 9–11] and artificial neural networks (ANN) are available for derivation of PTFs. In recent years, PTFs constructed by using artificial neural networks (ANN) (especially feed forward ANN) have proven popular with many researchers. ANN-PTFs have been developed by researchers such as Minasny et al., Minasny and McBratney, and Pachepsky et al. [5, 6, 11]. The overall conclusion made by these (and other) investigators was that when the number of input parameters is greater than three, ANN usually performs better than regression techniques, particularly when uncertainties in the quality of the data were small [12]. Multilayer perceptron (MLP) and radial basis function (RBF) are two of the most widely used neural network architecture. General difference between MLP and RBF is that RBF is a localist type of learning which is responsive only to a limited section of input space. RBF utilizes a local learning strategy versus MLP global learning and this leads to a higher rate of accuracy and faster training of RBF [13].

Adaptive neuro-fuzzy inference system (ANFIS) can be applied as a further alternative technique [14]. Similar to ANN, the neuro-fuzzy system constructs input-output membership function relationships. The fuzzy system is set to learn from the training data by adjusting the parameters of the membership functions using the least-squares method. The trained system can be used as a PTF to predict properties of an unknown soil sample [11]. A close spatial relationship between soil hydraulic properties and other easily available soil and terrain attributes, as auxiliary variables, can be exploited to predict these hydraulic characteristics, as target variables, with a reasonable accuracy at unobserved locations [15].

State of the art shows that, in most previous studies, $K_s$ was predicted using regression and ANN models [3, 16, 17]. Many comparisons of PTFs have been made with respect to different data sets, different mathematical procedures (regression versus ANN models), and different input parameters. However, there are few studies which have used ANFIS model for developing PTFs. Wieland and Mirschel compared a feed-forward neural network (NN), a radial basis function network (RBF), and a trained fuzzy algorithm for regional yield estimation of agricultural crops (winter rye, winter barley) [18]. As mentioned before, ANFIS model has more ability to develop PTFs than ANN and regression models. Moreover unlike the tradition, radial basis function (RBF) technique, as far as our knowledge goes, has not been used for this comparison purpose. Therefore the objectives of this study were (a) to employ the techniques of ANN (MLR and RBF methods), ANFIS, and multiple linear regression (MLR) to predict saturated hydraulic conductivity using easily available soil properties, (b) to evaluate the performance of these techniques and determine the best methodology for prediction of saturated hydraulic conductivity.

An ANN is a highly interconnected network of many simple processing units called *neurons*, which are analogous to the biological neurons in the human brain. Neurons having similar characteristics in an ANN are arranged in groups called *layers*. The neurons in one layer are connected to those in the adjacent layers but not to those in the same layer. The strength of connection between the two neurons in adjacent layers is represented by what is known as connection strength or weight. An ANN normally consists of three layers, an input layer, a hidden layer, and an output layer. In a feed-forward network, the weighted connections feed activations only in the forward direction from an input layer to the output layer. On the other hand, in a recurrent network additional weighted connections are used to feed previous activations back into the network. The structure of a feed-forward ANN is shown in Figure 1 [19].

In Figure 1, the circles represent neurons; the lines joining the neurons represent weights; sand, silt, clay, and $\rho_b$ are input variables; $K_s$ represents the output variables; $w_{kj}$ and $W_{ik}$ represent the weights between input and hidden and

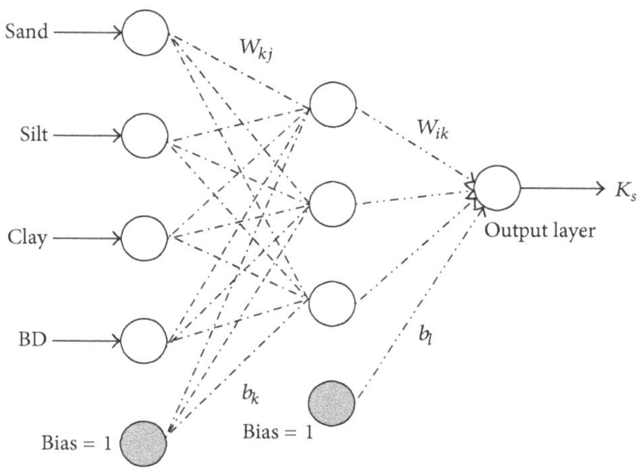

FIGURE 1: Structure of a feed-forward ANN.

hidden and output layers, respectively. Parameters $b_k$ and $b_l$ represent the bias of the corresponding hidden and output layer neurons. The role of bias in a neuron is to displace the original functional domain by a magnitude equal to that of the bias and thereby translate the area of influence to its activation state [19].

An important step in developing an ANN model is *training* of its weight matrix. The weights are initialized randomly in a suitable range and then updated using certain training mechanism. There are primarily two types of training mechanisms: supervised and unsupervised. A supervised training algorithm requires an external teacher to guide the training process. This typically involves a large number of examples (or patterns) of inputs and outputs for training. The inputs in an ANN are the cause variables and outputs are the effect variables of the physical system being modeled. The primary goal of training is to minimize the objective (error) function by searching for a set of connection strengths that cause the ANN to produce outputs that are equal to or closer to target data [19].

Multilayer perceptron (MLP) and radial basis function (RBF) are two of the most widely used neural network architecture in literature for classification or regression problems. They are robust classifiers with the ability to generalize for imprecise input data. Both artificial neural network and fuzzy logic are used in ANFIS's architecture. ANFIS is consisted of if-then rules and couples of input-output. ANFIS training uses learning algorithms of neural network [14, 20, 21]. An adaptive network, as its name implies, is a network structure consisting of nodes and directional links through which the nodes are connected. Moreover, parts or all of the nodes are adaptive, which means each output of these nodes depends on the parameters pertaining to this node and the learning rule specifies how these parameters should be changed to minimize a prescribed error measure [14]. ANFIS is a multilayer feed-forward network where each node performs a particular function on incoming signals. Both square and circle node symbols are used to represent different properties of adaptive learning (Figure 2). To perform desired input-output characteristics, adaptive learning parameters

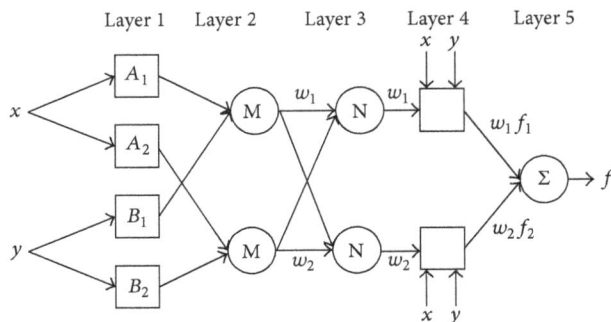

Figure 2: ANFIS architecture.

are updated based on gradient learning rules [14, 22, 23]. For simplicity, we assume that the fuzzy inference system under consideration has two inputs, $x$ and $y$, and one output $z$. suppose that the rule base contains two fuzzy if-then rules of Takagi and Sugeno's type.

*Rule 1.* If $x$ is $A_1$ and $y$ is $B_1$ then $f_1 = p_1 x + q_1 y + r_1$.

*Rule 2.* If $x$ is $A_2$ and $y$ is $B_2$ then $f_2 = p_2 x + q_2 y + r_2$,

where $A_i$ and $B_i$ are the fuzzy sets, $f_i$ are the outputs within the fuzzy region specified by the fuzzy rule, and $p_i$, $q_i$, and $r_i$ are the design parameters that are determined during the training process. The ANFIS architecture to implement these two rules is shown in Figure 2, in which a circle indicates a fixed node, whereas a square indicates an adaptive node.

*Layer 1.* Every node $i$ in this layer is a square node with a node function:

$$o_i^1 = \mu A_i(x), \quad i = 1, 2$$
$$o_i^1 = \mu B_{i-2}(y), \quad i = 3, 4, \tag{1}$$

where $x$ is the input to node $i$, and $A_i$ is the linguistic label (small, large, etc.) associated with this node function, and where $\mu A_i(x)$, $\mu B_{i-2}(y)$ can adopt any fuzzy membership function. Usually, we choose $\mu A_i(x)$ and $\mu B_{i-2}(y)$ to be bell-shaped with a maximum equal to 1 and a minimum equal to 0, such as

$$\mu A_i(x) = \frac{1}{1 + \left\{ \left( (x - c_i)/a_i \right)^2 \right\}^b}, \tag{2}$$

where $(a_i, b_i, \text{ and } c_i)$ is the parameter set. Parameters in this layer are referred to as *premise* parameters.

*Layer 2.* The nodes in this layer are fixed. These are labeled as M to indicate that they play the role of a simple multiplier. The outputs of these nodes are given by

$$o_i^2 = w_i = \mu A_i(x) \mu B_i(y), \quad i = 1, 2, \tag{3}$$

which are the so-called firing strengths of the rules.

*Layer 3.* Every node in this layer is a circle node labeled as N. The $i$th node calculates the ratio of the $i$th rule's firing strength to the sum of all rules' firing strengths:

$$o_i^3 = \overline{w} = \frac{w_i}{w_1 + w_2}, \quad i = 1, 2. \tag{4}$$

For convenience, outputs of this layer will be called normalized firing strengths.

*Layer 4.* In this layer, the nodes are adaptive nodes. The output of each node in this layer is simply the product of the normalized firing strength and a first-order polynomial (for a first-order Sugeno model). Thus, the outputs of this layer are given by

$$o_i^4 = \overline{w}_i f_i = w_i (p_i x + q_i y + r_i), \quad i = 1, 2. \tag{5}$$

Parameters in this layer will be referred to as consequent parameters.

*Layer 5.* The single node in this layer is circle node labeled as S that computes the overall output as the summation of all incoming signals; that is,

$$o_1^5 = \sum_{i=1}^{2} \overline{w}_i f_i = \frac{\sum_{i=1}^{2} w_i f_i}{w_1 + w_2}. \tag{6}$$

It is seen that there are two adaptive layers in this ANFIS architecture, namely, the first layer and the fourth layer. In the first layer, there are three modifiable parameters $\{a_i, b_i, c_i\}$, which are related to the input membership functions. These parameters are the so-called premise parameters. In the fourth layer, there are also three modifiable parameters $\{p_i, q_i, r_i\}$, pertaining to the first-order polynomial. These parameters are the so-called consequent parameters [14].

## 2. Materials and Methods

*2.1. Data Collection.* The data used in this study were obtained from the reports of basic soil science and land reclamation studies conducted by Khuzestan Water and Power Organization. Khuzestan province is located in the southwest of Iran from east $47°40'$ to $50°33'$ longitude and north of latitude. Khuzestan province has mild winters and very hot summers [24].

The data set which was used to develop the PTFs included the data from 175 soil profiles from different parts of Khuzestan province. Soil samples were taken from the 0–30 cm layer. The data consisted of five soil properties (four as independent variables and one as dependent variable). Independent variables were the percentages of clay (C), silt (Si), sand (S), and bulk density ($\rho_b$). Saturated hydraulic conductivity ($K_s$) was considered as dependent (output) variable. Soil samples were air-dried and sieved by a 2 mm sieve for physical analysis. Particle size distribution was determined using pipet method. $\rho_b$ was determined on undisturbed samples using cylinder method, being made of 10 cm and 100 cm$^3$ cylinders, after drying 24 h in 105°C ovens. Saturated hydraulic conductivity was measured using inversed auger hole method. The principle of the auger hole test above the water table consists of boring a hole to given depth, filling it with water and measuring the rate of fall of the water level [25].

MLP, RBF, ANFIS and MLR of pedotransfer functions were used to predict the $K_s$. The data set was divided into two separate data sets: the training one (80%) and the testing one (20%). The training data set was used to train the ANN, ANFIS, and MLR models, whereas the testing data set was used to verify the accuracy and the effectiveness of the trained models. A computer program was run in MATLAB (ver.7.6.0.323) to train and to test the data set using ANFIS and ANN structures for $K_s$ prediction. The performances of new techniques (ANN and ANFIS) were compared with MLR method.

*2.2. Regression Analysis.* In the multiple-linear regression (MLR) analysis, first, the most essential input variables were selected using backwards stepwise method, and then linear, quadratic, and possible interaction terms of these basic soil properties were investigated using the SPSS software. The general form of the regression equations was:

$$Y = a + b_1 X_1 + b_2 X_2 + \cdots + b_i X_i, \tag{7}$$

where $Y$ is the dependent variable representing $K_s$, $a$ is the intercept, $b_1, \ldots, b_i$ are regression coefficients, and $X_1$–$X_i$ are independent variables referring to basic soil properties.

The same data sets were used in the derivation ($N = 130$) and testing ($N = 45$) of PTFs developed using ANN, ANFIS, and regression methods for reliable comparison. The resulting functions were tested using data sets not included in the derivation procedure.

*2.3. Developing PTFs Using Artificial Neural Network Model (ANN).* First step in developing an ANN model involves identifying input and output variables and normalizing the data between 0 and 1. The method of *channelized normalization* was used to normalize the input and output data, wherein the data representing separate physical variables/parameters are normalized separately. Then the best ANN architecture is determined by finding the optimal number of hidden neurons through training of the various architectures using a trial and error method. Once the best ANN architecture is trained, it is validated using the testing data set.

In this study, two different types of ANN were developed. The first ANN model is multilayer perceptron (MLP) which is the most commonly used neural network structure in ecological modeling and soil science [26], whereas the second ANN model is radial basis function (RBF).

*2.4. Neural-Fuzzy Model Design.* For designing the neural-fuzzy system, the feed-forward multilayer neural networks with hybrid training algorithm and Takagi-Sugeno fuzzy inference system were used. In this model, the "gaussmf" membership function was used for input function, the "linear" membership function was used for output function, and defuzzification operation was performed by weight average function.

*2.5. Performance Evaluation Criteria.* Two standard statistics were used as evaluation criteria to evaluate the performance of ANN models (MLP and RBF) and ANFIS methods. These were mean square errors (MSE) and coefficient of determination ($R^2$) as calculated using the following equations:

$$R^2 = 1 - \frac{\sum_{i=1}^{N} (y_i - \widehat{y}_i)}{\sum_{i=1}^{N} (y_i - \overline{y}_i)^2}, \tag{8}$$

$$\text{MSE} = \frac{1}{N} \sum_{i=1}^{N} (y_i - \overline{y}_i)^2, \tag{9}$$

where $N$ is the number of data set, $y_i$ is the measured value of output variable, $\widehat{y}_i$ is the predicted value of output variable, and $\overline{y}_i$ is the average of predicted value of output variable.

## 3. Results and Discussion

Before developing PTFs using ANN, ANFIS, and MLR models, descriptive statistics of data were derived using SPSS (version 16). Descriptive statistics for the soil physical and hydraulic parameters which were used in the development (train) and validation (test) of PTFs using models are summarized in Table 1.

The studied soils have wide ranges of physical properties; for instance, the ranges of sand, silt, and clay contents, and $\rho_b$ are 4–70, 12–66, 9.4–62%, and 0.96–1.77 Mg m$^{-3}$ for training or derivation data set, respectively. Testing or validation data set had similar ranges. The studied soils are mostly originated from alluvial processes which probably cause such wide range in soil physical properties. The soils are on average with medium textures, where the mean silt percent and $\rho_b$ are 44.6 and 1.45 Mg m$^{-3}$, respectively.

The $K_s$ values of the studied soils are mostly very high, ranging from 6.4–207 (cm day$^{-1}$), with standard deviation of 29.4 and 10.7 cm day$^{-1}$ for training and testing data sets, respectively. A reason for high values of $K_s$ might be the large area of the studied region. Good structure and existence of macropores due to wide ranges of particle size distribution might be also responsible for high values of $K_s$.

*3.1. PTFs Development Using MLR Model.* In the regression analysis, normalizing the data distribution is one of the primary assumptions that have to be carried out. Therefore, the normality of the data was evaluated using the Kolmogrov-Smirnov method. $K_s$, $\rho_b$, sand, silt and clay data did not conform to normal distribution and were normalized using the natural-based logarithmic transformation. After normalizing data, in order to develop PTFs for predicting $K_s$ through MLR model, first the most essential input variables were selected using stepwise method, and then linear interaction terms of these basic soil properties were investigated by means of SPSS 16 software.

After training the regression model with training data set, the derived regression equation was as follows:

$$\text{Ln} K_s = 14.66 - 0.44 * \text{Ln sand} - 1.37 * \text{Ln silt} - 1.25 * \text{Ln clay} - 2.8 * \text{Ln} \rho_b. \tag{10}$$

TABLE 1: Descriptive statistics of the datasets used for training and testing (ANN, ANFIS, and MLR).

| Variable | Units | Training data ($N = 130$) | | | | Testing data ($N = 45$) | | | |
|---|---|---|---|---|---|---|---|---|---|
| | | Max. | Min. | Mean | S.D. | Max. | Min. | Mean | S.D. |
| $\rho_b$ | Mg m$^{-3}$ | 1.77 | 0.96 | 1.45 | 0.14 | 1.74 | 1.20 | 1.45 | 0.13 |
| Sand | % | 70.0 | 4.0 | 27.9 | 13.8 | 52.0 | 4.0 | 20.4 | 11.2 |
| Silt | % | 66.0 | 12.0 | 44.6 | 10.6 | 78.0 | 18.0 | 47.5 | 10.8 |
| Clay | % | 62.0 | 9.4 | 28.5 | 11.2 | 62.0 | 10.5 | 32.1 | 12.6 |
| $K_s$ | cm day$^{-1}$ | 207.0 | 6.9 | 27.4 | 29.4 | 60.0 | 6.4 | 18.2 | 10.7 |

TABLE 2: Summary of statistical analyses of training and testing the MLR for $K_s$ prediction.

| Model | Training | | Testing | |
|---|---|---|---|---|
| | $R^2$ | MSE | $R^2$ | MSE |
| MLR | 0.69 | 0.01 | 0.5 | 0.17 |

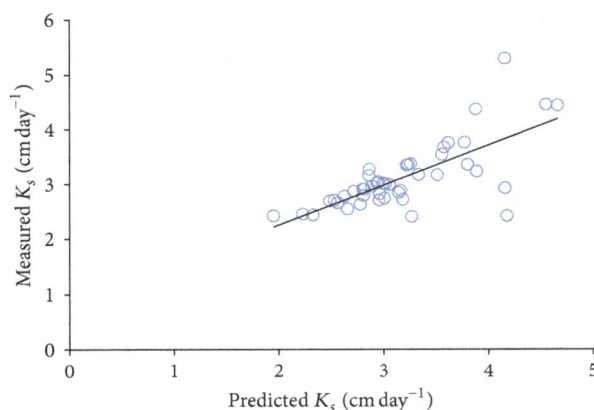

FIGURE 3: Measured versus predicted $K_s$ values using MLR.

In derived equation clay, sand, silt, and bulk density were chosen as the independent variables. After determining regression equation, the accuracy of MLR model was evaluated through comparing its predicted $K_s$ with experimental data. The obtained values of $R^2$ and MSE using MLR model for predicting $K_s$ are tabulated in Table 2. As observed from this table for test data set, the $R^2$ and MSE values that have been obtained are 0.5 and 0.17, respectively. Merdun et al. obtained higher $R^2$ and RMSE values varied from 0.637 to 0.979 and from 0.013 to 0.938 for regression method, respectively [3]. The scatter plot of the measured against predicted $K_s$ values obtained from the MLR model for the test data set with a poor correlation coefficient is illustrated in Figure 3.

### 3.2. PTFs Development Using ANN Models.

In current study two different algorithms of ANNs including radial basis function (RBF) and multilayer perceptron (MLP) models were investigated to predict $K_s$ through employing the same data set which was used by MLR. In order to reach this end, all data sets were first normalized between 0 and 1 to achieve effective network training. Luk et al. stated that

neural networks trained on normalized data achieve better performance and faster convergence in general, although the advantages diminish as network and sample size become large [27]. Normalizing the data set was done through

$$X_{\text{norm}} = 0.5\left(\frac{X_0 - \overline{X}}{X_{\text{max}} - X_{\text{min}}}\right), \quad (11)$$

where $X_{\text{norm}}$ is the normalized value, $X_0$ is the actual value, $\overline{X}$ is average value, for each parameter, $X_{\text{max}}$ is the maximum value and $X_{\text{min}}$ is the minimum value. A three-layered feed-forward ANN architecture with an input layer, one hidden layer, and an output layer was developed for predicting $K_s$ by means of both ANN models.

### 3.3. MLP Network.

Table 3 shows the results of statistical analyses between the observed and MLP ANN-simulated values of $K_s$ for training and testing stages. For MLP network, the best architecture consists of four neurons in the input layer, seven neurons in the hidden layer, and one neuron in the output layer with Tansig and Purelin threshold functions for hidden and output layers, respectively, gave the best results. The $R^2$ and MSE values among the observed and predicted $K_s$ are 0.66 and 0.02, respectively. The values are in accordance with the previous studies using ANN methods. Agyare et al. while estimating $K_s$ obtained $R^2$ and NMSE about 0.6 and 0.42, respectively [26]. Merdun et al. obtained $R^2$ ranges and RMSE varied from 0.444 to 0.952 and from 0.020 to 3.511, respectively [3]. Figure 4 shows the relationship between the measured and predicted $K_s$ values for testing stage indicating that MLP network can predict $K_s$ with acceptable accuracy. In order to employ RBF, Gaussian function that is the most widely used in applications was chosen as a threshold function for hidden layer. In the next step a regression analysis of the network response between ANN outputs and the corresponding targets was performed. Table 3 tabulated that the provided predictions between ANNs outputs and the corresponding targets using RBF model ($R^2 = 0.68$, MSE = 0.02) and MLP were approximately similar and differences were not significant.

The levels of $R^2$ and MSE derived by both ANN models had higher accuracy than those derived by multiple linear regressions for predicting $K_s$ which was a support for those previous studies conducted by Merdun et al., Tamari et al., Yilmaz et al., and other researchers [3, 28, 29]. This is due to that, unlike the traditional regression PTFs ANNs do not require a priori regression model which relates input and output data that in general is difficult because these

TABLE 3: Summary of statistical analyses of training and testing the MLP and RBF neural network for $K_s$ prediction.

| Network | Architecture | Threshold function | Spread | Training | | Testing | |
|---------|-------------|-------------------|--------|----------|-----|---------|-----|
| | | | | $R^2$ | MSE | $R^2$ | MSE |
| MLP | 4-7-1 | Tansig-Purelin | | 0.88 | 0.007 | 0.66 | 0.02 |
| RBF | | Radbas | 45 | 0.9 | 0.005 | 0.68 | 0.02 |

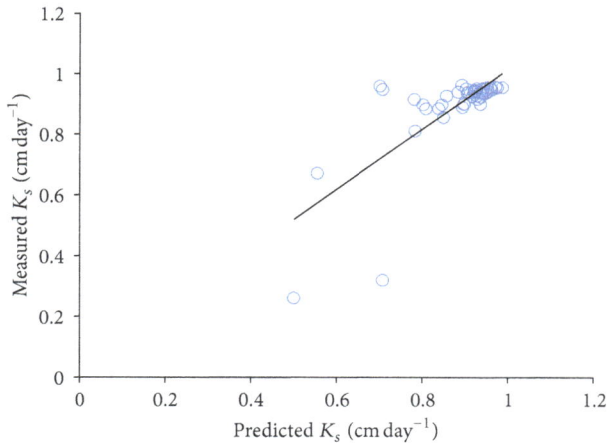

FIGURE 4: Measured versus predicted $K_s$ values using MLP network.

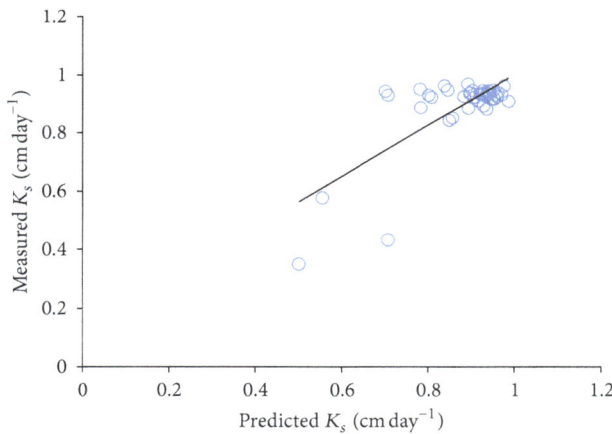

FIGURE 5: Measured versus predicted $K_s$ values using RBF network.

models are not known [30]. Also Minasny and McBratney, Pachepsky et al., and Tamari et al. stated that when the number of input parameters is greater than three, ANNs usually perform better than regression techniques, particularly when uncertainties in the quality of the data were small [11, 28, 31]. Additionally many investigations have indicated that a neural network with one hidden layer is capable of approximating any finite nonlinear function with very high accuracy [32, 33]. The scatter plots between measured and predicted $K_s$ using MLP and RBF models for testing stage with acceptable accuracy are indicated in Figures 4 and 5, respectively.

*3.4. PTFs Development Using ANFIS Model.* In order to develop PTFs for prediction $K_s$ through ANFIS model the

TABLE 4: Summary of statistical analyses of training and testing the ANFIS model for $K_s$ prediction.

| Training | | Testing | |
|----------|-----|---------|-----|
| $R^2$ | MSE | $R^2$ | MSE |
| 0.72 | 0.009 | 0.71 | 0.005 |

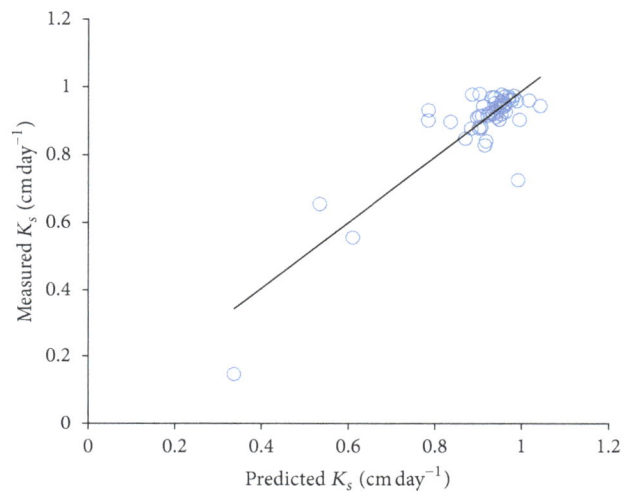

FIGURE 6: Results of running the ANFIS network in testing stage (actual versus predicted $K_s$ values).

same normalized data which employed by both ANN models were used as the input variables. In the ANFIS system, each input parameter might be clustered into several class values in layer 1 to build up fuzzy rules, and each fuzzy rule would be constructed using two or more membership functions in layer 2. Several methods have been proposed to classify the input data and to make the rules, among which the most widespread of them are grid partition and subtractive fuzzy clustering [34–36]. In this study subtractive fuzzy clustering was taken into consideration to develop the rule base relationship between the input and output variables. Sugeno fuzzy model, also known as Takagi-Sugeno-Kang (TSK) model [37], that is, one of the most widespread methodologies for developing fuzzy rules was used in the current study to predict soil $K_s$.

The performance of the ANFIS model for the test data set and the related statistical evolutionary results are given in Figure 6 and Table 4, respectively. The values of 0.71 and 0.005 for $R^2$ and MSE parameters, respectively, for ANFIS testing stage, show higher prediction accuracy of ANFIS model than the ANN models.

## 4. Conclusions

Predicting the saturated hydraulic conductivity ($K_s$) of soil is one of the important subjects in modeling water flow and solute transport processes in the vadose zone. This paper presents the development and validation of PTFs for prediction of $K_s$ from basic soil properties by using ANN and ANFIS models in Khuzestan province, southwest Iran. The predictive capabilities of these methods were also compared using some evaluation criteria of testing data. The results revealed that the ANFIS model had superiority to the ANN models for $K_s$ prediction. Both applied ANNs algorithms exhibited acceptable accuracy. The comparison of the RBF and MLP networks indicates that they had approximately similar performance for prediction $K_s$ as the obtained results using RBF were slightly better than MLP ones, but their differences were not found significant. Among the employed ANN models, RBF prediction was more accurate than MLP network prediction.

## References

[1] V. Bagarello, S. Sferlazza, and A. Sgroi, "Comparing two methods of analysis of single-ring infiltrometer data for a sandy-loam soil," *Geoderma*, vol. 149, no. 3-4, pp. 415–420, 2009.

[2] N. Islam, W. W. Wallender, J. P. Mitchell, S. Wicks, and R. E. Howitt, "Performance evaluation of methods for the estimation of soil hydraulic parameters and their suitability in a hydrologic model," *Geoderma*, vol. 134, no. 1-2, pp. 135–151, 2006.

[3] H. Merdun, Ö. Çinar, R. Meral, and M. Apan, "Comparison of artificial neural network and regression pedotransfer functions for prediction of soil water retention and saturated hydraulic conductivity," *Soil and Tillage Research*, vol. 90, no. 1-2, pp. 108–116, 2006.

[4] J. Tomasella, Y. Pachepsky, S. Crestana, and W. J. Rawls, "Comparison of two techniques to develop pedotransfer functions for water retention," *Soil Science Society of America Journal*, vol. 67, no. 4, pp. 1085–1092, 2003.

[5] B. Minasny, J. W. Hopmans, T. Harter, S. O. Eching, A. Tuli, and M. A. Denton, "Neural networks prediction of soil hydraulic functions for alluvial soils using multistep outflow data," *Soil Science Society of America Journal*, vol. 68, no. 2, pp. 417–429, 2004.

[6] Y. A. Pachepsky, W. J. Rawls, and H. S. Lin, "Hydropedology and pedotransfer functions," *Geoderma*, vol. 131, no. 3-4, pp. 308–316, 2006.

[7] J. H. M. Wösten, P. A. Finke, and M. J. W. Jansen, "Comparison of class and continuous pedotransfer functions to generate soil hydraulic characteristics," *Geoderma*, vol. 66, no. 3-4, pp. 227–237, 1995.

[8] J. Bouma and J. A. van Lanen, "Transfer functions and threshold values: from soil characteristics to land qualities," in *Proceedings of the International Workshop on Quantified Land Evaluation Procedures*, K. J. Beek, P. A. Burrough, and D. E. McCormack, Eds., pp. 106–111, International Inst. For Aerospace Survey and Earth Sciences, Washington, DC, USA, May 1987.

[9] H. S. Lin, K. J. McInnes, L. P. Wilding, and C. T. Hallmark, "Effects of soil morphology on hydraulic properties. II. Hydraulic pedotransfer functions," *Soil Science Society of America Journal*, vol. 63, no. 4, pp. 955–961, 1999.

[10] T. Mayr and N. J. Jarvis, "Pedotransfer functions to estimate soil water retention parameters for a modified Brooks-Corey type model," *Geoderma*, vol. 91, no. 1-2, pp. 1–9, 1999.

[11] B. Minasny and A. B. McBratney, "The neuro-$m$ method for fitting neural network parametric pedotransfer functions," *Soil Science Society of America Journal*, vol. 66, no. 2, pp. 352–361, 2002.

[12] L. Baker and D. Ellison, "Optimisation of pedotransfer functions using an artificial neural network ensemble method," *Geoderma*, vol. 144, no. 1-2, pp. 212–224, 2008.

[13] E. Z. Panagou, V. Kodogiannis, and G. J.-E. Nychas, "Modelling fungal growth using radial basis function neural networks: the case of the ascomycetous fungus Monascus ruber van Tieghem," *International Journal of Food Microbiology*, vol. 117, no. 3, pp. 276–286, 2007.

[14] J. S. R. Jang, "ANFIS: adaptive-network-based fuzzy inference system," *IEEE Transactions on Systems, Man and Cybernetics*, vol. 23, no. 3, pp. 665–685, 1993.

[15] H. R. Motaghian and J. Mohammadi, "Predictive infiltration rate mapping with improved soil and terrain predictors," *Journal of Applied Sciences*, vol. 9, no. 8, pp. 1562–1567, 2009.

[16] J. Tomasella, M. G. Hodnett, and L. Rossato, "Pedotransfer functions for the estimation of soil water retention in Brazilian soils," *Soil Science Society of America Journal*, vol. 64, no. 1, pp. 327–338, 2000.

[17] J. H. M. Wösten, Y. A. Pachepsky, and W. J. Rawls, "Pedotransfer functions: bridging the gap between available basic soil data and missing soil hydraulic characteristics," *Journal of Hydrology*, vol. 251, no. 3-4, pp. 123–150, 2001.

[18] R. Wieland and W. Mirschel, "Adaptive fuzzy modeling versus artificial neural networks," *Environmental Modelling and Software*, vol. 23, no. 2, pp. 215–224, 2008.

[19] K. Parasuraman, A. Elshorbagy, and B. C. Si, "Estimating saturated hydraulic conductivity in spatially variable fields using neural network ensembles," *Soil Science Society of America Journal*, vol. 70, no. 6, pp. 1851–1859, 2006.

[20] E. Avci and Z. H. Akpolat, "Speech recognition using a wavelet packet adaptive network based fuzzy inference system," *Expert Systems with Applications*, vol. 31, no. 3, pp. 495–503, 2006.

[21] J. Ryoo, Z. Dragojlovic, and D. A. Kaminski, "Control of convergence in a computational fluid dynamics simulation using ANFIS," *IEEE Transactions on Fuzzy Systems*, vol. 13, no. 1, pp. 42–47, 2005.

[22] A. Baylar, D. Hanbay, and E. Ozpolat, "Modeling aeration efficiency of stepped cascades by using ANFIS," *Clean—Soil, Air, Water*, vol. 35, no. 2, pp. 186–192, 2007.

[23] J. S. Jang and N. Gulley, *Fuzzy Logic Toolbox: Reference Manual*, The Mathworks, Natick, Mass, USA, 1996.

[24] M. H. Mahdian and R. S. Oskoee, "Developing pedotransfer functions to predict infiltration rate in flood spreading stations of Iran," *Research Journal of Environmental Sciences*, vol. 3, no. 6, pp. 697–704, 2009.

[25] H. P. Ritzema, *Drainage Principles and Applications*, ILRI, The Netherlands, 1994.

[26] W. A. Agyare, S. J. Park, and P. L. G. Vlek, "Artificial neural network estimation of saturated hydraulic conductivity," *Vadose Zone Journal*, vol. 6, no. 2, pp. 423–431, 2007.

[27] K. C. Luk, J. E. Ball, and A. Sharma, "A study of optimal model lag and spatial inputs to artificial neural network for rainfall forecasting," *Journal of Hydrology*, vol. 227, no. 1-4, pp. 56–65, 2000.

[28] S. Tamari, J. H. M. Wösten, and J. C. Ruiz-Suárez, "Testing an artificial neural network for predicting soil hydraulic conductivity," *Soil Science Society of America Journal*, vol. 60, no. 6, pp. 1732–1741, 1996.

[29] I. Yilmaz, M. Marschalko, M. Bednarik, O. Kaynar, and L. Fojtova, "Neural computing models for prediction of permeability coefficient of coarse-grained soils," *Neural Computing and Applications*, vol. 21, no. 5, pp. 957–968, 2012.

[30] M. G. Schaap and F. J. Leij, "Using neural networks to predict soil water retention and soil hydraulic conductivity," *Soil and Tillage Research*, vol. 47, no. 1-2, pp. 37–42, 1998.

[31] Y. A. Pachepsky, D. Timlin, and G. Varallyay, "Artificial neural networks to estimate soil water retention from easily measurable data," *Soil Science Society of America Journal*, vol. 60, no. 3, pp. 727–733, 1996.

[32] M. Kim and J. E. Gilley, "Artificial Neural Network estimation of soil erosion and nutrient concentrations in runoff from land application areas," *Computers and Electronics in Agriculture*, vol. 64, no. 2, pp. 268–275, 2008.

[33] I. Yilmaz, M. Marschalko, M. Bednarik, O. Kaynar, and L. Fojtova, "Neural computing models for prediction of permeability coefficient of coarse-grained soils," *Neural Computing and Applications*, pp. 1–12, 2011.

[34] M. Aqil, I. Kita, A. Yano, and S. Nishiyama, "A comparative study of artificial neural networks and neuro-fuzzy in continuous modeling of the daily and hourly behaviour of runoff," *Journal of Hydrology*, vol. 337, no. 1-2, pp. 22–34, 2007.

[35] H. M. Ertunc and M. Hosoz, "Comparative analysis of an evaporative condenser using artificial neural network and adaptive neuro-fuzzy inference system," *International Journal of Refrigeration*, vol. 31, no. 8, pp. 1426–1436, 2008.

[36] I. Yilmaz and O. Kaynar, "Multiple regression, ANN (RBF, MLP) and ANFIS models for prediction of swell potential of clayey soils," *Expert Systems with Applications*, vol. 38, no. 5, pp. 5958–5966, 2011.

[37] T. Takagi and M. Sugeno, "Fuzzy identification of systems and its applications to modeling and control," *IEEE Transactions on Systems, Man and Cybernetics*, vol. 15, no. 1, pp. 116–132, 1985.

# 13

# Variability of Soil Physical Properties in a Clay-Loam Soil and Its Implication on Soil Management Practices

**Samuel I. Haruna and Nsalambi V. Nkongolo**

*Center of Excellence for Geospatial Information Sciences, Department of Agriculture and Environmental Science, Lincoln University, Jefferson City, MO 65102-0029, USA*

Correspondence should be addressed to Nsalambi V. Nkongolo; nkongolo@lincolnu.edu

Academic Editors: R. K. Kolka and H. O. Liechty

We assessed the spatial variability of soil physical properties in a clay-loam soil cropped to corn and soybean. The study was conducted at Lincoln University in Jefferson City, Missouri. Soil samples were taken at four depths: 0–10 cm, 10–20, 20–40, and 40–60 cm and were oven dried at 105°C for 72 hours. Bulk density (BDY), volumetric (VWC) and gravimetric (GWC) water contents, volumetric air content (VAC), total pore space (TPS), air-filled (AFPS) and water-filled (WFPS) pore space, the relative gas diffusion coefficient (DIFF), and the pore tortuosity factor (TORT) were calculated. Results showed that, in comparison to depth 1, means for AFPS, Diff, TPS, and VAC decreased in Depth 2. Opposingly, BDY, Tort, VWC, and WFPS increased in depth 2. Semivariogram analysis showed that GWC, VWC, BDY, and TPS in depth 2 fitted to an exponential variogram model. The range of spatial variability ($A_0$) for BDY, TPS, VAC, WFPS, AFPS, DIFF, and TORT was the same (25.77 m) in depths 1 and 4, suggesting that these soil properties can be sampled together at the same distance. The analysis also showed the presence of a strong (≤25%) to weak (>75%) spatial dependence for soil physical properties.

## 1. Introduction

Characterizing the spatial variability and distribution of soil properties is important in predicting the rates of ecosystem processes with respect to natural and anthropogenic factors [1] and in understanding how ecosystems and their services work [2]. In agriculture, studies of the effects of land management on soil properties have shown that cultivation generally increases the potential for soil degradation due to the breakdown of soil aggregates and the reduction of soil cohesion, water content and nutrient holding capacity [3, 4]. Cultivation, especially when accompanied by tillage, has been reported to have significant effects on topsoil structure and thus the ability of soil to fulfill essential soil functions and services in relation to root growth, gas and water transport and organic matter turnover [5–7]. Soil properties vary considerably under different crops, tillage type and intensity, fertilizer types and application rates. Consequently, the physical properties of the soil are also affected by many factors that change vertically with depth, laterally across fields

and temporally in response to climate and human activity [8]. Since this variability affects plant growth, nutrient dynamics, and other soil processes, knowledge of the spatial variability of soil physical properties is therefore necessary. To study the spatial distribution of soil properties, techniques such as classical statistics and geostatistics have been widely applied [9–11]. Geostatistics provides the basis for the interpolation and interpretation of the spatial variability of soil properties [9, 12–14]. Information on the spatial variability of soil properties leads to better management decisions aimed at correcting problems and at least maintaining productivity and sustainability of the soils and thus increasing the precision of farming practices [1, 15]. A better understanding of the spatial variability of soil properties would enable refining agricultural management practices by identifying sites where remediation and management are needed. This promotes sustainable soil and land use and also provides a valuable base against which subsequent future measurements can be proposed [14]. Despite the importance of this topic in agriculture, the literature is not abundant on the variability of soil physical properties in

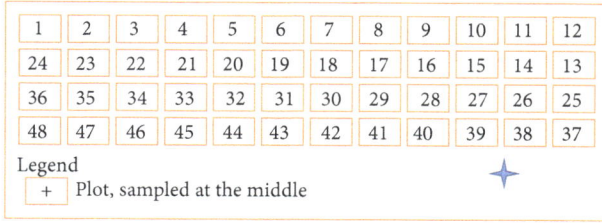

FIGURE 1: Study area (Lincoln University's Freeman farm) showing the plots.

Central Missouri. Furthermore, existing studies on the spatial variability of soil properties have focused on the top soil (0–20 cm) with less or no studies at deeper soil depths (30–100 cm). The objective of this study was therefore to assess the spatial variability of soil physical properties at various depths (0–10 cm, 10–20, 20–40 and 40–60 cm) in a clay-loam soil cropped to corn and soybean, and determine how knowledge on this variability can affect soil management practices.

## 2. Materials and Methods

*2.1. Experimental Site.* The study was conducted at Lincoln University's Freeman farm in Jefferson City, Missouri. The geographic coordinates of the study site are $38°58'116''$N latitude and $92°10'53''$W longitude. The soil of the experiment site is a Waldron clay-loam (Fine, smectitic, calcareous, mesic Aeric Fluvaquents). The study area is almost flat, with an average slope of 2%. The experimental field was made of 48 plots of 12.19 m width by 21.34 m length each. The 48 plots were arranged in a grid of 4 plots in the width by 12 plots in length as shown in Figure 1. One half of the plots was planted to corn (*Zea mays*) while the other half was planted to soybean (*Glycine max*). Soybean and corn plots all received 26.31 kg/ha of nitrogen, 67.25 kg/ha of phosphorus, and 89.67 kg/ha of potassium. Corn plots received 201.75 kg/ha of additional nitrogen in the form of urea.

*2.2. Soil Sampling.* Soil samples were collected in the middle of each plot after planting and full seeds emergence. Cylindrical cores of 3.15 cm radius and 10 or 20 cm height were used to collect soil samples at four depths: 0–10 cm, 10–20, 20–40 and 40–60 cm, corresponding to depths 1, 2, 3, and 4, respectively. The cylinders of 10 cm height were used for soil samples collection at depths 1 and 2 while the 20 cm height cylinders used for sampling at depths 3 and 4. A total of 576 soil samples were collected as follows: 48 plots × 4 depths × 3 replicates (at the middle of each plot). Collected samples were taken to the laboratory where they were weighed (fresh weight of sample; FWS) then oven dried at 105°C for 72 hrs. The weight was taken after oven drying (dry weight of soil; DWS). Soil physical properties were calculated as follows: Soil bulk density (BDY, g·cm$^{-3}$) = (DWS/V), where DWS is the dry weight of soil and $V$ the volume of cylinder (total volume of soil); Volumetric water content (VWC, cm$^3$·cm$^{-3}$) = (FWS – DWS)/$V$), with FWS being the

fresh weight of soil; gravimetric water content (GWC, g·g$^{-1}$) = [(FWS – DWS)/DWS] where FWS is the fresh weight of soil; total pore space (TPS, cm$^3$·cm$^{-3}$) = 1 – (BDY/PDY), where PDY is the soil particle density (taken as 2.65 g cm$^{-3}$); volumetric air content (VAC, cm$^3$·cm$^{-3}$) = TPS – VWC; water-filled pore space (WFPS, %) = 100 ∗ (VWC/TPS); air-filled pore space (AFPS, %) = 100 ∗ (VAC/TPS); relative gas diffusion coeffient (Diff., cm$^2$s$^{-1}$·cm$^{-2}$·s) = (VAC)$^2$, pore space tortuosity (Tort., m·m$^{-1}$) = (1/VAC) [16].

*2.3. Statistical and Geospatial Analysis.* After calculation, data on soil physical properties was first transferred to Statistix 9.0 to compute summaries of simple statistics, then to GS+ (Geostatistics for environmental science) 7.0 for semivariogram analysis. A semivariogram (a measure of the strength of statistical correlation as a function of distance) is defined by the following equation [17]:

$$\gamma(h) = \frac{1}{2m(h)} \sum_{i=1}^{m(h)} [z(x_i + h) - z(x_i)]^2, \qquad (1)$$

where $\gamma(h)$ is the experimental semivariogram value at a distance interval $h$, $m(h)$ is number of sample value pairs within the distance interval $h$, and $Z(X_i)$, and $Z(X_i + h)$ are sample values at two points separated by the distance $h$. Exponential and spherical models were the empirical semivariograms. The stationary models, that is, exponential (2) and spherical model (3) that fitted to experimental semivariograms were defined in the following equations [18]:

$$\gamma(h) = C_0 + C_1 \left[ 1 - \exp\left\{ -\left(\frac{h}{a}\right) \right\} \right], \qquad (2)$$

$$\gamma(h) = C_0 + C_1 \left[ \left(\frac{3h}{2a}\right) - \left(\frac{h^3}{2a^3}\right) \right], \quad \text{when } h \leq a,$$
$$= C_0 + C_1, \quad \text{when } h \geq a, \qquad (3)$$

where $C_0$ is the nugget, $C_1$ is the partial sill, and $a$ is the range of spatial dependence to reach the sill $(C_0 + C_1)$. The ratio $C_0/(C_0 + C_1)$ and the range are the parameters that characterize the spatial structure of a soil property. The $C_0/(C_0 + C_1)$ relation is the proportion in the dependence zone, and the range defines the distance over which the soil property values are correlated with each other [19]. A low value for the $C_0/(C_0 + C_1)$ ratio and a high range generally indicate that high precision of the property can be obtained by Parfitt et al. [19]. The classification proposed by Cambardella et al. [14], which considers the degree of spatial dependence (DSD = $C_0/(C_0 + C_1) \times 100$) as strong when DSD ≤ 25%; moderate when 25 < DSD ≤ 75%; and weak when DSD > 75%, was used in this study to classify the degree of spatial dependence of each soil property.

## 3. Results and Discussion

*3.1. Summaries of Statistics for Soil Physical Properties.* Overall, descriptive statistics for soil properties in this study showed moderate to high skewness for some of the properties

TABLE 1: Descriptive statistics for soil physical properties at four depths in a clay-loam soil.

| | AFPS | BDY | Diff. | GWC | TPS | Tort | VAC | VWC | WFPS |
|---|---|---|---|---|---|---|---|---|---|
| **D1 (0–10 cm)** | | | | | | | | | |
| Mean | 45.76 | 1.24 | 0.06 | 0.22 | 0.51 | 4.60 | 0.24 | 0.28 | 54.24 |
| SD | 10.61 | 0.11 | 0.03 | 0.03 | 0.04 | 1.50 | 0.07 | 0.04 | 10.61 |
| C.V | 23.18 | 8.99 | 56.29 | 15.25 | 8.36 | 32.60 | 29.64 | 15.26 | 19.56 |
| Median | 45.71 | 1.24 | 0.06 | 0.22 | 0.52 | 4.10 | 0.24 | 0.28 | 54.29 |
| **D2 (10–20 cm)** | | | | | | | | | |
| Mean | 26.54 | 1.47 | 0.02 | 0.21 | 0.42 | 12.46 | 0.12 | 0.31 | 73.46 |
| SD | 11.19 | 0.18 | 0.02 | 0.048 | 0.07 | 10.70 | 0.06 | 0.05 | 11.19 |
| C.V | 42.15 | 12.09 | 91.61 | 22.48 | 16.40 | 85.91 | 49.83 | 17.11 | 15.23 |
| Median | 27.32 | 1.46 | 0.01 | 0.22 | 0.43 | 9.04 | 0.11 | 0.31 | 72.68 |
| **D3 (20–40 cm)** | | | | | | | | | |
| Mean | 42.35 | 1.20 | 0.06 | 0.26 | 0.53 | 4.72 | 0.23 | 0.30 | 57.65 |
| SD | 8.19 | 0.12 | 0.03 | 0.04 | 0.05 | 1.29 | 0.06 | 0.03 | 8.19 |
| C.V | 19.33 | 9.75 | 53.22 | 13.92 | 8.68 | 27.30 | 26.27 | 11.48 | 14.20 |
| Median | 42.11 | 1.20 | 0.05 | 0.25 | 0.53 | 4.52 | 0.22 | 0.31 | 57.90 |
| **D4 (40–60 cm)** | | | | | | | | | |
| Mean | 39.34 | 1.18 | 0.05 | 0.28 | 0.54 | 4.94 | 0.21 | 0.32 | 60.66 |
| SD | 7.62 | 0.07 | 0.02 | 0.03 | 0.03 | 1.12 | 0.05 | 0.03 | 7.62 |
| C.V | 19.36 | 5.57 | 46.36 | 10.43 | 4.83 | 22.69 | 22.72 | 10.58 | 12.56 |
| Median | 39.74 | 1.18 | 0.05 | 0.27 | 0.54 | 4.67 | 0.22 | 0.33 | 60.27 |

AFPS: air-filled pore space (%); BDY: soil bulk density ($gcm^{-3}$); DIFF.: relative gas diffusion coefficient ($m^2 s^{-1} m^{-2} s$); GWC: gravimetric water content of soil ($g \cdot g^{-1}$); TPS: total pore spaces ($cm^3 cm^{-3}$); TORT: Pore tortuosity factor ($m \cdot m^{-1}$); VAC: volumetric air content ($cm^3 cm^{-3}$); VWC: volumetric water content ($cm^3 cm^{-3}$); WFPS: water-filled pore space (%).

(Table 1). The highly skewed soil parameters included soil bulk density (BDY), diffusivity (DIFF), and volumetric water content (VWC), whereas total pore space (TPS) was moderately skewed. Air-filled pore space (AFPS) had a low skewness. Highly skewed parameters indicate that these properties have a local distribution; that is, high values were found for these properties at some points, but most values were low [20]. The other soil physical properties were approximately normally distributed on the field. The underlying reason for soil properties being normally or nonnormally distributed may be associated with differences in management practices, land use, vegetation cover, and topographic effects on the variability of soil erosion across the landscape of the field. These factors can be the sources for a large or very small variation of soil properties in some of the samples, which leads to the nonnormal distribution [21]. A wide range of spatial variability was observed for soil physical properties (Table 1). For instance, soil bulk density (BDY) ranged from 1.01 to 1.23 $g \, cm^{-3}$ for depth 1, 1.15 to 1.46 $g \, cm^{-3}$ for depth 2, 0.96 to 1.19 $g \, cm^{-3}$ and 1.04 to 1.18 $g \, cm^{-3}$ for depths 3 and 4, respectively (Figure 2). Soil bulk density was also significantly higher in the second depth (1.4 $g \, cm^{-3}$) than all the other 3 depths, where it varied between 1.18 $g \, cm^{-3}$ and 1.24 $g \, cm^{-3}$. The mean value of AFPS was significantly lower in the second depth (26.5 $cm^3 \cdot cm^{-3}$) than in all other 3 depths, where it varied from 39.34 to 45.7 $cm^3 \cdot cm^{-3}$. Soil pore tortuosity factor (TORT) and water-filled pore space (WFPS) were also significantly higher in the second depth (12.46 $cm \cdot cm^{-1}$ and 73.46%, resp.). However, the relative gas diffusion

coefficient (DIFF), gravimetric water content (GWC), total pore space (TPS), and volumetric air content (VAC) were significantly lower in the second depth (0.02 $m^2 s^{-1} m^{-2} s$, 0.21 $g \cdot g^{-1}$, 0.42 $cm^3 \cdot cm^{-3}$, and 0.12 $cm^3 \cdot cm^{-3}$, resp.) (Table 1). The variability in soil physical properties is understandable since the soil of this site has a smectite layer (claypan) in the 10–20 cm, which corresponded to our second sampling depth. This layer of smectite is hard and compact, with very low pore space, high mass-volume ratio (bulk density) and high water retention capability (because of their large surface area). As a consequence of the presence of this smectite layer in depth 2, the mean of water-filled pore space (WFPS) was slightly lower in the first depth (54%) than in all four depths. In fact, air predominates the pore space in the first depth and cultivation loosened the soil, thereby allowing the water trapped in the pore space to evaporate. Higher GWC, VWC, and TPS at the lower depths (20–60 cm) mean that crops (especially corn and soybean grown in the field) were able to access water and dissolved nutrients through their roots. In fact, despite the claypan layer (10–20 cm), it has been reported by various researchers that crop roots were able to penetrate into and through this layer of smectitic clay [22–24] and that root growth may increase within the claypan layer [23] as a result of plant adaptation to water-limited soil layers. In general, the use of the coefficient of variation (CV) is a common procedure to assess variability in soil properties since it allows comparison among properties with different units of measurement. Overall, the coefficient of variation for all soil physical properties, in the four depths, ranged from 4.83 to

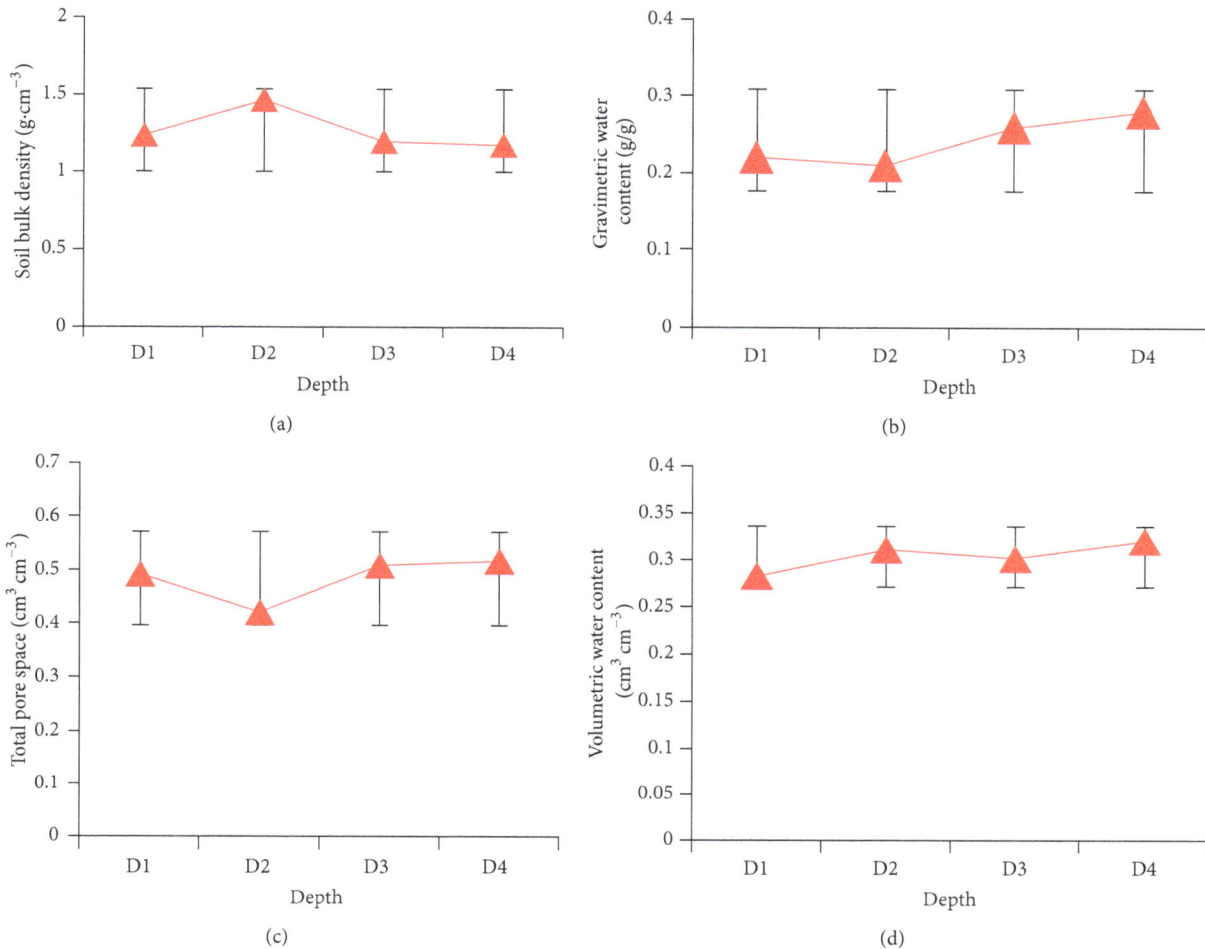

FIGURE 2: Variation of soil bulk density, gravimetric water content, total pore space, and volumetric water content with depth.

91.61% (Table 1). The pore tortuosity factor (TORT) showed the highest variation while soil bulk density (BDY) showed the least variation. The CV indicated that there was a strong spatial variability of the soil properties investigated. However, to have a better assessment of such spatial variability across the entire field, a geostatistical analysis was used.

3.2. Spatial Variability of Soil Properties. After computing summaries of simple statistics with Statistix 9.0, data on soil physical properties was transferred to GS+ (geostatistics for environmental science) 7.0 for semivariogram analysis. Semivariogram model fit was determined from the coefficient of determination ($R^2$) values, which range from 0 (very poor model fit) to 1 (very good model fit). Table 2 shows soil physical properties which mainly responded to exponential and linear variogram models, with the exponential model providing the best fit. In the 10–20 cm depth, exponential model provided the best fit for BDY ($R^2$ = 0.93), with the spherical model providing very poor model fit. Pore tortuosity also responded to an exponential variogram model in the 20–40 cm depth ($R^2$ = 0.57), although spherical model was noticed. Linear and exponential models were observed

in the 40–60 cm depth for TPS ($R^2$ = 0.46), with linear model providing a better fit (Table 2). In general, for all depths, model fit was not very strong with the exception of gravimetric water content and bulk density in the second depth. Overall, the exponential model provided the best fit with about 65% of the physical properties fitting this model. In geostatistical theory, the range of the spatial variability of the semivariogram is the distance between correlated measurements (the minimum lateral distance between two points before the change in property is noticed) and can be an effective criterion for the evaluation of sampling design and mapping of soil properties. The value that the semivariogram model attains at the range (the value on the $y$-axis) is called the sill. The partial sill is the sill minus the nugget [25, 26]. Theoretically, at zero separation distance (lag = 0), the semivariogram value is zero. However, at an infinitesimally small separation distance, the semivariogram often exhibits a nugget effect (the apparent discontinuity at the beginning of many semivariogram graphs), which is some value greater than zero. The nugget effect can be attributed to measurement errors or spatial sources of variation at distances smaller than the sampling interval (or both). Measurement error occurs because of the error inherent in measuring devices. To

kgo

TABLE 2: Variogram parameters for soil physical properties at four depths in a clay-loam soil.

| Depth (cm) | Model | Nugget ($C_0$) | Sill ($C_0 + C$) | Range ($A_0$) | $R^2$ | ($C/C_0 + C$) | DSD (%) |
|---|---|---|---|---|---|---|---|
| GWC | | | | | | | |
| 0–10 | Exponential | 0.00 | 0.00 | 6.42 | 0.12 | 0.94 | 0.01 |
| 10–20 | Exponential | 0.00 | 0.00 | 64.26 | 0.75 | 0.52 | 0.29 |
| 20–40 | Spherical | 0.00 | 0.00 | 7.17 | 0.11 | 1.00 | 0.00 |
| 40–60 | Exponential | 0.00 | 0.00 | 4.56 | 0.01 | 0.99 | 0.00 |
| VWC | | | | | | | |
| 0–10 | Exponential | 0.00 | 0.00 | 6.21 | 0.31 | 0.94 | 0.01 |
| 10–20 | Exponential | 0.00 | 0.00 | 11.82 | 0.44 | 0.89 | 0.04 |
| 20–40 | Exponential | 0.00 | 0.00 | 5.88 | 0.17 | 0.97 | 0.00 |
| 40–60 | Exponential | 0.00 | 0.00 | 0.45 | 0.00 | 1.00 | 0.00 |
| BDY | | | | | | | |
| 0–10 | Linear | 0.01 | 0.01 | 25.77 | 0.07 | 0.00 | 1.18 |
| 10–20 | Exponential | 0.02 | 0.04 | 40.17 | 0.93 | 0.50 | 0.04 |
| 20–40 | Spherical | 0.00 | 0.01 | 7.42 | 0.20 | 0.95 | 0.07 |
| 40–60 | Linear | 0.00 | 0.00 | 25.77 | 0.39 | 0.00 | 0.44 |
| TPS | | | | | | | |
| 0–10 | Linear | 0.00 | 0.00 | 25.77 | 0.07 | 0.00 | 0.16 |
| 10–20 | Exponential | 0.00 | 0.01 | 14.64 | 0.79 | 0.81 | 0.13 |
| 20–40 | Exponential | 0.00 | 0.00 | 7.50 | 0.21 | 0.90 | 0.03 |
| 40–60 | Linear | 0.00 | 0.00 | 25.77 | 0.47 | 0.00 | 0.07 |
| VAC | | | | | | | |
| 0–10 | Linear | 0.00 | 0.00 | 25.77 | 0.25 | 0.00 | 0.48 |
| 10–20 | Exponential | 0.00 | 0.00 | 2.73 | 0.00 | 0.86 | 0.06 |
| 20–40 | Exponential | 0.01 | 0.00 | 7.62 | 0.35 | 0.92 | 0.07 |
| 40–60 | Linear | 0.00 | 0.00 | 25.77 | 0.21 | 0.00 | 0.23 |
| WFPS | | | | | | | |
| 0–10 | Linear | 107.28 | 107.28 | 25.77 | 0.15 | 0.00 | 10727.60 |
| 10–20 | Spherical | 10.80 | 135.30 | 5.38 | 0.00 | 0.92 | 1173.91 |
| 20–40 | Exponential | 3.50 | 66.49 | 7.47 | 0.27 | 0.95 | 369.59 |
| 40–60 | Linear | 55.49 | 55.49 | 25.77 | 0.07 | 0.00 | 5549.30 |
| AFPS | | | | | | | |
| 0–10 | Linear | 107.28 | 107.28 | 25.77 | 0.15 | 0.00 | 10727.6 |
| 10–20 | Spherical | 10.80 | 135.30 | 5.38 | 0.00 | 0.92 | 1173.91 |
| 20–40 | Exponential | 3.50 | 66.49 | 7.47 | 0.27 | 0.95 | 369.588 |
| 40–60 | Linear | 55.49 | 55.49 | 25.77 | 0.07 | 0.00 | 5549.30 |
| DIFF | | | | | | | |
| 0–10 | Linear | 0.00 | 0.00 | 25.77 | 0.10 | 0.00 | 0.11 |
| 10–20 | Exponential | 0.00 | 0.00 | 4.71 | 0.00 | 0.85 | 0.00 |
| 20–40 | Exponential | 0.00 | 0.00 | 4.53 | 0.05 | 0.93 | 0.01 |
| 40–60 | Linear | 0.00 | 0.00 | 25.77 | 0.19 | 0.00 | 0.05 |
| TORT | | | | | | | |
| 0–10 | Linear | 2.12 | 2.12 | 25.77 | 0.15 | 0.00 | 211.80 |
| 10–20 | Exponential | 17.80 | 126.10 | 8.10 | 0.19 | 0.86 | 2072.20 |
| 20–40 | Exponential | 0.18 | 1.74 | 13.20 | 0.58 | 0.90 | 19.44 |
| 40–60 | Linear | 1.24 | 1.24 | 25.77 | 0.16 | 0.00 | 123.68 |

DSD: degree of spatial dependence: strong DSD (DSD ≤ 25%), moderate DSD (25 < DSD ≤ 75%), and weak DSD (DSD > 75%) according to Cambardella et al. (1994) [14].

eliminate this error, multiple samples were taken from each sampling point. Natural phenomena can vary spatially over a range of scales. Variation at microscales smaller than the sampling distances will appear as a part of the nugget effect. Table 2 shows that the spatial correlation (range) of soil properties widely varied from 1 m for volumetric water content (VWC) in depth four to 64 m for gravimetric water content (GWC) in depth 2. However, for the first and second depth (which are agriculturally more important), the range of spatial correlation varied from 3 m for volumetric air content (VAC) in depth 2 to 64 m for GWC in depth 2. Beyond these ranges, there is no spatial dependence (autocorrelation). The spatial dependence can indicate the level of similarity or disturbance of the soil condition. According to López-Granados et al. [27] and Ayoubi et al. [17], a large range indicates that the measured soil property value is influenced by natural and anthropogenic factors over greater distances than parameters which have smaller ranges. Thus, a range of about 64 m for GWC in this study indicates that the measured GWC values can be influenced in the soil over greater distances as compared to the soil parameters having smaller range (Table 2). This means that soil variables with smaller range such as VWC and VAC are good indicators of the more disturbed soils (the more disturbed a soil is, the more variable some soil properties become). The more variable properties have a shorter range of correlation. The different ranges of the spatial dependence among the soil properties may be attributed to differences in response to the erosion—deposition factors, land use-cover, parent material, and human interferences in the study area. The nugget, which is an indication of microvariability was significantly higher for water-filled pore space (WFPS) and air-filled pore space (AFPS) when compared to the others. This can be explained by our sampling distance which could not capture well their spatial dependence. The lowest nugget was for GWC (Table 2). This indicates that GWC had low spatial variability within small distances. Knowledge of the range of influence for various soil properties allows one to construct independent accurate datasets for similar areas in future soil sampling design to perform statistical analysis [17]. This aids in determining where to resample, if necessary, and design future field experiments that avoid spatial dependence. Therefore, for future studies aimed at characterizing the spatial dependency of soil properties in the study area and/or a similar area, it is recommended that the soil properties be sampled at distances shorter than the range found in this study. Cambardella et al. [14] established the classification of degree of spatial dependence (DSD) between adjacent observations of soil property > 75% to correspond to weak spatial structure. In this study, the semivariograms indicated strong spatial dependence (DSD $\leq$ 25%) for soil physical properties such as bulk density, gravimetric water content, volumetric water content, total pore space, and diffusivity. The rest of the soil physical properties (water-filled pore space, Air-filled pore space, and tortuosity) measured exhibited very weak spatial dependence (DSD > 75%) (Table 2). The strong spatial dependence of the soil properties may be controlled by intrinsic variations in soil characteristics such as texture and mineralogy, whereas extrinsic variations such as tillage and other soil and water

management practices may also control the variability of the weak spatially dependent parameters [14].

*3.3. Spatial Distribution of Soil Properties across the Field.* Interpolated maps portraying the distribution of soil physical properties in various depths are shown in Figure 3 for soil gravimetric (GWC) and volumetric (VWC) contents and water-filled pore space (WFPS). Gravimetric water content showed a good spatial distribution across the field with the highest values located around the southwestern portion of the field. Volumetric water content also showed good spatial distribution across the field with high values located in the northern, central, and southwestern portions of the field. Water-filled pore has a distribution similar to that of volumetric water content. The other soil properties, however, showed very poor spatial distribution in the field. This is most probably due to their poor sill ($C_0 + C$), model fit and coefficient of determination ($R^2$). Even though the spatial variability was not very pronounced, there were areas on the field that had slightly higher values of these physical properties than the rest of the field. In general, bulk density, total pore space, volumetric air content, Air-filled pore space, diffusivity, and tortuosity were very high in the field even though they did not exhibit very distinguishable variability. This lack of visible spatial variability is supported because the sampling distance (range) is 26 m for these properties.

*3.4. Implications of Spatial Variability of Soil Physical Properties on Soil Management.* Results of this study indicated that the spatial variability of soil water content (GWC and VWC) was high. This can be explained, among many other reasons, by soil type (clay-loam) which was able to hold more water. But with intensive tillage, this soil water content could be adversely affected. Studies have shown that tillage practices can alter soil physical properties and consequently the hydrological behavior of agricultural fields, especially when a similar tillage system has been practiced for a long period [15, 28–31]. Tillage intensity has also considerable effects on spatial structure and spatial variability of soil properties [15, 30]. Therefore, this study can help determine site-specific soil management and decision making. To do so, the spatial variability of soil properties developed through kriging will be an important tool. Different ranges of spatial dependence were noticed in the field. The different ranges of the spatial dependence among the soil properties may be attributed to differences in response to the erosion—deposition factors, land use-cover, parent material, and human interferences in the study area. The different ranges can also be used in future studies to determine the sampling distance of different soil physical properties on the field. Also, the sill ($C_0 + C$) can help determine where the variability or change in soil property stops. This will be useful especially for the irrigation purposes. Generally, with farmers facing the decision of whether or not to till and the intensity of tillage, a spatial variability study can help in this decision making. Maps produced in this study can also be used for irrigation purposes as they can clearly indicate which portion of the field needs irrigation (soil water content). To do this, soil water content

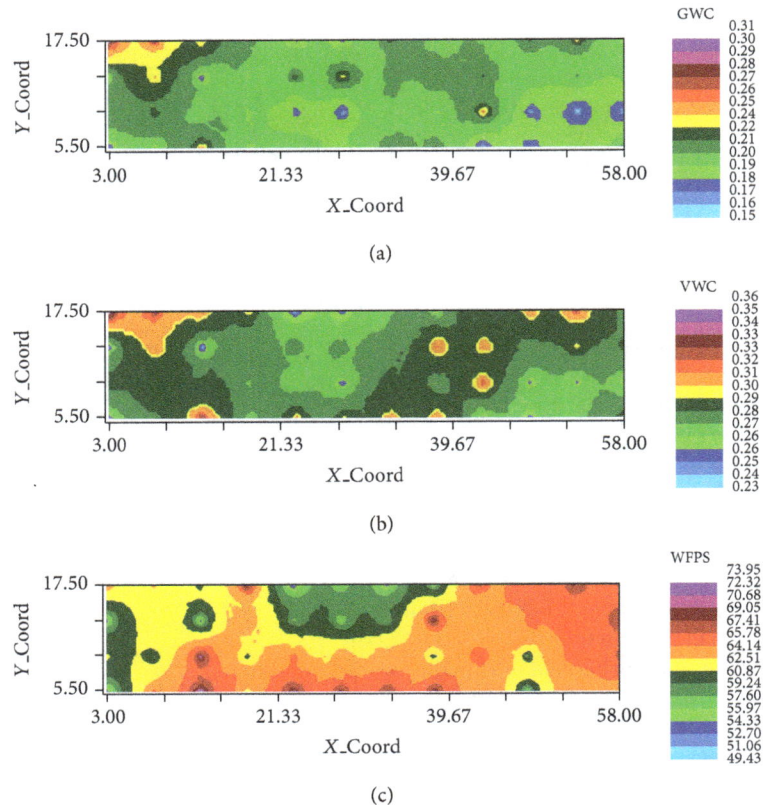

(a)

(b)

(c)

FIGURE 3: Spatial distribution of gravimetric (GWC) and volumetric (VWC) water contents in 0–10 cm depth and that of water-filled pore space (WFPS) in 20 cm depth.

information can be collected and analyzed geospatially to produce field maps. The process can be repeated frequently to obtain up-to-date soil water content information. To avoid frequent destructive sampling for water content analysis, equipment that allows insitu measurements such as TDR methods and water mark sensors can be used. Since a different range of spatial dependence among soil properties shows differences in response to human interferences and land use-cover, this will help reduce human activities that increase soil bulk density and cause soil compaction like the use of heavy equipment. It can also serve as a reference for the type of crop to be grown (cover crops for erosion susceptible areas).

## 4. Conclusion

We assessed the spatial variability of soil physical properties in a clay-loam soil cropped to corn and soybean. Results showed that soil physical properties either decreased or increased sharply in the second depth (due to the presence of a smectite layer) before leveling up or dropping off, but without reaching the first depth value in either case. In addition, depending on soil physical property, maps produced by kriging showed either good or poor spatial distribution. The semi-variogram analysis showed the presence of a strong ($\leq 25\%$) to weak ($>75\%$) spatial dependence of soil properties. Our understanding of the behavior of soil properties in this study provides new insights for soil site-specific management in

addressing issues such as "where to place the proper interventions" (tillage, irrigation, and crop type to be grown).

## Acknowledgment

This research is part of a regional collaborative project supported by the USDA-NIFA, Award no. 2011-68002-30190, Cropping Systems Coordinated Agricultural Project: Climate Change, Mitigation, and Adaptation in Corn-based Cropping Systems. (Project website: http://sustainablecorn.org/).

## References

[1] D. Schimel, J. Melillo, H. Tian et al., "Contribution of increasing $CO_2$ and climate to carbon storage by ecosystems in the United States," *Science*, vol. 287, no. 5460, pp. 2004–2006, 2000.

[2] C. Kosmas, S. Gerontidis, and M. Marathianou, "The effect of land use change on soils and vegetation over various lithological formations on Lesvos (Greece)," *Catena*, vol. 40, no. 1, pp. 51–68, 2000.

[3] J. Igbal, J. A. Thomasson, J. N. Jenkins, P. R. Owens, and F. D. Whisler, "Spatial variability analysis of soil physical properties of alluvial soils," *Soil Science Society of America Journal*, vol. 69, pp. 1–14, 2005.

[4] S. Zhang, X. Zhang, T. Huffman, X. Liu, and J. Yang, "Influence of topography and land management on soil nutrients variability in Northeast China," *Nutrient Cycling in Agroecosystems*, vol. 89, no. 3, pp. 427–438, 2011.

[5] L. J. Munkholm, A. Garbout, and S. B. Hansen, "Tillage effects on topsoil structural quality assessed using X-ray CT, soil cores and visual soil evaluation," *Soil & Tillage Research*, vol. 128, pp. 104–109, 2013.

[6] A. J. Franzluebbers, "Soil organic matter stratification ratio as an indicator of soil quality," *Soil and Tillage Research*, vol. 66, no. 2, pp. 95–106, 2002.

[7] L. J. Munkholm, P. Schjønning, K. J. Rasmussen, and K. Tanderup, "Spatial and temporal effects of direct drilling on soil structure in the seedling environment," *Soil and Tillage Research*, vol. 71, no. 2, pp. 163–173, 2003.

[8] A. Swarowsky, R. A. Dahlgren, K. W. Tate, J. W. Hopmans, and A. T. O'Geen, "Catchment-scale soil water dynamics in a mediterranean-type oak woodland," *Vadose Zone Journal*, vol. 10, no. 3, pp. 800–815, 2011.

[9] R. Webster and M. A. Oliver, *Statistical Methods in Soil and Land Resource Survey*, Oxford University Press, Oxford, UK, 1990.

[10] A. Saldaña, A. Stein, and J. A. Zinck, "Spatial variability of soil properties at different scales within three terraces of the Henares river (Spain)," *Catena*, vol. 33, no. 3-4, pp. 139–153, 1998.

[11] Z. M. Wang, K. S. Song, B. Zhang et al., "Spatial variability and affecting factors of soil nutrients in croplands of Northeast China: a case study in Dehui county," *Plant, Soil and Environment*, vol. 55, no. 3, pp. 110–120, 2009.

[12] R. Webster, "Quantitative spatial analysis of soil in the field," *Advances in Soil Science*, vol. 3, pp. 1–70, 1985.

[13] H. Pohlmann, "Geostatistical modelling of environmental data," *Catena*, vol. 20, no. 1-2, pp. 191–198, 1993.

[14] C. A. Cambardella, T. B. Moorman, J. M. Novak et al., "Field-scale variability of soil properties in central Iowa soils," *Soil Science Society of America Journal*, vol. 58, no. 5, pp. 1501–1511, 1994.

[15] E. Özgöz, "Long term conventional tillage effect on spatial variability of some soil physical properties," *Journal of Sustainable Agriculture*, vol. 33, no. 2, pp. 142–160, 2009.

[16] N. V. Nkongolo, R. Hatano, and V. Kakembo, "Diffusivity models and greenhouse gases fluxes from a forest, pasture, grassland and corn field in Northern Hokkaido, Japan," *Pedosphere*, vol. 20, no. 6, pp. 747–760, 2010.

[17] S. H. Ayoubi, S. M. Zamani, and F. Khomali, "Spatial variability of some soil properties for site-specific farming in northern Iran," *International Journal of Plant Production*, vol. 2, pp. 225–236, 2007.

[18] T. M. Burgess and R. Webster, "Optimal interpolation and isarithmic mapping of soil properties. I. The semi-variogram and punctual kriging," *Journal of Soil & Science*, vol. 31, no. 2, pp. 315–331, 1980.

[19] J. M. B. Parfitt, L. C. Timm, E. A. Pauletto et al., "Spatial variability of the chemical, physical and biological properties in lowland cultivated with irrigated rice," *Revista Brasileira de Ciencia do Solo*, vol. 33, no. 4, pp. 819–830, 2009.

[20] C. R. Grego, S. R. Vieira, and A. L. Lourenção, "Spatial distribution of Pseudaletia sequax Franclemlont in triticale under no-till management," *Scientia Agricola*, vol. 63, no. 4, pp. 321–327, 2006.

[21] G. B. Tesfahunegn, L. Tamene, and P. L. G. Vlek, "Catchment-scale spatial variability of soil properties and implications on site-specific soil management in northern Ethiopia," *Soil and Tillage Research*, vol. 117, pp. 124–139, 2011.

[22] S. J. Grecu, M. B. Kirkham, E. T. Kanemasu, D. W. Sweeney, L. R. Stone, and G. A. Milliken, "Root growth in a claypan with a perennial-annual rotation," *Soil Science Society of America Journal*, vol. 52, no. 2, pp. 488–494, 1988.

[23] D. B. Myers, N. R. Kitchen, K. A. Sudduth, R. E. Sharp, and R. J. Miles, "Soybean root distribution related to claypan soil properties and apparent soil electrical conductivity," *Crop Science*, vol. 47, no. 4, pp. 1498–1509, 2007.

[24] P. Jiang, N. R. Kitchen, S. H. Anderson, E. J. Sadler, and K. A. Sudduth, "Estimating palnt available water using the simple inverse yield model for claypan landscapes," *Agronomy Journal*, vol. 100, pp. 1–7, 2008.

[25] A. Utset, M. E. Ruiz, J. Herrera, and D. P. De Leon, "A geostatistical method for soil salinity sample site spacing," *Geoderma*, vol. 86, no. 1-2, pp. 143–151, 1998.

[26] W. Fu, H. Tunney, and C. Zhang, "Spatial variation of soil nutrients in a dairy farm and its implications for site-specific fertilizer application," *Soil and Tillage Research*, vol. 106, no. 2, pp. 185–193, 2010.

[27] F. López-Granados, M. Jurado-Expósito, S. Atenciano, A. García-Ferrer, M. Sánchez De La Orden, and L. García-Torres, "Spatial variability of agricultural soil parameters in southern Spain," *Plant and Soil*, vol. 246, no. 1, pp. 97–105, 2002.

[28] R. L. Hill, "Long-term conventional and no-tillage effects on selected soil physical properties," *Soil Science Society of America Journal*, vol. 54, no. 1, pp. 161–166, 1990.

[29] D. E. Buschiazzo, J. L. Panigatti, and P. W. Unger, "Tillage effects on soil properties and crop production in the subhumid and semiarid Argentinean Pampas," *Soil and Tillage Research*, vol. 49, no. 1-2, pp. 105–116, 1998.

[30] T. Tsegaye and R. L. Hill, "Intensive tillage effects on spatial variability of soil physical properties," *Soil Science*, vol. 163, no. 2, pp. 143–154, 1998.

[31] J. A. Gómez, J. V. Giráldez, M. Pastor, and E. Fereres, "Effects of tillage method on soil physical properties, infiltration and yield in an olive orchard," *Soil and Tillage Research*, vol. 52, no. 3-4, pp. 167–175, 1999.

# 14

# Significance of Some Soil Amendments and Phosphate Dissolving Bacteria to Enhance the Availability of Phosphate in Calcareous Soil

**Ahmed A. Khalil**

*Soils, Water, and Environment Research Institute, Agricultural Research Centre, P.O. Box 175 El-Orman, Giza 12112, Egypt*

Correspondence should be addressed to Ahmed A. Khalil; ahwafa@yahoo.com

Academic Editors: W. Ding and D. H. Phillips

A field experiment was carried-out on a private farm at the Salah El-Din village, El-Bostan district, Nobaria, El-Behera Governorate, Egypt. The aim of this study was to evaluate the best combination of rock phosphate (RP), sulphur (S), organic manure, and phosphate dissolving bacteria (PDB) inoculation to enhance the availability of phosphorous from rock phosphate and their effects on yield of broad bean plants (cv. *Luz doe Otono* L.). It was found that either sulphur application or PDB inoculation with RP had a significant effect on broad bean yield and its quality. Application of RP and different soil amendments individually or together increased N, P, and K contents in straw and seeds of broad bean plant. The highest contents of the studied nutrients were found when the plants were fertilized by a mixture of RP and different soil amendments. Results also showed the important role of organic matter, sulphur, and PDB for releasing phosphorus from rock phosphate. The combination of soil amendments with RP as a natural P-source, has the possibility of saving significant quantities of industrialized inorganic phosphate fertilizers.

## 1. Introduction

Broad bean (*Vicia faba* L.) is one of the most important legumes in Egypt. It is intensively used by both human and animals in many countries worldwide. It is considered as a cheap diet containing high protein and energy. Therefore, efforts to improve the quality and quantity of the vegetable crop are important. Under Egyptian soil conditions, phosphorus availability in soil is governed by many factors (pH, $CaCO_3$, organic matter, and clay contents). In spite of the considerable addition of P to these soils, the level of available phosphorus decreases sharply after a short period since application. Under alkaline soil conditions, the available phosphorus in the added fertilizer is rapidly transformed to tricalcium phosphate which is unavailable to the plants [1]. Rock phosphate is the main source for producing phosphate fertilizers. The direct application of apatite $Ca_5(PO_4,CO_3)_3(OH,F)$ instead of phosphate fertilizers is not suitable, especially in soils with a high pH. However, using acidic materials such as sulphur and sulphuric acid, or using rock phosphate combined with phosphate dissolving bacteria

(PDB) such as *Pseudomonas, Azospirillum, Burkholderia, Bacillus, Enterobacter, Rhizobium, Erwinia, Serratia, Alcaligenes, Arthrobacter, Acinetobacter,* and *Flavobacterium* which can produce some organic acids will release phosphorous from rock phosphate and can replace P-fertilizers. Gluconic acid was reported as the most frequent agent for mineral phosphate solubilization produced by *Pseudomonas sp.* and *Erwinia herbicola* [2, 3].

Therefore, rock phosphate is a good source of phosphorus if organic manure and powdered sulphur with phosphate dissolving bacteria are added [4]. The addition of compost to the soils improves their physical, chemical, and biological properties, which influence the growth and development of plants. Also, organic acids produced from decomposition of organic matter help to dissolve the rock phosphate and increase the availability of phosphorus. Subehia (2001) found that the use of rock phosphate in conjunction with different organic manures was similar to the use superphosphate [5]. Sulphur oxidation in soils is an effective process in the reclamation of sodic soils in addition to providing the sulphur needs for plants. More importantly, this process will lower the pH of the soil resulting in an increased activity of some plant nutrients near the root zone and consequently resulting in an improvement in the yield and quality of agricultural crops. Kumar et al. (1992) reported the superiority of rock phosphate and sulphur compared to rock. Phosphate alone in increasing macro- and micronutrients in soils and decreasing soil pH which may be due to the oxidation of sulphur to sulphuric acid [6]. El-Sayed (1999) revealed that PDB plays an important role in releasing P from rock, tri-calcium, or other difficult P-forms through the production of organic and inorganic acids, as well as $CO_2$ [7]. These substances convert the insoluble forms of P into soluble ones. PDB also affects other nutrients in addition to phosphorus. For example, it was reported that seed inoculation with PDB generally increased number of total bacteria PDB in the rhizosphere zone and number of nodules and released ammonia from bound complex nitrogen compounds [8].

Therefore, the aim of this study was evaluating the efficiency of PDB, sulphur, and (organic manure) on release of phosphorus from rock phosphate and their effects on vegetative growth, chemical composition, and yield of broad bean plants.

## 2. Materials and Methods

A field experiment was carried-out in a private farm of the Salah El-Din village, El-Bostan district (30° 43' 25.14" N 30° 17' 23.94" E elevation 14 m), Nobaria, El-Behera Governorate, Egypt, during two successive winter growing seasons of 2008/2009 and 2009/2010. Soil samples were collected, prior to tillage from surface layer (0–20 cm) for physical and chemical analyses (Table 1). Furthermore, organic manure (compost) was analyzed according to Cottenie et al. (1982) and Page et al. (1982) (Table 2) [9, 10].

Each experiment included nine treatments arranged in a complete randomized block design (CRBD) with three replicates. Data was collected separately for both years from the individual treatments in the CRBD, analyzed statistically through analysis of variance over the year's techniques, and tables of variance were constructed. Averages of significant treatments were compared in accordance with Duncan's Multiple Range Test (DMR) at 5% probability level. Seeds of broad bean (cv. *Luz de Otono* L.) were sown in hills spaced 20 cm apart on both sides of the ridge (planting line), under drip irrigation system.

For all treatments, phosphorus of two sources (super phosphate and rock phosphate) and sulphur were applied during seed bed preparation. Organic manure at a rate of 47.62 m$^3$/ha was uniformly incorporated into the soil layer at 20 cm depth with power tiller two weeks before planting. Plots received 47.62 kg N applied as a starter dose in the form of ammonium sulphate (20.6% N) after thinning as well as 57.14 kg $K_2O$ applied in the form of potassium sulphate (48% $K_2O$).

The treatments were as follows:

(1) 357.4 kg applied super phosphate (15% soluble $P_2O_5$) at a rate of 53.57 kg/ha $P_2O_5$;

TABLE 1: Some physical and chemical properties of the experimental soil.

| Property | Content |
| --- | --- |
| Moisture content (%) | 13.00 |
| EC (dS/m) | 3.80 |
| pH (1:10) | 7.50 |
| Total N (%) | 0.78 |
| Total P (%) | 0.52 |
| Total K (%) | 1.68 |
| C/N ratio | 1:22.70 |
| Organic matter (%) | 30.55 |
| Organic carbon (%) | 17.71 |

TABLE 2: Some properties of the used organic manure (compost).

| Characters | Value |
| --- | --- |
| Particle size distribution (%) | |
| Sand | 65.18 |
| Silt | 18.51 |
| Clay | 16.31 |
| Textural class | Sandy loam |
| $CaCO_3$ (%) | 22.30 |
| Chemical analysis | |
| pH (1:2.5) | 8.15 |
| EC (dS/m) 1:5 | 0.53 |
| Soluble ions (meq/100 g soil) | |
| $Ca^{++}$ | 0.37 |
| $Mg^{++}$ | 0.16 |
| $Na^+$ | 0.58 |
| $K^+$ | 0.08 |
| $CO_3^{--}$ | 0.00 |
| $HCO_3^-$ | 0.30 |
| $Cl^-$ | 0.59 |
| $SO_4^{--}$ | 0.30 |
| Organic matter (%) | 0.26 |
| Organic carbon (%) | 0.15 |
| Total (%) | |
| N | 0.04 |
| P | 0.12 |
| K | 1.21 |
| Available contents (mg/kg soil) | |
| N | 11.7 |
| P | 3.32 |
| K | 70.0 |
| Fe | 3.61 |
| Zn | 0.52 |
| Mn | 0.44 |

(2) 191.93 kg applied rock phosphate (RP) (28% total $P_2O_5$) at a rate of 53.57 kg/ha $P_2O_5$;

(3) RP + 476.19 kg S/ha;

(4) RP + PDB (seed inoculation with PDB just before sowing);

(5) RP + 47.62 m³/ha OM (organic manure);

(6) RP + PDB + 476.19 kg S/ha;

(7) RP + 476.19 kg S/ha + 47.62 m³/ha OM;

(8) RP + PDB + 47.62 m³/ha OM;

(9) RP + PDB + 476.19 kg S/ha + 47.62 m³/ha OM.

Broad bean plants were harvested and the following characteristics were recorded:

(1) total weight of green pods/plant (g);

(2) number of pods/plant;

(3) dry weight of seeds/plant (g);

(4) 100-seed weight (g) (the 100-seed weight is a measure of seed size. It is the weight in grams of 100-seeds. Seed size and the 100-seed weight can vary from one crop to another, between varieties of the same crop and even from year to year or from field to field of the same variety. Because of this variation in seed size, the number of seeds and, consequently, the number of plants in a pound or a bushel of seed are also highly variable. By using the 100-seed weight, a producer can account for seed size variations when calculating seeding rates, calibrating seed drills, and estimating shattering and combine losses);

(5) seed and straw yields (kg/ha).

From each plot, samples of both seeds and straw were oven dried (70°C); ground and wet digested using a 1:1 mixture of $H_2SO_4$ and $HClO_4$ acids ($V/V$). In the digested product, nitrogen content of the plants (straw and seeds) was determined by Kjeldahl method while potassium content was determined by flame photometer [11]. Phosphorus content of plant parts was determined by a colorimetric method [12]. Moreover, the biochemical constituents in broad bean seeds such as protein and carbohydrate were estimated using the methods outlined by A.O.A.C. (1990) [13]. All collected data were statistically analyzed according to K. A. Gomez and A. A. Gomez (1984) [14].

## 3. Results and Discussion

*3.1. Effect of the Applied Treatments on Yield and Its Components.* Results presented in Table 3 showed that total weight of green pods/plant, number of pods/plant, dry weight of seeds/plant, 100-seed weight, seed and straw yield, seed protein (%), protein yield, and total carbohydrate (%) were significantly affected ($P < 0.05$) by RP in combination with soil amendments (S and organic manure) and PDB either added individually or in combinations. However, the greatest values were associated with applied combined treatment of (PDB + compost + S). The corresponding relative percentages increases reached 34.03%, 22.45%, 23.26%, 6.64%, 41.99%, 41.34%, 7.95%, 53.28%, and 9.71% over the control treatment (SP%), respectively. These increases in yield and its components of broad bean plants may be attributed to the increases in both cell division and cell elongation. As reported earlier, phosphorus plays a major role in protein

synthesis and protoplasm formation. It may increase the proportion or protoplasm to cell wall, and it is a part of the molecular structure of nucleic acids DNA and RNA resulting in an increment in vegetative growth characters like increased cell size [15, 16]. In addition, phosphorus plays an important role in photosynthesis and respiration, and it is also essential for division and development of meristematic tissues. Similar results were previously reported [17, 18]. Seed inoculation with the PDB caused remarkable increases in most parameters as compared with rock phosphate fertilizer only.

These increases may be due to the inoculation with phosphate dissolving bacteria which solubilized unavailable forms of calcium bound phosphate by excreting organic acids such as gluconic acid, formic, acetic, lactic, propionic, fumaric, and succinic acids, those acids are lowering the pH which directly bring in insoluble phosphates in soil into soluble forms as well as the displacement of phosphorous on surfaces [8, 19–22]. There are indications that these bacteria may also produce growth promoting substances such as auxins, gibberellins, and cytokinins. Such substances could influence the plant growth by making the roots able to explore more soil and more zones where phosphate ions were chemically liberated from the P- source. Similar positive effects of PDB have been obtained on soybean, lentil, and mung bean [7, 8, 23].

Concerning the interactive effects of treatments, data showed that the application of two amendments was more effective than a single one, while the tricombinations had the most effect on enhancing the yield and yield components. Also, the same effect of elemental sulphur application and inoculation with PDB in the presence of RP on yield and its components had been observed. This could be attributed to the vital role of combined treatments in reducing soil pH, enhancing nutrient uptake, chlorophyll content, and photosynthetic rate which impacted growth and plant characteristics [24–26]. Sulphur has an important role in the formation of plant proteins and some hormones, as well as being necessary for enzymatic action, chlorophyll formation, and synthesis of certain amino acids and vitamins [19, 20]. Hence, sulphur is important for good vegetative growth leading to a high yield and increasing absorption of macro- and micronutrients. This is through its oxidation to sulphuric acid by soil microorganisms leading to the solubilization and availability of nutrients to plants [16, 27, 28]. Also, Table 3 showes the effect of soil amendments, that is, organic and sulphur on broad bean yield (seed and straw), yield components, as well crude as protein and carbohydrate (%). In general, the increase in (seed and straw) yield due to the addition of organic manure, which increase the availability of nutrients in soil during the decomposition process and produced $CO_2$, plays an important role in increasing phosphorus availability [2, 3].

Furthermore, the addition of organic manure significantly affected all studied characteristics of broad bean plants as being a source for all essential macro- and micronutrients. It plays a direct role for meliorating soil hydrophysical properties such as soil aggregation, bulk density, total porosity, aeration, hydraulic conductivity and available water range. Soil chemical characteristics such as soil pH, released organic

TABLE 3: Effect of different applied treatments on yield and its components of broad bean plants (combined data of two seasons).

| P-source and/or soil amendments | Total weight of green pods/plant (g) | Number of pods/plant | Dry weight of seed/plant (g) | 100-seeds weight (g) | Seed yield (kg/ha) | Straw yield (kg/ha) | Seed protein (%) | Seed constituents Total carbo-hydrates (%) | Seed protein (kg/ha) |
|---|---|---|---|---|---|---|---|---|---|
| SP | 388.89 | 29.67 | 183.93 | 172.96 | 5896.38 | 6373.33 | 21.13 | 51.47 | 1245.91 |
| RP | 326.16 | 25.33 | 140.90 | 147.51 | 4730.29 | 5370.67 | 20.00 | 45.49 | 946.05 |
| RP + ES | 372.13 | 28.67 | 170.22 | 170.28 | 5356.19 | 5852.57 | 20.88 | 49.16 | 1118.38 |
| RP + PDB | 365.42 | 27.33 | 148.53 | 151.45 | 5056.76 | 5688.76 | 20.50 | 46.25 | 1036.64 |
| RP + OM | 396.29 | 30.67 | 191.37 | 174.71 | 6484.57 | 7396.19 | 21.88 | 52.80 | 1418.83 |
| RP + PDB + ES | 418.22 | 31.67 | 200.51 | 175.95 | 7006.86 | 7675 | 21.56 | 51.47 | 1510.67 |
| RP + ES + OM | 460.65 | 34.33 | 215.00 | 179.38 | 8057.91 | 8565.05 | 22.56 | 55.80 | 1817.86 |
| RP + PDB + OM | 449.32 | 32.67 | 205.27 | 177.37 | 7528.38 | 7990.86 | 22.19 | 53.13 | 1670.55 |
| RP + PDB +OM + ES | 521.23 | 36.33 | 226.72 | 184.45 | 8372.48 | 9008.38 | 22.81 | 56.47 | 1909.76 |
| L.S.D at 5% | 20.84 | 3.70 | 5.26 | 3.44 | 156.38 | 137.62 | 0.23 | 3.76 | 15.05 |

SP: super phosphate; RP: rock phosphate; OM: organic manure; ES: elemental sulphur; PDB: phosphate dissolving bacteria.

TABLE 4: Effect of different applied treatments on NPK total content (kg/ha) of broad bean plants (combined data of two seasons).

| P-source and/or soil amendments | Nitrogen total content (kg/ha) | | | Phosphorous total content (kg/ha) | | | Potassium total content (kg/ha) | | |
|---|---|---|---|---|---|---|---|---|---|
| | Seed | Straw | Total | Seed | Straw | Total | Seed | Straw | Total |
| SP | 199.29 | 93.69 | 292.98 | 21.81 | 14.02 | 35.83 | 34.19 | 141.48 | 175.67 |
| RP | 151.38 | 75.19 | 226.57 | 9.93 | 7.52 | 17.45 | 23.64 | 114.93 | 138.57 |
| RP + ES | 178.9 | 84.86 | 263.76 | 18.74 | 12.29 | 31.02 | 29.45 | 128.76 | 158.21 |
| RP + PDB | 165.86 | 81.93 | 247.79 | 14.67 | 10.24 | 24.90 | 26.81 | 124.02 | 150.83 |
| RP + OM | 226.95 | 114.64 | 341.6 | 25.29 | 17 | 42.29 | 42.14 | 169.38 | 211.52 |
| RP + PDB + ES | 241.74 | 121.26 | 363 | 31.52 | 19.19 | 52 | 43.45 | 172.69 | 216.14 |
| RP + ES + OM | 290.88 | 140.93 | 431.81 | 39.48 | 23.07 | 62.55 | 54.79 | 200.71 | 255.5 |
| RP + PDB + OM | 267.26 | 129.45 | 396.71 | 35.38 | 20.79 | 56.17 | 52.69 | 194.98 | 247.67 |
| RP + PDB + OM + ES | 305.6 | 152.24 | 457.83 | 46.05 | 25.21 | 71.26 | 60.29 | 224.31 | 284.57 |
| L.S.D at 5% | 2.45 | 1.74 | 3.12 | 1.57 | 1.67 | 2.36 | 1.86 | 3.31 | 3.29 |

constituents of active groups such as fulvic and humic acids able to retain the essential plant nutrients in complex and available chelated forms are also impacted by the organic manure. Organic manure affects soil biological conditions. That is, a source of energy for the microorganism activities which enhance the release of necessary nutrients in available forms throughout their mineralization, in return improves soil fertility status (i.e., slow release of nutrients) which support root development among the different growth stages, and finally leads to higher yield and its content of broad bean plants. Similar results were gained previously [27, 29, 30].

### 3.2. Effect of Different Applied Treatments on NPK Contents of Broad Bean Plant.

The data presented in Table 4 shows broad bean seeds and straws differ significantly in their uptake of N, P, and K as a result of the applied treatments. The highest N, P, and K contents of broad bean plants were recorded from plants received soil amendments combined with RP compared with RP only. Phosphorus uptake was higher in plants fertilized with RP in combination with biofertilizer (PDB) or fertilized with organic manure as compared to those that received RP only. This is because since organic manure and PDB inoculation increased the efficiency of phosphorus released in low phosphate source, this is due to the release of organic acids, pH reduction, and/or dissolution the mineral phosphate through the anion exchange of $(PO4)-3$ or due to the chelating property of the organic acid produced by PDB such as acetate, lactate, oxalate, and citrate [31]. Moreover, application of P increases its concentration in the vicinity of plant roots and its availability in the soil solution as well as reduces its fixation by soil factors resulting from the introduction of most P requirements at preplanting in a limited zone, where root growth is highly concentrated.

Application of PDB increased the uptake of N and K for building new tissues [8]. The positive effect of P application on N contents in different parts of broad bean plants can be attributed to increase of the nodular number, size, and mass, which in turn increases $N_2$-fixation by bacteria [32].

Marschner (1986) described the enhancing effect of P on K uptake to the energy rich phosphates (in the form ATP) and the close relationship between K-uptake and the ATP-ase activity [33].

The maximum increase percentage was obtained by mixing soil amendments combined with RP, which resulted in increments of 53.35%, 111.14%, and 76.32%, respectively, in N, P and K-uptake of broad bean seeds and 62.49%, 79.80%, and 58.55%, respectively, in N, P and K-uptake of broad bean straw as compared with the control treatment (SP). Also, the data showed that a combination of two amendments was more effective than a single one, while a combination of all three had the most effect, for enhancing the nutrients taken up by seeds and straw of broad bean. Such effect of sulphur amendment on nutrients uptake may be due to its important role in reducing the soils' pH, oxidation to sulphuric acid by soil microorganisms. Subsequently, this results in solubilization and availability of nutrients to plants. Also, under those experimental conditions, it was concluded that application of organic manure to sandy loam soil increased the efficiency of P mineral from the RP fertilizers used. It improves the physical and chemical properties of sandy soil through its ability to adsorb nutrients on active groups or colloidal surfaces and increased the efficiency of nutrients uptake by plants reflected on growth and productivity. These results are in agreement with those of S. N. Sahu and B. B. Jana (2000) and Evans et al. (2006) [34, 35].

In conclusion, the addition of soil amendments and bio fertilizer to sand loamy soil improved the estimated parameters reflecting improved soil available nutrients, chemical, and physical soil properties. Applied soil amendments gave a significantly positive effect than the control. The combined treatments of OM and S and PDB gave highly significant increases in yield and nutrient contents of broad bean as compared to their individual application. The most effective treatment was RP and S and OM or RP and PDB and S and OM which achieved the highest yield parameters and N, P, and K contents of seeds and straw of broad bean as compared to the control.

Significance of Some Soil Amendments and Phosphate Dissolving Bacteria to Enhance the Availability of Phosphate in Calcareous Soil

157

# References

[1] R. W. Miller, R. L. Danhaue, and J. U. Miller, *An Introduction to Soil and Plant Growth*, Prentice Hall International, London, UK, 6th edition.

[2] P. Illmer and F. Schinner, "Solubilization of inorganic phosphates by microorganisms isolated from forest soils," *Soil Biology and Biochemistry*, vol. 24, no. 4, pp. 389–395, 1992.

[3] S.-T. Liu, L.-Y. Lee, C.-Y. Tai et al., "Cloning of an *Erwinia herbicola* gene necessary for gluconic acid production and enhanced mineral phosphate solubilization in *Escherichia coli* HB101: nucleotide sequence and probable involvement in biosynthesis of the coenzyme pyrroloquinoline quinone," *Journal of Bacteriology*, vol. 174, no. 18, pp. 5814–5819, 1992.

[4] M. Lotfollahi, M. J. Malakout, K. Khavazi, and H. Besharat, "Effect of different methods of direct application of rock phosphate on the yield of feed corn in Karaj region," *Journal of Soil and Water*, vol. 12, pp. 11–15, 2001.

[5] S. K. Subehia, "Direct and residual effect of Udaipur rock phosphate as a source of P to wheat soybean cropping System in a Western Himalayan soil," *Research on Crop*, vol. 2, pp. 297–300, 2001.

[6] V. Kumar, R. J. Gilkes, and M. D. A. Bolland, "The residual value of rock phosphate and superphosphate from field sites assessed by glasshouse bioassay using three plant species with different external P requirements," *Fertilizer Research*, vol. 32, no. 2, pp. 195–207, 1992.

[7] S. A. M. El-Sayed, "Influence of *Rhizobium* and phosphate solubilization bacteria on nutrient uptake and yield of lentil in New Valley," *Egyptian Journal of Soil Science*, vol. 39, pp. 175–186, 1999.

[8] K. Nassar, M. Y. Gebrail, and K. M. Khalil, "Efficiency of phosphate dissolving bacteria (PDB) combined with different forms and rates of P fertilization on the quantity and quality of Faba bean (*Vicia Faba L.*)," *Menoufya Journal of Agricultural Research*, vol. 25, pp. 1335–1349, 2000.

[9] A. Cottenie, M. Verloo, L. Kiekens, G. Velghe, and R. Camerlynck, "Chemical Analysis of Plants and Soils," Lab. Anal. Agrochem. Faculty of Agriculture, State University Gent, Gent, Belgium, 1982.

[10] A. L. Page, R. H. Miller, and D. R. Keeny, *Methods of Soil Analysis—part 2: Chemical and Microbiological Properties*, American Society of Agronomy, Madison, Wis, USA, 2nd edition, 1982.

[11] H. D. Chapman and P. F. Pratt, "Methods of Analysis for Soils, Plants and Waters," The University of California's Division of Agriculture Sciences, Davis, Calif, USA, 1961.

[12] E. Troug and K. H. Mayer, "Improvements in the Denige; chloromeric method for phosphorous and srsenic," *Industrial and Engineering Chemistry*, vol. 1, pp. 136–139, 1949.

[13] A.O.A.C., *Official Methods of Analysis*, Association of Official Agricultural Chemists, Washington, DC, USA, 15 edition, 1990.

[14] K. A. Gomez and A. A. Gomez, *Statistical Procedures for Agricultural Research*, John Wiley & Sons, New York, NY, USA, 1984.

[15] K. Mengel and E. A. Kirkby, "Principles of Plant Nutrition," International Potash Institute, Bern, Switzerland, 1987.

[16] H. Marschner, *Mineral Nutrition of Higher Plants*, Harcourt Brace and Company, London, UK, 1998.

[17] A. A. Abdul-Galil, E. M. El-Naggar, H. A. Awad, and T. S. Mokhtar, "Response of two faba bean cultivars to different N, P and K levels under sandy soil conditions," *Zagazig Journal of Agricultural Research*, vol. 5, pp. 1787–1808, 2003.

[18] F. A. Abdo, "Effect of bio-fertilizer with phosphate dissolving bacteria under different levels of phosphorus fertilization on mung bean plant," *Journal of Agricultural Research*, vol. 30, pp. 187–211, 2003.

[19] B. S. Kundu and A. C. Gaur, "Rice response to inoculation with N2-fixing and P-solubilizing microorganisms," *Plant and Soil*, vol. 79, no. 2, pp. 227–234, 1984.

[20] M. Monib, I. Hosny, and Y. B. Besada, "Seed inoculation of castor oil plant (*Ricinus communis*) and effect on nutrient uptake," *Soil Biology and Conservation of the Biosphere*, vol. 2, pp. 723–732, 1984.

[21] S. A. Azer, A. M. Awad, Sadek et al., "A comparative study on the effect of elements and bio-phosphatic fertilizers on the response of Faba bean (*Vicia Faba L.*) to P fertilization," *Egyptian Journal of Applied Science*, vol. 18, pp. 324–363, 2003.

[22] M. A. Ewais, "Response of vegetative growth, seed yield and quality of peanut grown on a sandy soil to application of organic manure, inoculation with rhizobium and phosphate dissolving bacteria," *Egyptian Journal of Applied Science*, vol. 21, pp. 794–816, 2006.

[23] E. M. Abd El Lateef, M. M. Selim, and T. G. Behairy, "Response of some oil crops to bio-fertilization with phosphate dissolving bacteria associated with different levels of phosphatic fertilization," *Bulletin of National Research Center (NRC)*, vol. 23, pp. 193–202, 1998.

[24] A. M. Hewedy, "Effect of sulphur application and biofertilizer phosphorein on growth and productivity of tomato," *Minufiya Journal of Agricultural Research*, vol. 24, pp. 1063–1078, 1999.

[25] H. A. El-Shamma, "Effect of chemical and bio-fertilizer on growth, seeds and quality of new cv. of dry bean," *Annals of Agricultural Science*, vol. 38, pp. 461–468, 2000.

[26] T. G. A. Ali, *Effect of some agriculture treatments on growth and dry seeds yield of bean [M. S. thesis]*, Faculty of Agriculture, Minia University, Minia Governorate, Egypt, 2002.

[27] F. S. Salem, "Effect of some soil amendment on the clayey soil properties and some crops production," *Menoufya Journal of Agricultural Research*, vol. 28, pp. 1705–1715, 2003.

[28] F. S. Salem, M. Y. Gebrail, M. O. Easa, and M. Abd El-Warth, "Raising the efficiency of nitrogen fertilization for wheat plants under salt affected soils by applying some soil amendments," *Menoufya Journal of Agricultural Research*, vol. 29, pp. 1059–1073, 2004.

[29] S. M. Abdel Aziz, F. S. Salem, M. A. Reda, and L. A. Hussien, "Influence of some amendments on the clayey soil properties and crop production," *Fayoum Journal of Agricultural Research and Development*, vol. 12, pp. 196–204, 1998.

[30] S. E. A. Mohamed and S. S. S. El-Ganaini, "Effect of organic, mineral and bio-fertilizers on growth, yield and chemical constituents as well as the anatomy of broad bean (*Vicia Faba L.*) in reclaimed soil," *Egyptian Journal of Applied Science*, vol. 18, pp. 38–63, 2003.

[31] A. Abd El Latif, M. Maged, A. Ewais, A. A. Mahmoud, and M. M. Hanna, "The response of maize to organic manure, bio-fertilizer and foliar spray with citric acid under sandy soil condition," *Egyptian Journal of Applied Science*, vol. 20, pp. 661–681, 2005.

[32] B. Y. El-Koumey, E. S. Abou Husien, and F. S. El-Shafie, "Influence of phosphatic fertilizers on NPK and some heavy metals in soil and plant," *Zagazig Journal of Agricultural Research*, vol. 20, pp. 2029–2044, 1993.

[33] H. Marschner, *Mineral Nutrition of Higher Plants*, Harcourt Brace Javanovich, London, UK, 1986.

[34] S. N. Sahu and B. B. Jana, "Enhancement of the fertilizer value of rock phosphate engineered through phosphate-solubilizing bacteria," *Ecological Engineering*, vol. 15, no. 1-2, pp. 27–39, 2000.

[35] J. Evans, L. McDonald, and A. Price, "Application of reactive phosphate rock and sulphur fertilisers to enhance the availability of soil phosphate in organic farming," *Nutrient Cycling in Agroecosystems*, vol. 75, no. 1–3, pp. 233–246, 2006.

# 15

# Passive and Active Restoration Strategies to Activate Soil Biogeochemical Nutrient Cycles in a Degraded Tropical Dry Land

**Manuel F. Restrepo, Claudia P. Florez, Nelson W. Osorio, and Juan D. León**

*Universidad Nacional de Colombia, Calle 59A No. 63-20, Oficina 14-225 050034, Medellín, Colombia*

Correspondence should be addressed to Juan D. León; jdleon@unal.edu.co

Academic Editors: J. A. Entry and D. Lin

The potential use of two restoration strategies to activate biogeochemical nutrient cycles in degraded soils in Colombia was studied. The active model was represented by forest plantations of neem (*Azadirachta indica*) (FPN), while the passive model by successional patches of native plant species was dominated by mosquero (*Croton leptostachyus*) (SPM). In the field plots fine-litter traps and litter-bags were established; samples of standing litter and surface soil samples (0–10 cm) were collected for chemical analyses during a year. The results indicated that the annual contributions of fine litterfall in FPN and SPM were 557.5 and 902.2 kg ha$^{-1}$, respectively. The annual constant of decomposition of fine litter ($k$) was 1.58 for neem and 3.40 for mosquero. Consequently, the annual real returns of organic material and carbon into the soil from the leaf litterfall decomposition were 146 and 36 kg ha$^{-1}$ yr$^{-1}$ for FPN and 462 and 111 kg ha$^{-1}$ yr$^{-1}$ for SPM, respectively. Although both strategies showed potential to activate soil biogeochemical cycles with respect to control sites (without vegetation), the superiority of the passive strategy to supply fine litter and improve soil properties was reflected in higher values of soil organic matter content and cation exchange capacity.

## 1. Introduction

Land degradation in arid and semiarid lands increases as a result of soil misuse or mismanagement, which, together with climatic variations, may promote desertification and reduces soil productivity [1, 2]. In Colombia, 78.9% of dry lands show some degree of desertification, mainly due to soil erosion by overgrazing and soil salinity [3]. Passive and active restoration strategies have been proposed to restore the functioning of ecological processes [4]. Passive restoration strategies imply minimal human intervention and are based on natural succession process, and in this way the restorer has a passive role regarding the process. On the other hand, active restoration strategies include planting trees at high density and their respective management [5]; this strategy implies a more active role of the restorer. Although passive restoration strategies are simple, inexpensive, and based on natural regeneration processes, they are not always successful [6, 7]. Alternatively, active restoration strategies accelerate the restoration of ecosystem functioning through the activation of soil biogeochemical cycling of nutrients and carbon sequestration [4].

The hypothesis of this study is that the activation of soil biogeochemical nutrient cycles and soil quality improvement of degraded dry land depend on the strategy of restoration (active and passive). Thus, the objective of this study was to evaluate the potential use of both active and passive strategies to restore soil biogeochemical nutrient cycles in fine litterfall and soil quality in tropical degraded dry lands by overgrazing. The active restoration strategy consisted of a plantation of neem (*Azadirachta indica*) established six years ago for restoration purposes in soils severely eroded. The passive restoration strategy consisted of six-year-old successional patches dominated by native species, where mosquero (*Croton leptostachyus*) is the most abundant plant species that grow in the same eroded soils. To this purpose, we characterized several processes related to fine litterfall dynamics that control the flow of organic matter and nutrients and evaluated some soil physic-chemical parameters.

TABLE 1: Mean values of structural parameters of the successional patches of mosquero (SPM) and forest plantations of neem (FPN) studied in Antioquia (Colombia); standard deviation in parentheses.

| Parameters | FPN | SPM |
|---|---|---|
| DBH (cm) | 3.55 (0.85) | 1.88 (0.97) |
| $D_q$ (cm) | 3.89 (0.92) | 2.91 (0.83) |
| $H$ (m) | 3.81 (0.46) | 1.84 (0.53) |
| $G$ (m$^2$ ha$^{-1}$) | 1.36 (0.76) | 0.76 (0.24) |

DBH: diameter at breast height (1.3 m), $D_q$: mean square diameter, $H$: height, $G$: stand basal area.

## 2. Material and Methods

### 2.1. Research Study Area.
This study was conducted in Santa Fe de Antioquia, northeastern Colombia (6°54'N, 75°81'W, 560 m of altitude). The annual average temperature, sunlight, precipitation, and evaporation of the region are 26.6°C, 2172 h, 1034 mm, and 1637 mm, respectively, characterized by a pronounced annual water deficit derived from the Precipitation/Evaporation ratio of 0.63. The landscape consists of hills with a low to medium slope formed from sediments from the Tertiary. The soils are alkaline and classified as Typic Ustorthents (USDA soil taxonomy); the most dominant soil use is grassland and unfortunately it is degraded by overgrazing. The neem plantations studied here (active strategy) were established in 2004 on hillsides severely eroded by overgrazing; since then, forestry management practices have not been carried out. In the middle of these plantations have grown natural successional patches (passive strategy), which are constituted by native plant species heavily dominated by mosquero.

### 2.2. Field and Laboratory Research Methods.
To evaluate the biogeochemical cycle processes we established 20 circular plots of 250-m$^2$ in forest plantations (FPN) and 13 similar plots in successional patches (SPM) (Table 1). In each plot three circular litter traps (fine netting of 0.5-m$^2$) were placed 1-m above the soil surface. Every 15 days for 1 year, we collected the litter material in each plot; it means 60 samples in FPN and 39 samples in SPM per sampling time.

The litterfall was separated in the following fractions: (a) neem leaves (NL), (b) mosquero leaves (ML), (c) leaves from other species (OL), (d) wood material (WM) from branches of <2-cm diameter and small bark pieces, (e) reproductive material (RM), and (f) other materials (OM). The divided material was oven dried at 65°C and weighed. The samples of NL and ML were then separately combined and homogenized for every two 15-day periods (1 month) and a subsample was taken for chemical analysis. At the end of the year, samples from the accumulated litterfall layer or standing litter layer on the soil of the plantations and successional patches (a 50 × 50 cm sample per plot) were collected. Each sample was separated in the same fractions than fine litterfall. Each fraction was dried at 65°C until reaching constant mass (72 h) and weighed. A composite and homogenized sample from each fraction was prepared for chemical analysis.

The decomposition of neem and mosquero leaf litter was studied by installing 18 litter-bags (20 × 20 cm, 2 mm pores) with 3 g of senescent dry leaves per plant species. Three litter-bags were randomly retrieved monthly. The residual leaf material was air-dried, cleaned with a brush to remove soil particles, dried at 65°C and weighed and named "residual dry matter" (RDM). The samples from the three collected litter-bags in plantations and successional patches were then separately combined and homogenized for chemical analysis. Additionally, in each plot of successional patches and forest plantations, surface soil samples (0–10 cm depth) were collected. Soil samples were also collected from 13 control sites where there was not vegetation; soil samples were transported to laboratory for physical and chemical analyses.

### 2.3. Physical and Chemical Analyses.
Leaf nutrient contents were analyzed by different methods: carbon (C) by the Walkley and Black method, nitrogen (N) by the Kjeldahl method, phosphorus (P) by the molybdate-blue method; calcium (Ca), magnesium (Mg), and potassium (K) by atomic absorption spectroscopy.

In soil samples the methods used were pH (1:2, water), total N (Nt) (Kjeldahl method), available P (Bray-Kurtz's method), exchangeable Ca, Mg, and K (1 M ammonium acetate; atomic absorption spectroscopy). We also determined soil bulk density (BD) and aggregate stability (AS) (Yoder method). Details about soil and plant analysis methods are available in Westerman [8].

### 2.4. Statistical Analysis and Calculations.
The rate of potential nutrient return (PNR) by the leaf litterfall was calculated as the product of the nutrient concentration and the dry mass of the leaf litterfall. The retention of nutrients in the standing litter (RNSL) was obtained by multiplying the dry weight of the leaves present in the standing litter and their respective concentration. We calculated nutrient release from the decomposition coefficient $k_j$ as proposed by Jenny et al. [9]: [$k_j$ = PNR/(PNR + RNS)]. The mean residence time (MRT) of litterfall and nutrients was calculated as the inverse of $k_j$. The real rate of nutrient return (RNR) from the leaf litterfall was calculated as the product of PNR and the corresponding $k_j$ coefficient [10].

The weight loss in the litter-bags was expressed by a simple exponential model [11]:

$$\frac{X_t}{X_0} = e^{-kt}, \tag{1}$$

where $X_t$ is the weight of the remaining material at moment $t$, $X_0$ is the weight of the initial dry material, $e$ is the basis of natural logarithm, and $k$ is the decomposition rate. The time required to obtain losses of 50% and 99% of the dry material was calculated as $t_{50} = -0.693/k$ and $t_{99} = -4.605/k$, respectively.

Regression models were fitted using nonlinear regression for the weight loss of leaf material deposited in the litter-bags. Linear coefficient of determination ($R^2$), the Durbin-Watson (D-W) coefficient, and the sum of the squares of the error were employed to select the models. Correlation analyses

TABLE 2: Mean values of fractions for fine litter production (FLP) (kg ha$^{-1}$ yr$^{-1}$) and standing litter (SL) (kg ha$^{-1}$) in successional patches (SP) and forest plantations (FP). Standard deviation is in parentheses. ML and NL: leaf litter of mosquero and neem in their respective ecosystem; OL: other leaves; RM: reproductive material; WM: woody material; OR: other rests unidentified.

| Fraction | FLP | | | | | SL | | |
| | (kg ha$^{-1}$ per two weeks) | | | (kg ha$^{-1}$ yr$^{-1}$) | | (kg ha$^{-1}$ yr$^{-1}$) | | |
| | SPM | FPN | $t$ value | SPM | FPN | SPM | FPN | $t$ value |
| --- | --- | --- | --- | --- | --- | --- | --- | --- |
| ML, NL | 15.4 (10.6) | 8.0 (6.2) | 2.85** | 477.9 | 184.8 | 238.9 | 73.1 | 4.87*** |
| OL | 3.1 (3.1) | 7.3 (3.3) | 4.43*** | 116.9 | 176.9 | 46.7 | 181.3 | 2.35 |
| RM | 8.7 (6.9) | 5.8 (6.0) | 1.07 | 227.1 | 135.8 | ND | ND | ND |
| WM | 1.7 (1.6) | 1.9 (0.9) | 0.74 | 52.7 | 48.1 | 64.3 | 111.8 | 0.17 |
| OR | 0.5 (1.0) | 0.5 (0.5) | 0.28 | 27.6 | 11.9 | 19.11 | 84.9 | 0.0002 |
| Total | 29.3 (18.3) | 23.4 (11.5) | 1.09 | 902.2 | 557.5 | 369.0 | 451.1 | 1.08 |

**, *** Denote significant differences between means at $P$ values $\leq 0.01$ and $\leq 0.001$, respectively ($t$-test).
ND: not determined.

TABLE 3: Fitted models for residual dry matter (RDM) as a function of time for neem and mosquero leaf litter.

| Plant species | Model | $t_{0.5}$ | $t_{0.99}$ | $k$ | $R^2$ | SSR | D-W |
| --- | --- | --- | --- | --- | --- | --- | --- |
| Mosquero (C. leptostachyus) | RDM = $1.29 * e^{(-0.00919934 * t)}$ | 0.21 | 1.37 | 3.36 | 0.97 | 0.06 | 1.31 |
| Neem (A. indica) | RDM = $3 * e^{(-0.00433362 * t)}$ | 0.44 | 2.91 | 1.58 | 0.84 | 0.01 | 0.18 |

$t_{0.5}$: decomposition time for half of the leaf litter, $t_{0.99}$: decomposition time for 99% of the leaf litter, $k$: yearly decomposition rate, $R^2$: coefficient of determination, SSR: sum of squared error, D-W: Durbin-Watson statistics.

(r-Pearson, $P \leq 0.05$) were also used to determine associations between weight loss rates and precipitation. Because litter data were normally distributed, a $t$-test was used to determine the differences in fine litter production, standing litter, and potential nutrient return between SPM and FPN. To compare soil parameters between control sites and both restoration strategies the Mann-Whitney (Wilcoxon) comparison test ($P \leq 0.05$) was employed since these data were not normally distributed. The analyses were performed with Statgraphics Centurion XV (StatPoint Technologies, Inc.).

## 3. Results and Discussion

*3.1. Fine Litter Production, Accumulation, and Decomposition.* The results of this study clearly showed that there were significant differences between the active and passive strategies characterized here to activate soil biogeochemical nutrient cycles in tropical dry lands. The annual production of leaf litter and total fine litter per unit area was 2.6-times and 1.6-times higher in SPM than in FPN (Table 2). The higher fine litter production found in SPM represents a greater potential return of organic matter to these degraded soils than that of FPN. The clear dominance of the leaf fraction in fine litter found in both types of ecosystems has been reported in other studies [12]. This also determines a potential source of nutrients for soil recovery because the higher decomposition rate of this fraction represents a faster nutrient return path [13]. By comparing the fine litterfall values with other studies, the FPN were lower than those of tropical lowland forest plantations with *ca.* of 5–10 Mg ha$^{-1}$ y$^{-1}$ [14–16]. The fine litterfall values found in the SPM coincided with other tropical dry successional forests (0.3 to 4.2 Mg ha$^{-1}$ y$^{-1}$) as reported by Descheemaeker et al. [17].

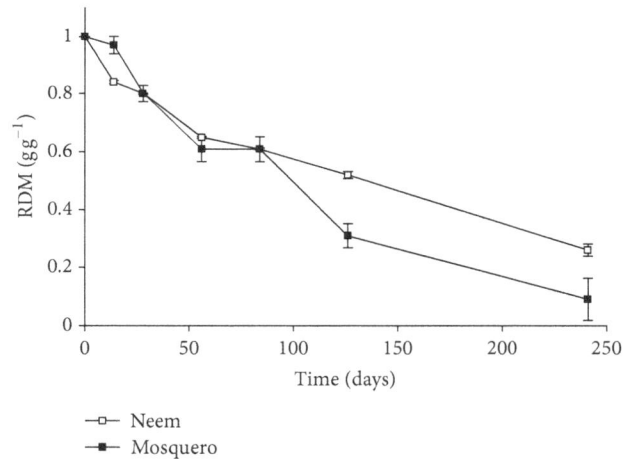

FIGURE 1: Residual dry matter (RDM) from leaf litter of neem (*A. indica*) and mosquero (*C. leptostachyus*). Each point represents the average of three litter-bags. The bars indicate the standard deviation.

In the results we saw a contradiction about which type of leaf litter decays faster. The constant of decomposition ($k$) showed that mosquero leaves are decomposed faster than neem leaves. In fact, it was noteworthy that the $k$ constant of mosquero (3.4) was more than twice that of neem (1.6), and the models predicted that the estimated time for 99% decomposition of leaf litter is 1.4 yr and 2.9 yr, respectively (Table 3). Furthermore, at the end of the litter-bags experiment, the RDM was 9% and 26% for mosquero and neem, respectively (Figure 1). However, if we consider the changes in the participation of neem and mosquero leaves collected in the litterfall traps (32 and 53%, resp.) with respect

TABLE 4: Mean monthly values (±SD) for leaf litter nutrient concentration (LLC, %) and potential nutrient return (PNR, kg ha$^{-1}$) via leaf litter in successional patches of mosquero (SPM) and forest plantations of neem (FPN) studied. Coefficient of variation (%) in parentheses.

| Nutrient[a] | LLC (%) | | PNR (kg ha$^{-1}$) | | $t$ value | $e$-value | $n$ |
|---|---|---|---|---|---|---|---|
| | SPM | FPN | SPM | FPN | | | |
| N | 1.08 ± 0.13 (11.9) | 1.29 ± 0.31 (24.2) | 0.44 ± 0.35 (78.9) | 0.20 ± 0.18 (114.4) | 2.105 | 0.047* | 12 |
| P | 0.05 ± 0.01 (15.6) | 0.03 ± 0.01 (19.0) | 0.02 ± 0.02 (79.2) | 0.005 ± 0.004 (86.7) | 3.052 | 0.006** | 12 |
| Ca | 1.75 ± 0.33 (18.9) | 2.16 ± 0.65 (30.1) | 0.76 ± 0.60 (79.7) | 0.39 ± 0.44 (82.0) | 1.721 | 0.099[NS] | 12 |
| Mg | 0.59 ± 0.08 (13.9) | 0.46 ± 0.06 (13.9) | 0.25 ± 0.21 (83.4) | 0.07 ± 0.06 (64.5) | 2.829 | 0.010** | 12 |
| K | 0.28 ± 0.13 (44.9) | 0.29 ± 0.14 (49.3) | 0.09 ± 0.08 (80.2) | 0.04 ± 0.03 (89.0) | 2.226 | 0.036* | 12 |

[a]Analytical methods in Westerman [8]. *, **Denote significant differences between means of PNR at $P$ values ≤ 0.05 and ≤ 0.01, respectively ($t$-test).

to their participation in the standing litter (SL) (16 and 64%, Table 2), this suggests that neem leaves decomposed faster. It is necessary to consider that the neem leaves are fragile and can be easily fragmented, and for this reason they can be part of the OR fraction (plant remains unidentified). Thus, the OR fraction (OR/total) was only 2% in the fine litter (FLP) collected in the traps, while this was 19% in the standing litter (SL). Consequently, we accepted that the $k$ is a better indicator and mosquero leaves decay faster. In the litter-bags experiment the RDM of mosquero was only 9% which is lower than that of neem 26%. Also, in the field the SL of FPN is higher suggesting that the decomposition of neem leaves is slower and for that reason they tend to accumulate on the soil surface. It has been reported that forest plantations with exotic species usually generate significant accumulations of litter on the ground [18–20]. Despite this accumulation, the values found in the SL are relatively low in comparison to those reported by other authors [10, 20, 21]. Perhaps, this is a result of the young age of these forest plantations, lack of canopy closure, and low fine litterfall. The $k$ values of mosquero litter are comparable to those reported in tropical arid ecosystems by several authors [15, 22, 23].

The high rate of decomposition of organic debris in SPM and therefore their low residence time are aspects of special significance to the reactivation of biogeochemical nutrient cycle in these degraded soils [24]. The inverse relationship between the RDM in the litter-bags and the precipitation in SPM (r-Pearson = −0.59, $P < 0.05$) indicates the favorable effect of the precipitation as a source of moisture for the decomposer microorganisms; perhaps, this was not observed in the FPN because of the limitations of degrader microbes to decompose nonnative leaf material.

3.2. Return, Accumulation, and Release of Nutrients. In both SPM and FPN, nutrient concentrations in the leaf litter (LLC) followed the sequence Ca > N > Mg > K > P (Table 4), with the highest temporal variability for K. The high concentrations of Ca and Mg in the leaf litter found in this study (1.75 and 2.16%) differ from those found in other tropical dry lowland forests [23], likely due to the high soil availability of both nutrients (Table 6). By contrast, the low concentration of K in the leaf litter (0.28–0.29%) is near the lowest end of the pantropical interval (0.27 ± 0.11%) [25] despite the availability of this nutrient in the soil. Likely, the low soil Ca/Mg ratio <1 caused an abnormally high Mg plant uptake and altered the K uptake [26]. In both ecosystems, the most restrictive nutrient was P likely as a result of its extreme scarcity in the soil (Table 6), which was reflected in the high values of the leaf litter N/P ratio (FPN: 43, SPM: 20) [10]. These values of the N/P ratio are much higher than the critical value of 11.9 suggested by [27].

The potential nutrient return (PNR) through leaf litter followed the same sequence of concentrations (Table 4) and was significantly higher in SPM because of the higher leaf litter production than in the FPN. Furthermore, the retention of nutrients in the standing litter (RNSL) was higher in SPM, which represents an important source of energy for the micro- and mesofauna, whose participation is a key functional aspect to reactivate the biogeochemical cycling in these degraded soils [28].

The real nutrient return (RNR) via leaf litter was higher in SPM for all nutrients (Table 5). This situation was primarily determined by the higher production of this fraction (ML), despite the fact that in FPN the $K_j$ for some elements (C, P, and K) was higher or similar (N) to those obtained in SPM. The low RNR of P also reflected the restrictive nature of this nutrient for the productivity of both ecosystems [10].

In terms of the return and incorporation of organic matter and C into the soil by leaf decomposition in the SL, the superiority of SPM was notorious (Table 5). Thus, the real return of C (ML) in SPM was more than twice higher than in the FPN (NL). In fact, the highest rate of leaf litter decomposition of mosquero (higher values $K_j$ and $k$) coincided with the highest contents of soil organic matter in SPM (Table 6). The times needed to achieve a decomposition degree of 99% of leaves (1.4 yr for mosquero and 2.9 yr for neem) were close to those obtained from the inverse $K_j$ in the leaves of the SL (1.5 and 2.0 years, resp.).

TABLE 5: Indexes calculated for return, retention, and release of nutrients via leaf litter in successional patches of mosquero (SPM) and forest plantations of neem (FPN) ($kg\,ha^{-1}\,yr^{-1}$).

| Indexes | SPM | | | | | | FPN | | | | | |
|---|---|---|---|---|---|---|---|---|---|---|---|---|
| | C | P | Ca | Mg | K | N | C | P | Ca | Mg | K | N |
| PNR | 114.4 | 0.22 | 8.4 | 2.8 | 1.3 | 5.2 | 45.9 | 0.06 | 4.6 | 0.9 | 0.5 | 2.4 |
| RNSL | 67.9 | 0.18 | 1.6 | 0.2 | 0.9 | 3.9 | 18.0 | 0.03 | 1.5 | 0.3 | 0.1 | 1.8 |
| $k_j$ | 0.6 | 0.55 | 0.8 | 0.9 | 0.6 | 0.57 | 0.7 | 0.68 | 0.7 | 0.7 | 0.9 | 0.6 |
| MRT | 1.6 | 1.81 | 1.2 | 1.1 | 1.6 | 1.76 | 1.4 | 1.47 | 1.3 | 1.3 | 1.1 | 1.8 |
| RNR | 71.8 | 0.12 | 7.04 | 2.6 | 0.8 | 2.94 | 33.0 | 0.04 | 3.5 | 0.6 | 0.5 | 1.4 |

PNR: potential nutrient return rate ($kg\,ha^{-1}\,yr^{-1}$), RNSL: retention of nutrients in the standing litter ($kg\,ha^{-1}\,yr^{-1}$), $k_j$: decomposition coefficient [$k_j$ = PNR/(PNR + RNS)], MRT: mean residence time [MRT = $1/k_j$], RNR: real nutrient return rate [RNR = PNR $*$ $k_j$].

TABLE 6: Mean values (±SD) for some soil parameters (0–10 cm) in successional patches of mosquero (SPM), forest plantations of neem (FPN), and control sites without vegetation studied in Santafe de Antioquia (Colombia).

| Parameter[a] | Control sites | SPM | PCI | FPN | PCI |
|---|---|---|---|---|---|
| pH | 6.3 ± 0.6 | 6.3 ± 0.3 | 0.99 | 6.4 ± 0.4 | 1.01 |
| SOM (%) | 2.0 ± 0.9 | 4.2 ± 0.4* | 2.19 | 3.4 ± 1.3* | 1.72 |
| Nt (%) | 0.21 ± 0.09 | 0.25 ± 0.06 | 1.19 | 0.27 ± 0.04* | 1.27 |
| P ($mg\,kg^{-1}$) | 3.3 ± 0.9 | 1.8 ± 0.5* | 0.53 | 4.3 ± 1.3* | 1.31 |
| Ca ($cmol_c\,kg^{-1}$) | 6.2 ± 2.5 | 11.6 ± 3.8* | 1.87 | 7.1 ± 3.6 | 1.15 |
| Mg ($cmol_c\,kg^{-1}$) | 6.7 ± 2.8 | 14.1 ± 3.4* | 2.11 | 7.3 ± 2.9 | 1.10 |
| K ($cmol_c\,kg^{-1}$) | 0.23 ± 0.05 | 0.25 ± 0.07 | 1.14 | 0.36 ± 0.14* | 1.61 |
| ECEC ($cmol_c\,kg^{-1}$) | 13.0 ± 4.6 | 25.9 ± 6.5* | 1.98 | 14.8 ± 6.0 | 1.14 |
| BD ($Mg\,m^{-3}$) | 1.35 ± 0.11 | 1.25 ± 0.15* | 0.93 | 1.25 ± 0.09* | 0.93 |
| AE (%) | 72.8 ± 10.6 | 68.5 ± 12.9 | 0.94 | 80.1 ± 9.5 | 1.10 |

[a]Analytical methods available in Westerman [8].
PCI: parameter change index (FPN/control or SPM/control), SOM: soil organic matter, ECEC: effective cation exchange capacity, BD: bulk density, AS: aggregate stability. *Indicates significant difference with control sites (Mann-Whitney, $P \leq 0.05$).

N and P were released more slowly (lower values of $k_j$) than other nutrients and they are expected to remain longer in the above ground leaf litter, as indicated by MRT values (Table 5). P was released faster in FPN ($K_j$ = 0.68, MRT = 1.47 years), while N was released at similar rates in both ecosystems ($K_j$ = 0.57, MRT = 1.76 years). In both ecosystems, Ca had the highest release. The time necessary for the effective release of all elements considered in both ecosystems was 1.1–1.81 years.

*3.3. Soil Reclamation.* Soils of both SPM and FPN showed changes of some properties with respect to soil of control sites (without vegetation) (Table 6). In SPM significant increases were detected with respect to the control sites on parameters such as soil organic matter the content (SOM), exchangeable Mg and Ca, and effective cation exchange capacity (ECEC). On the other hand, in FPN were observed significant increases in SOM, total N (Nt), available-P, and exchangeable-K and significant reduction in bulk density (BD).

Despite their short period of time for both strategies, the contributions of fine litter and its decomposition have improved various soil properties of these degraded lands. The sharp increases of SOM observed (compared to control sites) also increased soil moisture retention capacity and soil cation exchange, key aspects in the reclamation of soils of degraded dry land. Although FPN showed a significant increase of P,

its very low concentration in the soil determined a severe constraint on ecosystem primary productivity.

## 4. Conclusions

From the perspective of land restoration, both models showed different advantages. The passive model represented by the SPM showed a higher dynamics in the reactivation of soil biogeochemical cycles. It is expected that as the successional process continues the consequently greater complexity of the ecosystem will lead to an effective improvement not only on the soil, but also on ecosystem functions. On the other hand, the active model represented by the FPN showed significant improvements in soil parameters, even though the returns of litter and nutrients were lower. Likely, this situation is the result of differences in litter contributions, whose potential effect on soil rehabilitation has not been fully evaluated. These are issues to consider in selecting a restoration model and the degree and speed expected of the degradation process. Thus, an active model should be considered when the rate of degradation of the area of interest is high, because the planted species can be established quickly and create better conditions for a more diverse biological community as pointed by [5]. When the state and rate of degradation are not severe, the most appropriate model might be the passive restoration, allowing the ecosystem a natural recover [29], which had advantages from ecological and economic perspectives.

## Acknowledgments

The authors thank the Direction of Research of the Universidad Nacional de Colombia for financial support of the Project "Restoration of lands in a process of desertification with neem plantations (*Azadirachta indica*) in Western Antioquia." Juan D. León was supported by Convocatoria Nacional de Investigación y de Creación Artística de la Universidad Nacional de Colombia 2010–2012. They also thank the Biogeochemistry Laboratory of the Universidad Nacional de Colombia at Medellin campus. The authors are grateful to A. N. Marín, L. F. Osorio, J. C. Guingue, G. E. Mazo, and N. Alvarez for their technical collaboration.

## References

[1] Y. Zha and J. Gao, "Characteristics of desertification and its rehabilitation in China," *Journal of Arid Environments*, vol. 37, no. 3, pp. 419–432, 1997.

[2] J. F. Reynolds and D. M. Stafford Smith, *Global Desertification: Do Humans Cause Deserts? Vol. 88*, University Press, Berlin, Germany, 2002.

[3] *Plan de Acción Nacional de Lucha Contra la Desertificación y la Sequía en Colombia (PAN)*, Ministerio de Ambiente, Vivienda y Desarrollo Territorial, Bogotá, Colombia, 2004.

[4] D. Celentano, R. A. Zahawi, B. Finegan, R. Ostertag, R. J. Cole, and K. D. Holl, "Litterfall dynamics under different tropical forest restoration strategies in Costa Rica," *Biotropica*, vol. 43, no. 3, pp. 279–287, 2011.

[5] S. D. Reay and D. A. Norton, "Assessing the success of restoration plantings in a temperate New Zealand forest," *Restoration Ecology*, vol. 7, no. 3, pp. 298–308, 1999.

[6] K. D. Holl, "Tropical moist forest restoration," in *Handbook of Ecological Restoration*, M. R. Perrow and A. J. Davy, Eds., pp. 539–558, Cambridge University Press, Cambridge, UK, 2002.

[7] J. Schrautzer, A. Rinker, K. Jensen, F. Muller, P. Schwartze, and C. Dier Ben, "Succession and restoration of drained fens: perspectives from northwestern Europe," in *Linking Restoration and Ecological Succession*, L. R. Walker, J. Walker, and R. J. Hobbs, Eds., pp. 90–120, Springer, New York, NY, USA, 2007.

[8] R. L. Westerman, *Soil Testing and Plant Analysis*, Soil Science Society of America, Madison, Wis, USA, 1990.

[9] H. Jenny, S. Gessel, and F. Bingham, "Comparative study of decomposition of organic matter in temperate and tropical regions," *Soil Science*, vol. 68, pp. 419–432, 1949.

[10] J. D. León, M. I. González, and J. F. Gallardo, "Ciclos biogeoquímicos en bosques naturales y plantaciones de coníferas en ecosistemas de alta montaña de Colombia," *Revista Biología Tropical*, vol. 59, pp. 1883–1894, 2011.

[11] J. Olson, "Energy storage and balance of producers and decomposer in ecological systems," *Ecology*, vol. 44, pp. 322–331, 1963.

[12] V. Meentemeyer, E. O. Box, and R. Thompson, "World patterns and amounts of terrestrial plant litter production," *Bioscience*, vol. 32, pp. 125–128, 1982.

[13] C. Strojan, F. Turner, and R. Castetter, "Litter fall from shrubs in the northern Mojave desert," *Ecology*, vol. 60, pp. 891–900, 1979.

[14] J. A. Parrotta, "Productivity, nutrient cycling, and succession in single- and mixed-species plantations of Casuarina equisetifolia, Eucalyptus robusta, and Leucaena leucocephala in Puerto Rico," *Forest Ecology and Management*, vol. 124, no. 1, pp. 45–77, 1999.

[15] J. Goma-Tchimbakala and F. Bernhard-Reversat, "Comparison of litter dynamics in three plantations of an indigenous timber-tree species (*Terminalia superba*) and a natural tropical forest in Mayombe, Congo," *Forest Ecology and Management*, vol. 229, no. 1–3, pp. 304–313, 2006.

[16] J. Barlow, T. A. Gardner, L. V. Ferreira, and C. A. Peres, "Litter fall and decomposition in primary, secondary and plantation forests in the Brazilian Amazon," *Forest Ecology and Management*, vol. 247, no. 1–3, pp. 91–97, 2007.

[17] K. Descheemaeker, B. Muys, J. Nyssen et al., "Litter production and organic matter accumulation in exclosures of the Tigray highlands, Ethiopia," *Forest Ecology and Management*, vol. 233, no. 1, pp. 21–35, 2006.

[18] J. Sawyer, *Plantations in the Tropics: Environmental Concerns*, IUCN, Gland, Switzerland, 1993.

[19] A. E. Lugo, "The apparent paradox of reestablishing species richness on degraded lands with tree monocultures," *Forest Ecology and Management*, vol. 99, no. 1-2, pp. 9–19, 1997.

[20] J. F. Dames, M. C. Scholes, and C. J. Straker, "Litter production and accumulation in Pinus patula plantations of the Mpumalanga Province, South Africa," *Plant and Soil*, vol. 203, no. 2, pp. 183–190, 1998.

[21] J. F. Dames, M. C. Scholes, and C. J. Straker, "Nutrient cycling in a Pinus patula plantation in the Mpumalanga Province, South Africa," *Applied Soil Ecology*, vol. 20, no. 3, pp. 211–226, 2002.

[22] S. E. Attignon, D. Weibel, T. Lachat, B. Sinsin, P. Nagel, and R. Peveling, "Leaf litter breakdown in natural and plantation forests of the Lama forest reserve in Benin," *Applied Soil Ecology*, vol. 27, no. 2, pp. 109–124, 2004.

[23] A. N. Singh, A. S. Raghubanshi, and J. S. Singh, "Comparative performance and restoration potential of two Albizia species planted on mine spoil in a dry tropical region, India," *Ecological Engineering*, vol. 22, no. 2, pp. 123–140, 2004.

[24] D. L. Moorhead and R. L. Sinsabaugh, "A theoretical model of litter decay and microbial interaction," *Ecological Monographs*, vol. 76, no. 2, pp. 151–174, 2006.

[25] J. M. Duivenvoorden and J. F. Lips, *A Land-Ecological Study of Soils, Vegetation, and Plant Diversity in Colombian Amazonia*, Tropenbos, Series 12, The Tropenbos Foundation, Wageningen, The Netherlands, 1995.

[26] H. Marschner, *Mineral Nutrition of Higher Plants*, Academic Press, London, UK, 1995.

[27] R. Aerts, "Climate, leaf litter chemistry and leaf litter decomposition in terrestrial ecosystems: a triangular relationship," *Oikos*, vol. 79, no. 3, pp. 439–449, 1997.

[28] F. J. Stevenson, *Cycles of Soil*, John Wiley & Sons, New York, NY, USA, 1986.

[29] D. Lamb and D. Gilmour, *Issues in Forest Conservation. Rehabilitation and Restoration of Degraded Forests*, International Union for Conservation of Nature and Natural Resources and World Wide Fund, Cambridge, UK, 2003.

# 16

# Application of Digital Image Cross Correlation to Study Sinkhole Collapse

**Mahmoud Ahmed**

*Department of Transportation, Route 9A Project, Suite 1701, 115 Broadway, New York, NY 10006, USA*

Correspondence should be addressed to Mahmoud Ahmed; mahmoud.ahmed@dot.ny.gov

Academic Editors: J. A. Entry, D. Hui, D. Lin, and W. R. Roy

This paper presents the results of a study using a transparent soil experimental technique and numerical modeling to detect 3D deformations resulting from submerged cavities that lead to a sinkhole. Excessive deformations from underground activity beneath highway pavements could lead to sinkhole collapse. The formation of a sinkhole is often sudden and can lead to extensive damage and loss of life, especially in urban areas. The use of transparent soils permitted the visualization of internal ground deformations which allowed for comprehensive evaluation of the extension of failure. A series of finite element analyses have also been carried out for the tests conditions. The observed sinkhole, at the surface, is found to be a small indicator of the final size and magnitude of the internal deformations as a subsequent funnel-shaped depression developed with a hole at the center. The modeling results emphasized the need to extend the repair zone following sinkhole collapse by a minimum distance that equals twice the cavity diameter away and ahead of the developed hole. Results of this study are believed to be of practical interest for predicting surface and internal ground deformations following sinkhole collapse which could be useful for the stability assessment of underground utilities and the development of a restoration plan after collapse occurred. The results also provided approximate bounds to areas affected by the sinkhole allowing for collapse risk to be assessed.

## 1. Introduction

Sinkholes are depressions that develop at the soil surface due to stress release in the soil beneath. They could develop naturally like in karst terrains, due to dissolution of limestone and the subsequent subsidence above the depression, or as a result of human underground activities such as horizontal directional drilling (HDD) or tunneling operation. They could also be formed due to leaking sewer or culvert by washing away surrounding soils. Risk of sinkhole collapse will continue to increase due to increased urbanization which intensifies the need to utilize underground space to accommodate large utility systems, subways, and highways, thereby decreasing traffic congestion and allowing preservation of aboveground space.

A sinkhole may develop on a roadway when an opening is formed, either beneath or within the subgrade layer, followed by the flowing of layers above into this opening. Such sinkholes require immediate and costly repairs to avoid further traffic delays and closures. The current maintenance approach to this problem is to quickly remove and/or repave damaged pavement. This approach may temporarily fix the surface problem. However, this may accelerate the underlying problem by increasing loads above the sinkholes and does not properly fix the foundation soils. Additionally, the problem is only fixed in areas that have currently failed, but nearby areas may still be on the verge of failure. If these pending failures can be detected, a more extensive and economical repair solution may be obtained and implemented. Nowadays, most of transportation agencies would not permit HDD or tunneling operation without proper assessment of the opening stability. Therefore, prediction of sinkhole collapse is important to insure the safe construction and protection of nearby utilities. However, the extension of the developed sinkhole and the occurrence of soil flow into the opening is complex phenomena and have yet to be well-understood. The relationship between ground movement, soil properties, and geometry of the opening is neither simple nor linear which made it difficult to achieve a comprehensive theoretical solution. Majority of previous studies on the subject of sinkhole formation are empirical in nature and usually associated with a specific ground condition. Other types of studies include analytical solutions, laboratory scale models, and numerical models. Major limitation in some of these methods widely used is that they neglect the possible soil-structure interactions. The objective of this research is to predict 3D deformations for a sinkhole collapse and develop a relationship between the geometry of the cavity and final deformations.

Sinkhole collapse is generally an instantaneous event which makes it difficult to get field measurement during the collapsing processes. Therefore, small-scale or full-scale experimental modeling may be a good practice to predict the size and shape of the sinkhole and to collect information about influential factors and triggering mechanism of a specific collapsing event. Full-scale experiments are expensive and difficult. Small-scale experiments have been reported by some authors such as Craig [1] and Abdulla and Goodings [2]. These authors focused on the stability of soils over cavities. Atkinson and Potts [3], Davis et al. [4], Mühlhaus [5], and Leca and Dormieux [6] used the underground cavity conditions in centrifuge testing to demonstrate the way in which soil around a circular cavity deforms as the overburden pressure increased and/or in situ stress released. The ultimate goal of these authors was to study the stability of lined and unlined tunnels by using circular cavity deformations. A similar approach has been adopted in this study by using transparent soil modeling technology and associated image processing techniques to investigate the aspects of sinkhole formation. A cylindrical cavity was preplaced inside a transparent soil model, which represents sand. The face of the cylinder was simulated using an internal pressure ($\sigma_T$) equal to the effective stress at the center of the cavity and applied inside the cylinder. Tests were conducted by reducing $\sigma_T$ in stages until collapse of the soil occurred when the transparent soil caved in or flowed inside the cavity. The transparent soil permits the use of simple optical techniques to visualize 3D deformation patterns within the soil mass by slicing the model using a laser light sheet. In this study, the model was sliced parallel to and perpendicular to the cylinder axis in order to obtain 3D deformation patterns. Images of the soil, illuminated by a laser light sheet, were captured after each decrement of $\sigma_T$ and used to obtain the corresponding deformation fields at the precollapse and collapse stages. The technique might not model the exact construction sequencing of HDD or failure due to leaking sewer. Nevertheless, it demonstrates the way in which soil around a circular cavity deforms due to in-situ stress release which ultimately leads to sinkhole collapse. Transparent soil modeling has been used by the author before to study ground movements associated with tunneling [7] and tunnel face stability [8].

## 2. Development of Transparent Soil Model

Sinkhole failures usually result from reduction of in-situ stress at the face of excavated subsurface cavity. This process can be experimentally simulated by a contracting cylindrical cavity from the in-situ stress state to collapse. A simplified

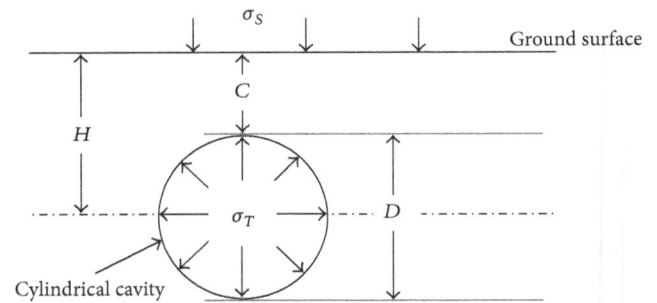

FIGURE 1: Simplified layout of collapsing cavity problem.

layout of the problem implemented in this study is shown in Figure 1. The submerged cavity is assumed to be cylindrical of diameter $D$ and cover $C$. The cavity was subject to an internal normal stress represented by internal pressure $\sigma_T$. The ground surface was horizontal and subject to a vertical surcharge $\sigma_S$. Plane strain was assumed.

*2.1. Transparent Synthetic Soil.* The transparent soil was made by matching the refractive indices of silica gel and pore fluids. For the preparation of the models, silica gel was immersed in the pore fluid. At the same time, the mix was stirred to release gas bubbles entrapped during the pouring of silica gel. Saturation of the particles was accomplished by soaking and mixing the particles in pore fluid under vacuum for several hours. Vacuum was applied to deair the mix until the mixture turned to transparent. Time for vacuum depends on the depth of silica gel and the power of the vacuum pump. Approximately 24 hours was required to fully saturate the silica gel. One advantage of transparent soil is that saturation can be readily viewed when the particles become transparent. The clarity of transparent soil depends on the perfect matching of the refractive indices of silica and pore fluid which was made by a blend of Drakeol35 mineral oil and Norpar 12 paraffinic solvent. The blend is 1:1 by weight. The refractive index, viscosity, and density of the oil blend at room temperature (24°C) were 1.447, 5.0 cp, and 800 kg/m$^3$, respectively. This mix resulted in a material which was completely transparent, thereby permitting the use of simple optical techniques to visualize 3D deformation patterns within the soil mass. Silica gel materials used in this study are produced by Multisorb Technologies located in Buffalo, New York, with particle size 0.5–1.5 mm. The specific gravity of silica gel is 2.2 [9], which is approximately 80% of the specific gravity of natural silicate sands. The minimum density of silica gel determined in accordance to the American Society for Testing and Materials (ASTM) standard D4254 was $\gamma_{min}$ = 7 kN/m$^3$. Because of the porous structure of the silica gel, ASTM standard D698 was not applicable. A maximum density $\gamma_{max}$ = 8 kN/m$^3$ was obtained using a vibrator compaction. The saturated unit weight depends on the pore fluid used, and it was 11–14 kN/m$^3$, for the mineral oil and solvent pore fluid used in this research. The volume change of the silica was consistent with that of natural sands. The hydraulic conductivity ranged from $1 * 10^{-2}$ to $1 * 10^{-4}$ cm/sec (12–45 Darcys), depending

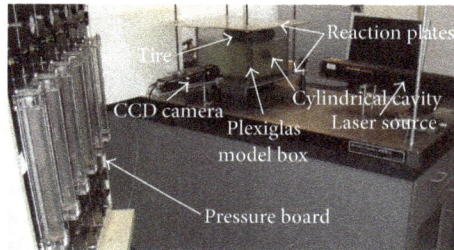

FIGURE 2: Test setup.

on grain size. The angel of friction is 30°–36°, and Young's modulus is 24–84 MPa depending on density and size. These values were similar to the values reported for the natural sands. The average stress strain behavior of silica gel is consistent with typical stress strain behavior of sand for both dense and loose conditions [10].

*2.2. Experimental Modeling Technique.* A plexiglas model (Figure 2) 30.50 cm long, 25.40 cm wide, and 20.30 cm high was used to contain the transparent soil. The model dimensions have been chosen based on parametric analysis (cavity size and depth), performed by empirical formulae, in such a way that the influence of the boundaries was minimized. Transparent soil was placed into the plexiglas model box at approximately 25 mm lifts. The box was shaken by hand while being packed to a density $\gamma = 7.5$ kN/m$^3$. The contracting cylindrical cavity was modeled by a PVC tube of 2.50 cm diameter and 25.40 cm long preinstalled inside the model. A latex membrane (0.3 mm thick), of negligible strength, was attached to the end of the cylinder to represent the contracting cavity. The membrane was left slack to prevent mechanical influence on the displacement of the face. The cylinder was then filled with water under pressure ($\sigma_T$) which can be read and controlled by a pressure board. A surcharge was needed to counter the effects of the small size model and the low-unit weight of the transparent soil. The surcharge may affect the scaling of stress-based problems such as scaling of liner deformations or soil reinforcement, to natural soils. For the deformational problems such as the one presented here, it is believed that the role of geometry is more significant than surcharge, but small differences in failure geometry may occur due to the presence of surcharge. For application of surcharge or surface pressure $\sigma_S$, the plexiglas model container was placed between two identical metal plates (Figure 2) connected by four threaded rods. A rubber tire with internal pressure $\sigma_S$ was placed on top of the transparent soil and connected to the pressure board. The cylinder pressure $\sigma_T$ was increased such that $\sigma_T = \sigma_S$ at the beginning of the test. In addition to the tunnel container, the setup also included a Cohu 2622 black and white Charged Coupled Device (CCD) camera, 35 mW Melles Griot laser light source, a line generator lens, a loading frame, a test table, and a PC for image processing (Figure 2). The camera has a resolution of 640 × 480 pixels and is controlled by the PC through a Matrox Meteor 2/4 frame grabber. A macrozoom lens with a variable focus length from 18 to 108 mm was mounted on the CCD camera.

The tests were conducted by reducing the cylinder pressure $\sigma_T$ in stages until collapse occurred. At some point, the cavity becomes sufficiently large and the remaining overburden is no longer able to arch across the cavity and collapsed. This process was found to be similar to in-situ cavity collapsing event. After each decrement of pressure, the model was sliced optically using laser light sheet to illuminate plane of measurements inside the model and an image was taken by CDD camera. Later, these images were processed to obtain corresponding deformations relative to pressure drop and volume loss in the soil mass induced by the cavity contraction. Complete strain and deformation fields were obtained from the set of images taken during the test.

*2.3. Deformation Measurement Using Digital Image Processing.* The measurement technique described in this paper operates by processing digital images, which can be captured directly from digital camera. The interaction between laser light and transparent soil produces a distinctive specklepattern. This speckle pattern manifests the interaction between the transparent soil matrix, impurities, entrapped air, and the laser. Small particle movement will result in a change in the speckle distribution in the plane of measurement. If the deformation is small, the contrast distribution resulting from the speckle effect will follow the particle movement. Images captured before and after deformations are analyzed using digital image correlation (DIC). The captured images are divided into a large number of interrogation areas or windows. It is then possible to calculate a displacement vector for each window with the help of DIC techniques. DIC has proven to be a valuable flow and deformation measurement tool that provides enhanced measurement capabilities not possible with other techniques and has been widely used for experimental measurements in solid mechanics and in fluid mechanics [11–13].

An advanced form of DIC which employs window shifting and window sizing, called adaptive cross correlation (ACC) has been adapted for this study. Comparison between conventional DIC and ACC demonstrated that ACC gives a better performance than conventional DIC. ACC not only reduces the prediction errors but also relieves the limit of the maximum reliable displacement, which depends on the interrogation window size. ACC uses both variable window sizing and window shifting. ACC permitted measuring deformation with a resolution of 0.1 pixels, which corresponded to a displacement on the order of 0.01 mm based on the CCD camera and the optical settings in the test. This accuracy can be improved by increasing the resolution of the camera or adjusting the optical setting, for example, using a smaller focus area. ACC is implemented in Flow Manager Software [14], which is the software used in this research.

## 3. Finite Element Modeling

PLAXIS 2D (V.8) has been used in this study. Soil properties, cavity size, and depth were chosen similar to the transparent soil models conditions. The pore pressure distribution was assumed to be hydrostatic and phreatic level was located at

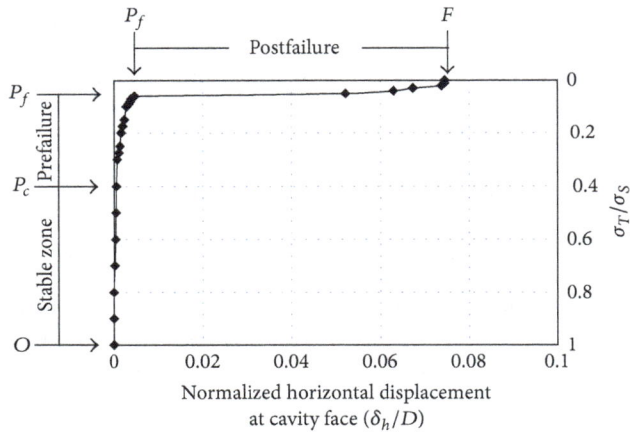

FIGURE 3: Plot of incremental cylindrical cavity contraction versus stress release.

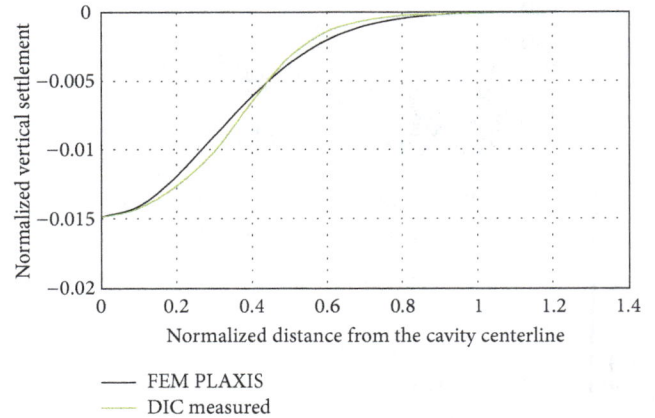

— FEM PLAXIS
— DIC measured

FIGURE 4: Transverse surface depression at pre-collapse.

the surface. The cylinder is modeled by curve plates with PVC properties and curved interface. All models were analyzed by assuming drained behavior for transparent soil, rigid interface strength, and elastic-plastic Mohr-Coulomb model. The input parameters used in PLAXIS were back-calculated from drained triaxial tests performed on transparent soil except the dilatancy angle which was assumed (unit weight = 9 kN/m$^3$; cohesion = 0, permeability = $1.5 * 10^{-4}$ cm/sec, Young's modulus = 21,000 kN/m$^2$, Poisson's ratio = 0.45, friction angle = 36°, and dilatancy angle = 1°). Staged construction with plastic calculation option was used to simulate the contracting cavity. The first stage included activation of the cylinder, deactivating of soil inside the cavity, and generation of the pore water pressure. The cavity development was simulated in the following stages by applying contraction measured from transparent soil models to the cylinder.

# 4. Results and Discussion

Multiple tests were performed at the same soil density for a single cavity size located at various depths. For a constant cavity size, the depth was found to be of little or no effect on final deformations. Therefore, only one set of results has been presented in this paper.

*4.1. Deformations Due to Stress Release.* Plot of incremental cylindrical cavity contraction versus stress release is presented in Figure 3. The stress release, represented by reduced support pressure $\sigma_T$ normalized by the vertical effective stress $\sigma_V$ at the cavity axis, is plotted against horizontal displacement $\delta_h$, is measured at the center of the cavity, and normalized by cavity diameter, $D$. As $\sigma_T$ is reduced, no movement has observed down to pressures that equal is 50% of the initial support pressure. Further reduction in support pressure, beyond point $P_c$ down to point $P_f$, resulted in small displacements, which are approximately proportional to the reduction in $\sigma_T$. Further reduction in supporting pressure results in a sudden collapse where larger deformations were measured without any further reduction in the supporting pressure.

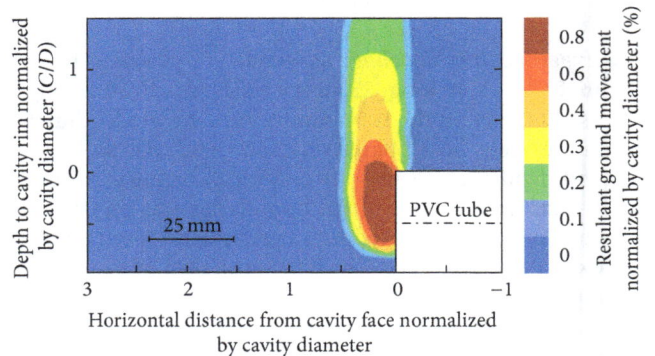

FIGURE 5: Contour plots of resultant ground movement normalized by cavity diameter.

*4.2. Deformations at Precollapse Stage.* Measured and computed transverse surface depression at precollapse, point $P_f$, is shown in Figure 4. The plot visualizes the extension of ground movement above the cavity in a plane perpendicular to the cylinder axis. The measured deformation was found to be consistent with the numerical modeling. A surface depression approximately equal to 1.5% of the cavity diameter $D$ was reported at the center of the cavity. The deformation extended horizontally to a distance equal to 1D from the cavity centerline.

Deformations in a plane parallel to the cylinder axis are presented in Figure 5 where contour plots of resultant ground movement normalized by cavity diameter are shown. It appears that deformations extend approximately 1D in front of the cavity. Unlike the deformations presented in Figure 4, these internal deformations (Figure 5) cannot be achieved by optical survey or surface monitoring instrumentation. The innovation of transparent soil allowed for measuring subsurface deformations in this experiment. Deformations were found to be much larger near the cavity and only minimally transferred to the surface. These findings highlight the importance of conducting a thorough impact assessment on nearby subsurface utilities and foundations for all subsurface construction projects. In practice, this issue seems to be overlooked with the application of grouting and other ground

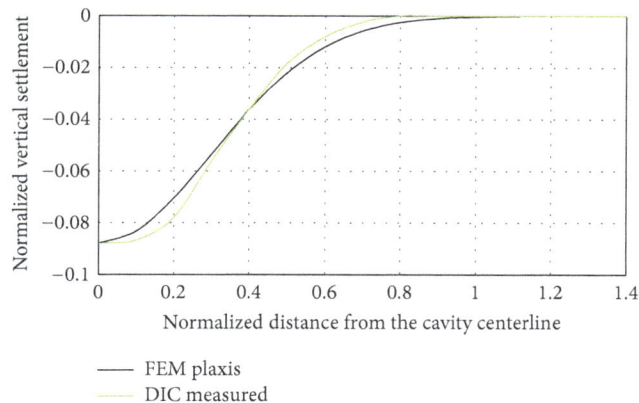

FIGURE 6: Transverse surface depression at collapsing stage.

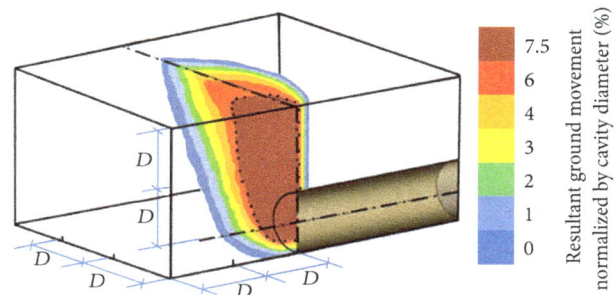

FIGURE 7: 3D deformations at collapsing stage.

improvement methods to control subsurface deformations. However, the quality of these underground techniques very much depends on the experience of the contractor and familiarity with the ground conditions which is questionable sometimes.

*4.3. Postcollapse Deformations and Development of Sinkhole.* After point $P_f$ was reached (Figure 3), further decrease of the supporting pressure led to sudden collapse (point $F$) as transparent soil tended to flow into the cavity and the failure envelope propagated progressively upward. The effect of these deformations on the surface was generating excessive surface disruption as shown in Figure 6. Again, the measured deformation was found to be consistent with the numerical modeling. A larger surface depression was reported at collapsing stage approximately. Surface depression at the center of the cavity was found to be equal to 9% of the cavity diameter $D$ and 8% larger than the values reported in Figure 4. Similar to precollapse, the deformation extended horizontally up to a distance equal to 1D from the cavity centerline. These findings indicate that the horizontal extension of deformation is independent of the magnitude of the cavity internal deformations. A similar conclusion was reported by Peck [15] and many other authors who studied settlement trough induced by tunneling.

3D contour plots of resultant ground movements at collapsing stage are shown in Figure 7 for circular cavity of diameter $D$ with cover $C$ equal to 1.5D. The figure visualizes the extension of ground movement above and ahead of the cavity. The dashed line in Figure 7 indicates the failed zone at the moment of collapse. Observed deformations were substantially larger than those observed at pre-collapse. Deformations were largest near the cavity but propagated upward to form a narrow chimney. The failed zone was relatively narrow and reached approximately 1D above the cavity crown, when arching of the soil prevented further deformation. The failed soil mass outcropped at the ground surface. The sinkhole appearance at the surface was found to be a small indicator of the final size of the deformations, as shown by the 3D plot (Figure 7). The failure of the overhanging soil around the cavity has led to a subsequent

funnel-shaped depression with a hole at the center. These results emphasize the need to extend the repaired zone following sinkhole formation by a minimum distance equal to 2D away and ahead of the cavity. Also some investigation should be conducted to evaluate the stability of buried utilities, foundation, and subgrade layers within this zone following sinkhole collapse.

## 5. Conclusions

Experimental technique and procedures were developed to analyze sinkhole pre-collapse and collapsing deformations. The use of transparent soil permitted visualization of actual internal ground deformations under the test conditions. Measured deformations parallel to the cavity axis were compared to numerical modeling and were found to be consistent. The observed failure mechanisms resembled a prismatic wedge in front of the cavity face extending upward in the form of a vertical chimney confirming the mechanisms proposed in the literature by various authors. The sinkhole appearance at the surface was found to be a small indicator of the final size and magnitude of the internal deformations as a subsequent funnel-shaped depression with a hole at the center developed. Therefore, the impact on nearby of utilities should be thought whenever underground construction is performed. Prediction of internal ground deformations is particularly important for ensuring stability of underground utilities and development of restoration plan after collapse occurred. Results of this study are believed to be of practical interest for predicting surface and internal ground deformations following sinkhole collapse. The modeling results emphasized the need to extend the repair zone following sinkhole collapse by a minimum distance that equals twice the cavity diameter away and ahead of the developed hole to cover pending failures illustrated by areas that experienced 1% to 6% deformations (resultant ground movement normalized by cavity diameter) as indicated in Figures 5 and 7. These results can also be used to provide approximate bounds to areas affected by the sinkhole allowing for collapse risk to be assessed.

The good agreement between transparent soil and numerical modeling confirms that transparent soil is a valid tool for studying sinkhole collapse. The results presented here can be used to provide models for predicting soil behavior around collapsing cavities. It can also assist the transportation agencies and contractors to predict unseen or buried unstable areas of a roadway after sinkhole collapse and develop a comprehensive restoration plan.

It is important to mention that these results were controlled by specific tests conditions (i.e., geotechnical properties, cavity depth, diameter, surcharge pressure, surface pressure, and boundary conditions). Further studies continuing from this research could look at different ground conditions, such as different densities, layered soil, and interaction with nearby utilities and foundations.

## Acknowledgments

Research on transparent soils is currently supported by the Defense Threat Reduction Agency under Grant no. HDTRA1-10-1-0049. Transparent soils were originally developed with the National Science Foundation funding under Career Grant no. CMS9733064. Continued NSF funding under Grants nos. DGE741714 and DGE0337668 is gratefully acknowledged.

## References

[1] W. H. Craig, "Collapse of cohesive overburden following removal of support," *Canadian Geotechnical Journal*, vol. 27, no. 3, pp. 355–364, 1990.

[2] W. A. Abdulla and D. J. Goodings, "Modeling of sinkholes in weakly cemented sand," *Journal of Geotechnical and Geoenvironmental Engineering*, vol. 122, no. 12, pp. 998–1005, 1996.

[3] J. H. Atkinson and D. M. Potts, "Subsidence above Shallow circular tunnels in soft ground," *Journal of Geotechnical Engineering Division*, vol. 103, no. 4, pp. 307–325, 1977.

[4] E. H. Davis, M. J. Gunn, R. J. Mair, and H. N. Seneviratne, "The stability of shallow tunnels and underground openings in cohesive material," *Geotechnique*, vol. 30, no. 4, pp. 397–416, 1980.

[5] H.-B. Mühlhaus, "Lower bound solutions for circular tunnels in two and three dimensions," *Rock Mechanics and Rock Engineering*, vol. 18, no. 1, pp. 37–52, 1985.

[6] E. Leca and L. Dormieux, "Upper and lower bound solutions for the face stability of shallow circular tunnels in frictional material," *Geotechnique*, vol. 40, no. 4, pp. 581–606, 1990.

[7] M. Ahmed and M. Iskander, "Analysis of tunneling-induced ground movements using transparent soil models," *Journal of Geotechnical and Geoenvironmental Engineering*, vol. 137, no. 5, pp. 525–535, 2011.

[8] M. Ahmed and M. Iskander, "Evaluation of tunnel face stability by transparent soil models," *Tunnelling and Underground Space Technology*, vol. 27, no. 1, pp. 101–110, 2012.

[9] M. Iskander, *Modelling With Transparent Soils, Visualizing Soil Structure Interaction and Multi Phase Flow, Non-Intrusively*, Springer, Dordrecht, The Netherlands, 2010.

[10] S. Sadek, M. G. Iskander, and J. Liu, "Geotechnical properties of transparent silica," *Canadian Geotechnical Journal*, vol. 39, no. 1, pp. 111–124, 2002.

[11] R. Taylor, R. Grant, S. Robson, and J. Kuwano, "An image analysis system for determining plane and 3D displacements in soil models," in *Proceedings of Centrifuge '89*, pp. 73–78, Taylor and Francis, London, UK, 1998.

[12] S. G. Paikowsky and F. Xi, "Particle motion tracking utilizing a high-resolution digital CCD camera," *Geotechnical Testing Journal*, vol. 23, no. 1, pp. 123–134, 2000.

[13] D. R. Gill and B. M. Lehane, "An optical technique for investigating soil displacement patterns," *Geotechnical Testing Journal*, vol. 24, no. 3, pp. 324–329, 2001.

[14] Dantec Dynamics, *Flow Manager Software User Guide*, Tonsbakken, Denmark, 2001.

[15] R. B. Peck, "Deep excavations and tunneling in soft ground," in *Proceedings of the 7th International Conference on Soil Mechanics and Foundation Engineering, Mexico City: State of-the-Art Volume*, pp. 225–290, 1969.

# 17

# Effects of Unburned Lime on Soil pH and Base Cations in Acidic Soil

**Athanase Nduwumuremyi,[1,2] Vicky Ruganzu,[1] Jayne Njeri Mugwe,[2] and Athanase Cyamweshi Rusanganwa[1]**

[1] *Department of Natural Resource Management, Rwanda Agriculture Board (RAB), P.O. Box 5016, Kigali, Rwanda*
[2] *Department of Agricultural Resource Management, Kenyatta University, P.O. Box 43844-00100, Nairobi, Kenya*

Correspondence should be addressed to Athanase Nduwumuremyi; nduwatha@gmail.com

Academic Editors: R. Ciccoli, W. Ding, W. Robarge, and J. Thioupouse

Sustainable agriculture is threatened by the widespread soil acidity in many arable lands of Rwanda. The aim of this study was to determine the quality of unburned limes and their effects on soil acidity and base cations in acidic soils of high land of Buberuka. The lime materials used were agricultural burned lime and three unburned lime materials, Karongi, Musanze, and Rusizi. The test crop was Irish Potato. All lime materials were analyzed for Calcium Carbonate Equivalent (CCE) and Fineness. A field trial in Randomized Complete Block Design was established in 2011 at Rwerere research station. The treatments comprised of the four lime materials applied at four levels: 0, 1.4, 2.8, and 4.3 t ha$^{-1}$ of CCE. Soil cations (Ca$^{2+}$, Mg$^{2+}$, K$^{+}$, and Na$^{+}$) were determined by extraction method using atomic absorption spectrophotometer for Ca and Mg and flame photometer for K and Na. The Al$^{3+}$ was determined using potassium chloride extraction method. Experimental soil baseline showed that the soil was very strongly acidic (2.8 cmol kg$^{-1}$ Al$^{3+}$). The unburned limes were significantly ($P < 0.001$) different in terms of CCE and fineness. A higher CCE was recorded in agricultural burned and Rusizi unburned limes (86.36% and 85.46%, resp.). In terms of fineness, agricultural burned and Musanze unburned lime were higher (70.57 and 63.03%, resp.). Soil acidity significantly affected from 4.8 to 5.6 pH and exchangeable Al reduced from 2.8 cmol kg$^{-1}$ to 0.16 cmol kg$^{-1}$ of Al$^{3+}$. Similarly all cations affected by unburned limes application, significantly ($P < 0.001$) Ca saturation increased from 27.44 to 71.81%, Mg saturation from 11.18 to 36.87% and significantly ($P < 0.001$) Al saturation reduced from 58.45 to 3.89%. The increase of Mg saturation was observed only with Karongi unburned lime application. This study recommends therefore, the use of 2.8 t ha$^{-1}$ of CaCO$_3$ of Rusizi or Musanze unburned lime as alternative to the agricultural burned lime for improving soil acidity and base cations in acidic soils.

## 1. Introduction

The constraints of sustainable agriculture can be partly attributed to continuous cropping, soil acidity [1], and inadequate soil fertility management [2]. The sustainable agriculture is threatened by widespread acidity in many parts of the tropical region, and applications of lime [3] to these soils have been reported to significantly improve soil fertility. Acidity affects the fertility of soils through nutrient deficiencies (P, Ca, and Mg) and the presence of phytotoxic nutrient such as soluble Al [4].

The population pressure in Rwanda triggers subsistence agriculture and is being continuously done on hills and mountains, while soil acidity is covering about one third of arable soils [5]. To feed the growing population, exploitation of all agricultural resources for sustainable agriculture and soil fertility improvement are the most important interventions to rely on.

The effect of lime is long lasting but not permanent [6]. When values of exchangeable Ca$^{2+}$, Mg$^{2+}$, and pH fall below optimum levels for a given crop species, liming should be repeated. The base enrichment especially of Ca$^{2+}$ ions in soil will neutralize exchangeable Al [7] thus enhancing root growth. The base cations include K, Ca, Mg, and Na, and the base saturation is the proportion of the CEC (cations exchange capacity) occupied by these base cations. A relatively high base saturation of CEC (70 to 80%) should be maintained for most cropping systems, since the base saturation determines in large measure the availability of bases for plant uptake and strongly influences soil pH as well. Low base saturation levels results in very acid soils and potentially toxic cations such as Al and Mn in the soil. A high base saturation (>50%) enhances Ca, Mg, and K availability and prevents soil pH decline. Low base saturation (<25%) is indicative of a strongly acidic soils that may maintain $Al^{3+}$ activity high enough to cause phytotoxicity [8]. Highly weathered tropical soils such as Oxisols have very low levels of exchangeable Ca and crops grown on such soils exhibit Ca deficiency when exchangeable Ca is <1 cmol kg$^{-1}$ [9]. The application of limestone (calcium carbonate) and or dolomitic lime (Ca and Mg bicarbonate) increases soil exchangeable Ca and Mg, respectively. The improvement of plant growth in acidic soil is not due to addition of basic cations (Ca and Mg) but is caused by the increasing pH which reduces toxicity of phytotoxic levels of Al [10]. In acidic soils, most of the Ca present would exist in soluble form, but both soluble and exchangeable Ca decreases with decreasing soil pH [11]. When $Ca^{2+}$, $K^+$, and $H^+$ concentration increase in the soil, they induce Mg uptake to be decreased in plant due to competitive inhibition [12]. Mg is also a poor competitor with Al and Ca for the exchange sites; it tends to accumulate in the solution phase and is therefore prone to leaching [13]. Thus, a greater attention has to be made when liming to prevent cations imbalance in the soil.

Locally available carbonates are relatively common in many countries of sub-Saharan Africa and are well suited for small-scale mining and processing [14]. The lime production through burning in vertical kilns [15] consume large amounts of firewood (energy) and cause environmental hazard such as the release of greenhouse gasses (GHG) including carbon dioxide ($CO_2$) to the atmosphere. The production of one ton of lime emit around 0.785 t of $CO_2$ due to mineralogical transformation [16]. Contrary, the production of unburned lime is environment friendly because they do not require burning energy and the $CO_2$ emission is null. In Rwanda, there is three main limestone deposits in western region (Karongi and Rusizi districts) and northern (Musanze and Gakenke districts) region of Rwanda. However, the production of lime in many parts of limestone mines in the country is targeting construction purpose and not for agriculture benefit. In addition, all of the available limestone materials have not been evaluated and compared to determine their effects on soil acidity and base cations. The objective of this study was therefore to determine quality of unburned limes and to evaluate their effects on the improvement of soil base cations saturation, soil acidity and available phosphorus, and yield of potato in acidic soils.

## 2. Materials and Methods

### 2.1. Experiment Details and Lime Application.
The study was carried out at Rwerere Research station located in Burera District in Northern Province of Rwanda. Rwerere Research

TABLE 1: Soil properties of experimental site before trial establishment in 2012A season, 2011.

| Soil properties | |
| --- | --- |
| $pH_W$ | 4.8 |
| $pH_{KCl}$ | 3.7 |
| Exchangeable Al (cmol kg$^{-1}$) | 2.8 |
| Total exchangeable acidity (cmol kg$^{-1}$) | 8.2 |
| Organic Carbon % | 1.3 |
| Organic matter % | 2.24 |
| Total nitrogen % | 0.11 |
| Available P (mg kg$^{-1}$) | 3.63 |
| Base saturation % | 42.5 |
| Exchangeable Ca (cmol kg$^{-1}$) | 1.3 |
| Exchangeable Mg (cmol kg$^{-1}$) | 0.5 |
| Exchangeable K (cmol kg$^{-1}$) | 0.12 |
| Exchangeable Na (cmol kg$^{-1}$) | 0.01 |
| ECEC (cmol kg$^{-1}$) | 4.8 |
| Clay % | 8.24 |
| Silt % | 11.9 |
| Sand % | 79.8 |

station lies in the agro-bio-climatic zone of highlands of Buberuka in northern parts of Rwanda. It has an altitude ranging from 2060 up to 2312 meters above sea level. The relief is characterized by steeply sloping hills connected either by steep sided valleys or by flooded marshes. Annual rainfall ranges from 1400 to 1800 mm and Annual average minimum and maximum temperature is 9°C and 25°C, respectively. Population density is 522 per km$^2$ with farm land holding ranging from 0.15 to 0.2 ha per household [17]. This implies conversion of degraded land into arable land and continuous farming on unsuitable hills and mountains. Before establishment of trial, soil fertility analysis showed that the soil was very strongly acidic [8] with soil texture of loamy sand, soil pH of 4.8, exchangeable $Al^{+3}$ of 2.8 cmol kg$^{-1}$, ECEC of 4.8 cmol kg$^{-1}$ and 42.5% of base saturation. The level of organic matter was 2.2% while Nitrogen was 0.11%, available P (BrayII) was 3.6 mg kg$^{-1}$ (Table 1).

The field trial had 13 treatments arranged in randomized complete block design (RCBD) and was established in September, 2011. The treatments comprised of four lime materials applied at three levels (1.4, 2.8 and 4.2 t ha$^{-1}$ of $CaCO_3$ equivalent) and control. Each experimental unit was 2.4 × 3 m in size. The treatments were replicated three times and the randomization was done within each block. Application of limes was done two weeks before planting by broadcast method and tilled in immediately after application. Lime requirement (LR) was determined following the method described by Kamprath [18] due to its ability to neutralize all extractable Al in soil. This method neutralizes exchangeable

TABLE 2: Limes quality (CCE, Fineness).

TABLE 2: Limes quality (CCE, Fineness).

| Limes sources | CCE (%) | Fineness (%) |
|---|---|---|
| Agricultural burned lime | 86.67 | 70.57 |
| Musanze unburned lime | 66.67 | 63.03 |
| Karongi unburned lime | 73.33 | 55.63 |
| Rusizi unburned lime | 86.00 | 56.90 |
| LSD | **17.926** | **9.132** |
| *P* value | **0.018** | **0.003** |

LSD: Least significant differences of means (5% level).

Al in the soil at the rate of 85–90% [5] and has been applied successfully in different countries [19]. The calculation of unburned lime rates (ULR) needed was done using (1)

$$\text{Unburned lime rate} = \frac{\text{Rate of pure lime}}{\text{CCE (unburned lime)}} * 100. \quad (1)$$

The unburned lime requirement rate ($t\,ha^{-1}$) depends on its quality in terms of CCE. Unburned lime with low CCE implies its higher quantity in reducing soil acidity compared to pure lime. Taking into consideration of resources poor farmers, this study evaluated three rates (0.5: economic rate, 1: normal rate and 1.5: high rate) which were equivalent to 1.4, 2.8 and $4.2\,t\,ha^{-1}$, respectively, of pure lime (100% of CCE).

Kirundo (Irish potato variety) was used in this study as test crop. Planting was done with an intrarow spacing of 0.3 m and inter-row spacing of 0.8 m. A $300\,kg\,ha^{-1}$ blanket application of (Nitrogen, Phosphorus and Potassium) NPK was done following the recommended rates by Rwanda Agriculture Board (RAB).

*2.2. Measurements.* The potassium chloride extraction method was used to determine exchangeable $Al^{3+}$ [20]. All base cations ($Ca^{2+}$, $Mg^{2+}$, $K^+$ and $Na^+$) were determined by extraction method using atomic absorption spectrophotometer for Ca and Mg and flame photometer for K and Na [21]. Available P was determined using Bray and Kurtz P-II method.

The yield was determined by weighing total fresh tubers per plot. The relative agronomic efficiency (RAE) of the unburned limes was then calculated to determine more effective local lime relatively to agricultural lime. RAE was calculated as the ratio [22] using the following equation:

$$\text{RAE} = \frac{\text{Yield (unburned lime)}}{\text{Yield (agricultural burned lime)}} * 100. \quad (2)$$

*2.3. Statistical Analysis.* The data were subjected to analysis of variance (ANOVA) using GenStat 14th edition. Means separation was performed using Turkey's test at 0.05 level of significance and mean comparisons were done using least significant difference (LSD).

# 3. Results and Discussion

*3.1. Selected Chemical and Physical Property of Lime Quality.* The CCE (as chemical property determining lime quality) of

unburned and agricultural burned limes were significantly different ($P = 0.01$ at $\alpha = 5\%$), it varies from 66.6% to 86.6% (Table 2). The agricultural burned lime and Rusizi unburned lime had similar and higher CCE compared to Musanze and Karongi unburned limes. The fineness (as physical property determining lime quality) was significantly different ($P = 0.003$) among lime types and ranged from 55.6 to 70.6 (Table 2). The agricultural burned lime was the finest compared to the three unburned limes.

The observed CCE in this study agree with the findings of Crawford and Su [23], who reported CCE of Rwandan travertine (unburned lime) to have CCE varying from 59.7 to 126%. Similarly, Beernaert [5] reported CCE variation within and between mining sites of local limes in Rwanda. The variability was attributed to the quantity of Ca, Mg, impurities and treatments (burning) of limestone as reported by several authors [5, 23–26].

The highest fineness factor of agricultural burned lime compared to the three unburned limes was attributed to the effect of heating limestone at high temperature (900–1200 kcal), crushing and sieving during its manufacture. Similar observations have been made by Millar et al. [27] who reported that fineness through various treatments (calcination, crushing, and sieving) of limestone increases the solubility of limes. The results of fineness factor observed in this study were similar to those reported by Crawford and Su [23]. These authors reported the fineness factor of Rwandan local limes to vary between 28.4 and 97.7%.

*3.2. Effect of Lime on Soil Acidity and Available Phosphorus.* The application of lime significantly increased soil pH and available phosphorus. The soil pH was significantly different ($P = 0.004$) among plots. The highest pH was recorded in plots that had Rusizi unburned lime and burned lime (Figure 1). The application of lime decreased amounts of soil exchangeable Al. The decrease of Al in the soil varied significantly ($P = 0.001$) from the types and rates of limes (Figure 1). The agricultural burned, Rusizi, and Musanze unburned limes at 4.2 and $2.8\,t\,ha^{-1}$ had relatively similar effects in reducing exchangeable Al. At lime rate of $1.4\,t\,ha^{-1}$, both agricultural burned and Musanze unburned lime had similar effects in reducing Al and they reduced 2.16 and $2.06\,cmol\,kg^{-1}$, respectively. In general, Karongi unburned lime had the least effects in reducing soil acidity based on low effect on exchangeable Al (Figure 1). The effects of limes on available phosphorus also were significantly different ($P < 0.001$). However, soil available P decreased in the control treatments, while increased in other plots (Figure 1).

The findings observed on soil pH changes in soil agree with the findings of many authors [28–30] who reported the increase of 0.4 to 0.9 units of soil pH after unburned limes application in acidic soils of Rwanda. The effects observed on exchangeable Al are corroborated by the findings of Fox [31] who reported reduction of exchangeable Al and Aluminium saturation to adequate levels following application of lime in acidic soil. Other authors such as Oates and Kamprath

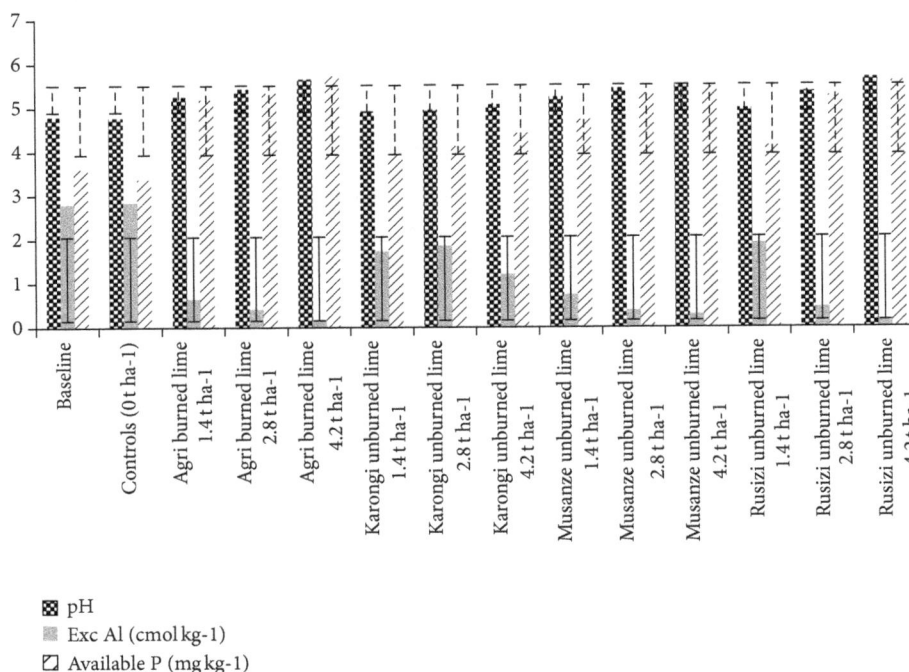

FIGURE 1: Effects of agricultural burned and unburned limes on soil pH, exchangeable Al, and available P.

[32], Conyers et al. [33], Synder and Leep [25], Caires et al. [34], and Crawford and Su [23] and Awkes [35] have reported a decrease of exchangeable Al following liming of acidic soils. Agricultural burned lime, Rusizi, and Musanze unburned limes were more effective than Karongi unburned lime in increasing available P possibly because of their effects in raising soil pH and reducing exchangeable Al. Similarly, Fageria [36] reported an increase of soil phosphorus as pH increased from 5.0 to 6.5, due to release of P ions from Al and Fe oxides, which are responsible of P fixation. Furthermore, Nurlaeny et al. [37] reports that, acidic soils are naturally deficient in available P and significant portions of applied P are immobilized due to precipitation of P as insoluble Al phosphates. The range of available P increase recorded is in agreement with the findings of Clements and McGrowen [38] who reported an increase ranging from 0 to $8 \, mg \, kg^{-1}$ of Bray P in acidic loam sandy soils of New South Wales in the United State of America (USA). Ruganzu [30] also reported an increase of available phosphorus from 3 to $13 \, mg \, kg^{-1}$ after application of travertine (local lime) combined with fertilizer in acidic soils of Crete Zaire-Nile (Karongi district) and central plateau (Huye district) in Rwanda.

### 3.3. Effects of Limes on Soil Exchangeable Cations Saturation.

Ca, Mg, and Al saturation were significantly ($P \leq 0.001$) affected by both application of agricultural burned and unburned limes. The lime rate of $2.8 \, t \, ha^{-1}$ of agricultural burned, Rusizi, and Musanze unburned limes had similar effects in reducing Al saturation. Karongi unburned lime applied at all rates was the lowest in reducing Al saturation

than other limes. However, potassium and sodium saturation were not significantly affected by lime application (Table 3).

The high increase of Ca saturation observed with agricultural burned, Rusizi and Musanze unburned limes than in Karongi unburned lime could be attributed to the fact that calcite lime releases more Ca in soil solution that dolomitic lime as reported by Fageria and Stone [39]. The application of 4.2 and $2.8 \, t \, ha^{-1}$ of agricultural burned lime, Rusizi unburned lime were able to bring Ca saturation at adequate level in the soil which is estimated at 65 to 85% by Hazelton and Murphy [40]. The highest Mg saturation recorded in the plots with Karongi unburned lime compared to the other limes could be attributed to its dolomitic nature. The results are in accordance with the findings of Fageria and Stone [39] who reported increase of Mg content in acidic soils as result of liming. Beernaert [5] also reported the increase of Mg in soil following application of dolomitic lime in acidic soil of Rwanda. The reduction of Al saturation recorded in this study were in accordance with the findings of Ruganzu [30] who reported 49% reduction of Al saturation after application of travertine combined with *Tithonia diversifolia* in Rubona acidic soils. According to Abbott [41], the adequate level of Al saturation in soil should be <5%. Markedly, only application of $4.2 \, t \, ha^{-1}$ of agricultural burned and Rusizi unburned lime were able to reduce Al saturation to 3.1 and 3.9%, respectively.

Despite the fact that K and Na saturation were not affected by limes, according to Abbott [41] and Hazelton and Murphy [40], the baseline of K and Na saturation were in adequate range where it was ranged from 1 to 5% for K and at 0 to 1% for Na.

TABLE 3: Exchangeable cations saturation (%) as affected by limes.

| Treatments | Ca (%) | Mg (%) | K (%) | Na (%) | Al (%) |
|---|---|---|---|---|---|
| Baseline | 27.44 | 11.18 | 2.64 | 0.28 | 58.45 |
| Controls(0 t ha$^{-1}$) | 27.98 | 11.73 | 2.30 | 0.07 | 57.91 |
| Agricultural burned lime 1.4 t ha$^{-1}$ | 64.11 | 15.3 | 3.55 | 0.09 | 16.96 |
| Agricultural burned lime 2.8 t ha$^{-1}$ | 62.6 | 25.33 | 3.08 | 0.07 | 8.96 |
| Agricultural burned lime 4.2 t ha$^{-1}$ | 75.09 | 18.03 | 3.67 | 0.09 | 3.1 |
| Karongi unburned lime 1.4 t ha$^{-1}$ | 35.04 | 31.11 | 2.71 | 0.07 | 31.06 |
| Karongi unburned lime 2.8 t ha$^{-1}$ | 30.12 | 36.04 | 3.06 | 0.03 | 30.74 |
| Karongi unburned lime 4.2 t ha$^{-1}$ | 34.21 | 36.87 | 7.66 | 0.07 | 21.19 |
| Musanze unburned lime 1.4 t ha$^{-1}$ | 59.94 | 18.07 | 3.19 | 0.09 | 18.71 |
| Musanze unburned lime 2.8 t ha$^{-1}$ | 68.34 | 18.81 | 3.24 | 1.00 | 9.51 |
| Musanze unburned lime 4.2 t ha$^{-1}$ | 69.31 | 20.45 | 3.33 | 0.08 | 6.83 |
| Rusizi unburned lime 1.4 t ha$^{-1}$ | 41.24 | 14.22 | 3.15 | 0.04 | 41.34 |
| Rusizi unburned lime 2.8 t ha$^{-1}$ | 64.84 | 19.97 | 3.65 | 0.05 | 11.48 |
| Rusizi unburned lime 4.2 t ha$^{-1}$ | 71.81 | 20.27 | 3.93 | 0.08 | 3.89 |
| P value | **<0.001** | **0.001** | **0.463** | **0.278** | **<0.001** |
| LSD | **11.552** | **6.568** | **3.236** | **0.043** | **14.580** |

LSD: Least significant differences of means (5% level).

TABLE 4: Effects of limes on potato yield (P < 0.001 and LSD 3.28) and RAE (%).

| Lime rates | Yield (kg ha$^{-1}$) | | | | RAE (%) | | |
|---|---|---|---|---|---|---|---|
| | Agricultural lime | Karongi lime | Musanze lime | Rusizi lime | Karongi lime | Musanze lime | Rusizi lime |
| Control (0 t ha$^{-1}$) | 14.32 | 14.32 | 14.32 | 14.32 | — | — | — |
| 1.4 t ha$^{-1}$ (1/2 rate) | 18.88 | 17.88 | 19.63 | 18.06 | 80.43 | 113.04 | 82.61 |
| 2.8 t ha$^{-1}$ (full rate) | 22.47 | 19.62 | 22.15 | 21.08 | 65.85 | 95.12 | 82.93 |
| 4.2 t ha$^{-1}$ (1.5 rate) | 24.82 | 19.23 | 24.9 | 21.85 | 47.62 | 100 | 72.38 |

RAE = 100: equal efficiency of local lime and agricultural lime; RAE > 100: more efficiency than agricultural lime; RAE < 100: less efficiency than agricultural lime.

*3.4. Effects of Limes on Potato Tuber Yield and Relative Agronomic Efficiency (RAE).* The yield of Irish potato was significantly (P = 0.01 at α = 5%) affected by the application of limes. Notably, agricultural burned lime, Musanze, and Rusizi unburned limes applied at lime rate of 2.8 t ha$^{-1}$ had relatively similar effects. However, the lowest yield was obtained in the control plots. The unburned limes were different in their RAE where Musanze unburned lime applied at 1.4 t ha$^{-1}$ had the highest RAE (113.04%) compared to application of the other unburned lime rates (Table 4). The application of 4.2 t ha$^{-1}$ of Musanze unburned lime had RAE (100%), which means it had the same effectiveness as agricultural burned lime (Table 4).

The high yield obtained in plots that were limed was probably due to the positive effects of liming on soil properties. The agricultural burned and unburned limes improved the overall soil base saturation. Markedly, when a little amount of lime or unburned lime applied in acidic soils, it results to the changes in soil properties and other nutrients [5] which in turn affect potato production positively. These findings are in agreement with Harelimana [42] and Kayitare [29] who reported that unburned limes increased Irish potato yield in acidic soils of Rwanda. The high RAE of Musanze unburned lime compared to the other unburned limes could be attributed to its fineness and CCE.

## 4. Conclusion

The agricultural burned lime, Rusizi and Musanze unburned limes had different quality. A higher calcium carbonate equivalent (CCE) was observed in agricultural burned and Rusizi unburned limes (86.6% and 86%, resp.). This indicates that the two lime types are comparable. In terms of fineness, agricultural burned and Musanze unburned limes were higher (70.57 and 63.03%, resp.) compared to the other two limes. This could be an indication of their higher effectiveness observed in this study. The application of 4.2 and 2.8 t ha$^{-1}$ of agricultural burned lime, Rusizi and Musanze unburned limes had similar effects in increasing Ca saturation and reducing Al saturation. These imply that unburned limes could be used in alternative to agricultural burned lime which is very expensive to farmers. In addition, the unburned lime do not possess burning effects, hence provides farmers with safer material to work with and with easy manipulation (handling practices). Furthermore, the production of agricultural burned lime was reported to cause environmental hazard through the emission of greenhouse gasses. Therefore this study recommends the use of unburned limes instead of relying on agricultural burned lime in improving base cations and available phosphorus in acidic soils.

## Conflict of Interests

The authors declare that there is no conflict of interests regarding the publication of this paper.

## Acknowledgments

The authors are grateful to the Alliance for Green Revolution in Africa (AGRA) for financial support. Gratitude is also expressed to the Rwanda Agriculture Board (RAB), Kenyatta University (KU), and Higher Institute of Agriculture and Animal Husbandry (ISAE) for facilities provided during this research work.

## References

[1] W. W. Kiiya, S. W. Mwwoga, R. K. Obura, and A. O. Musandu, "Soil acidity ameriolation as a method of sheep sorrel (rumex acetosella) weed management in Potato (solanum tuberosum L.) in cool highlands of the north Rift, Kenya," in *Proceedings of the KARI Biannual Scientific Conference*, KARI, Nairobi, 2006.

[2] L. Berga, D. Siriri, and P. Ebanyat, "Effect of soil amendments on bacterial wilt incidence and yield of potatoes in southernwestern Uganda," *African Crop Science Journal*, pp. 267–278, 2001.

[3] C. Yamoah, N. M. C. Ngong, and D. K. W. Dias, "Reduction of P fertilizer requirement using lime and mucuna on high P-sorption soil of NW Cameroon," *African Crop Science Journal*, vol. 4, pp. 441–451, 1996.

[4] A. S. Awad, D. G. Edwards, and P. J. Milham, "Effect of pH and phosphate on soluble soil aluminium and on growth and composition of kikuyu grass," *Plant and Soil*, vol. 45, no. 3, pp. 531–542, 1976.

[5] F. R. Beernaert, "Feasibility study of production of lime and or ground travertine for manegement of acidic soils in Rwanda," in *Pro-Inter Project Consultants*, p. 250, 1999.

[6] N. K. Fageria and V. C. Baligar, "Ameliorating soil acidity of tropical oxisols by liming for sustainable crop production," in *Advances in Agronomy*, D. L. Sparks, Ed., pp. 345–389, Academic Press, São Paulo, Brazil, 2008.

[7] L. C. Bell and T. Bessho, "Assessment of aluminium detoxification by organic materials in ultisol using soil solution characterization and plant response," in *Soil Organic Matter Dynamics and Sustainability of Tropical Agriculture*, pp. 317–330, Wiley-Sayce co-publication, 1993.

[8] Soil Survey Division Staff, "Soil survey manual," in *US Departement of Agriculture Handbook*, US Govertement Printing Office, Washington, DC, USA, 1993.

[9] D. D. Cregan, J. R. Hirth, and M. K. Conyers, "Amelioration of soil acidity by liming and other amendemnts," in *Soil Acidity and Plant Growth*, pp. 206–264, Academic press, Sydney, Australia, 1989.

[10] H. Marschaner, *Mineral Nutrition of Higher Plants*, Academic Press, New York, NY, USA, 1995.

[11] R. J. Haynes and T. E. Ludecke, "Effect of lime and phosphorus applications on concentrations of available nutrients and on P, Al and Mn uptake by two pasture legumes in an acid soil," *Plant and Soil*, vol. 62, no. 1, pp. 117–128, 1981.

[12] R. B. Clark, "Physiological aspects of calcium and magnbesium and molybdenum deficiencies in plants," in *Soil Acidicty and Liming*, pp. 99–170, 1984.

[13] J. A. Myers, E. O. McLean, and J. M. Bigham, "Reductions in exchangeable magnesium with liming acidic Ohio soils," *Soil Science Society of America Journal Journal*, vol. 52, no. 1, pp. 131–136, 1988.

[14] P. van Straaten, *Rocks for Crops: Agrominerals of Sub-Saharan Africa*, ICRAF, Nairobi, Kenya, 2002.

[15] J. A. H. Oates, *Lime and Limestone: Chemistry and Technology, Production and Uses*, Wiley-VCH Verlag GmbH, Weinheim, Germany, 1998.

[16] Y. d. Lespinary, "The EU greenhouse gas emission trading scheme: is the discrimination for process $CO_2$ ignored?" in *International Lime Association*, ILA, Prague, Czech Republic, 2006.

[17] CIIT, *Energy Baseline for the UNEP-GEF Pilot Project on Reducing the Vulnerability of the Energy Sector to the Impacts of Climate Change in Rwanda*, International Institute for Sustainable Development, Manitoba, Canada, 2006.

[18] E. J. Kamprath, "Exchangeable Al as criterion for liming leached minal soils," *Soil Science Society of America Journal*, pp. 252–254, 1970.

[19] P. Sanchez, *Properties and Management of Soils in Tropics*, Wiley-Interscience, New York, NY, USA, 1976.

[20] J. R. Page, R. H. Miller, D. R. Keeney, D. E. Baker, J. R. Ellis, and J. D. Rhoades, *Methods of Soil Analysis. II Chemical and Microbiology Properties*, 1982.

[21] IITA, *Selected Methods for Soil and Plant Analysis*, IITA Manual Services, Ibadan, Nigeria, 1979.

[22] O. A. Mercy and A. A. Ezekiel, "Lime effectiveness of some fertilizers in a tropical acid alfisol," *Journal of Central European Agriculture*, vol. 8, no. 1, pp. 17–24, 2007.

[23] T. W. Crawford and B. H. Su, *Solving Agricultural Problems Related to Soil Acidity in Central Africa's Great lakes Region*, International Center for Soil Fertility and Agriculture Development, Auburn, Ala, USA, 2008.

[24] J. Munyengabe, *Qualitative Study of Travertin Deposits in Rwanda on Productivity Effects of Acid Soils*, National University of Rwanda, Agriculture Faculty, Butare, Rwanda, 1993.

[25] C. S. Synder and R. H. Leep, "Fertilization forages," *Science of Grassland Agriculture*, pp. 355–379, 2007.

[26] M. Verhaeghe, *Inventory of Deposit of Limestone, Dolomite and Travertine in Kivu, Rwanda and Burundi*, Ministry of Agriculture and Economic Affairs, Geology Service, Ruhengeri, Rwanda, 1963.

[27] C. E. Millar, L. M. Turk, and H. D. Forth, *Fundamentals of Soils Sciences*, John Wiley & Sons, New York, NY, USA, 1958.

[28] M. Hartmann, *Possibilities and Limits of Boron Improvement to the Degraded Soils by Liming and Application of Colcanic Ash in Rwanda*, Geographical Intitute of J.Gutenberg University, Mainz, Germany, 1993.

[29] L. Kayitare, *Potential Increase of Yield and Economic Utilization of Mineral Fertilizers in Rwanda*, MINAGRI, Kigali, Rwanda, 1989.

[30] V. Ruganzu, *Potential of Improvement of Acid Soils Fertility by Incorporation of Natural Fresh Plant Biomass Combined with Travertine in Rwanda*, Agricultural University, Gembloux, Belgium, 2009.

[31] R. H. Fox, "Soil pH, aluminium saturation and corn grain yield," *Soil Science*, pp. 330–335, 1979.

[32] K. M. Oates and E. J. Kamprath, "Soil acidity and liming: I effect of the extracting solution cation and pH on the removal

of aluminium from acid soils," *Soil Science Society of America Journal Journal*, vol. 47, no. 4, pp. 686–689, 1983.

[33] M. K. Conyers, D. P. Heenan, W. J. McGhie, and G. P. Poile, "Amelioration of acidity with time by limestone under contrasting tillage," *Soil and Tillage Research*, vol. 72, no. 1, pp. 85–94, 2003.

[34] E. F. Caires, P. R. S. Pereira Filho, R. Zardo Filho, and I. C. Feldhaus, "Soil acidity and aluminium toxicity as affected by surface liming and cover oat residues under a no-till system," *Soil Use and Management*, vol. 24, no. 3, pp. 302–309, 2008.

[35] M. M. Awkes, *Comparison of Calcium Ameliorants and Coal Ash in Alleviating the Effects of Subsoil Acidity on Maize Root Development Near Middelburg, Mpumalanga*, Faculty of Agrisciences, Stellenbosch University, Stellenbosch, South Africa, 2009.

[36] N. K. Fageria, "Effect of phosphorus on growth, yield and nutrient accumulation in the common bean," *Journal of Tropical Agriculture*, pp. 249–255, 1989.

[37] N. Nurlaeny, H. Marschner, and E. George, "Effects of liming and mycorrhizal colonization on soil phosphate depletion and phosphate uptake by maize (Zea mays L.) and soybean (Glycine max L.) grown in two tropical acid soils," *Plant and Soil*, vol. 181, no. 2, pp. 275–285, 1996.

[38] B. Clements and I. McGrowen, *Strategic Fertilizer Use on Pastures*, NSW Agriculture, Orange, Australia, 1994.

[39] N. K. Fageria and L. F. Stone, "Yield of common bean in no-tillage system with application of lime and zinc," *Pesquisa Agropecuaria Brasileira*, vol. 39, no. 1, pp. 73–78, 2004.

[40] P. A. Hazelton and B. W. Murphy, *Interpreting Soil Test Results: What Do All Numbers Mean?* CSIRO, Collingwood, Australia, 2007.

[41] T. S. Abbott, *BCRI Soil Testing Method and Interpretation*, NSW Agriculture and Fisheries, Rydalmere, Australia, 1989.

[42] B. Harelimana, *Comparative Study of Burned Lime and Travertin Application in Acidic Soils of High Altitude of Gikongoro*, National University of Rwanda, Butare, Rwanda, 1990.

# 18

# Diversity of *Rhizobium leguminosarum* from Pea Fields in Washington State

**Rita Abi-Ghanem,[1] Jeffrey L. Smith,[2] and George J. Vandemark[3]**

[1] Department of Crop and Soil Sciences, Washington State University, Pullman, WA 99164-6420, USA
[2] Land Management and Water Conservation Research Unit, USDA-ARS and Washington State University, Pullman, WA 99164-6421, USA
[3] Grain Legume Genetics Physiology Research, USDA-ARS and Washington State University, Pullman, WA 99164-6421, USA

Correspondence should be addressed to Rita Abi-Ghanem; rita ag@wsu.edu

Academic Editors: G. Benckiser, J. A. Entry, H. K. Pant, and A. P. Schwab

Rhizobia-mediated biological nitrogen (N) fixation in legumes contributes to yield potential in these crops and also provides residual fertilizer to subsequent cereals. Our objectives were to collect isolates of *Rhizobium leguminosarum* from several pea fields in Washington, examine genetic diversity among these isolates and several commercial isolates of *R. leguminosarum*, and compare genetically distinct isolates for their ability to fix N in a range of pea hosts. Seventy-nine isolates were collected from pea root from four noninoculated pea fields. Sequence-related amplified polymorphism (SRAP) markers generated by PCR were used to discriminate among isolates. Isolates fell into 17 clusters with robust bootstrap support values. Nearly half of the isolates fell into a single large cluster, but smaller clusters were also detected for isolates from all four field locations. The majority of commercial isolates fell into a distinct cluster. Four genetically distinct isolates were compared for their efficiency in fixing N in a greenhouse experiment. Host plant variety effects were significant for plant biomass due to N fixation and also for the quantity of N fixed per variety. Significant effects of *R. leguminosarum* isolates were observed for the quantity of N fixed per isolate, plant biomass, and the quantity of N per plant.

## 1. Introduction

Global demand for food will increase commensurately with a world population that may grow to reach 8.3 billion by 2025 [1]. Such increased demand will most likely occur in developing countries, many of which suffer from limited access to fertilizers and other exogenous farm inputs [1]. Furthermore, in the US, there are competing energy demands for limited supplies of natural gas required by the Haber-Bosch process in the nitrogen- (N-) fertilizer industry.

Rhizobia-mediated biological N fixation, which occurs on legume roots, can provide residual fertilizer to subsequent small grain crops. Specifically, rhizobiaceae are free-living diazotrophic saprophytes able to form nitrogen- (N-) fixed symbiotic associations with legumes by forming root nodules. Pea plants (*Pisum sativum* L.) are nodulated by *Rhizobium leguminosarum* bv. *viciae* [2]. The efficiency of biological

N fixation is also influenced by many environmental factors including soil conditions, such as acidity, temperature, mineral nutrients, salinity, alkalinity [1], high nitrogen and phosphorus levels [3], and soil type [4, 5]. The influence of these environmental factors can be seen in the differences in the amount of N fixed in nodules of pea roots observed across different regions of pea production. For instance, the average N fixed by pea plants is 200 kg ha$^{-1}$ in Europe [6] and 83 kg ha$^{-1}$ in Australia, where peas are grown under rain-fed conditions [7].

Legume-rhizobial symbiosis is species-specific, but some recent studies have found that this symbiosis may even be cultivar-strain-specific. That is, different varieties within the same plant species may prefer different optimal rhizobial strains for maximum N fixation. In several legumes, including pea (*Pisum sativum* L.) [8] and chickpea (*Cicer arietinum* L.) [9], significant plant host, strain, and host x strain interaction effects have been observed on N fixation. Abi-Ghanem et al. [10] examined N fixation in growth chamber experiments using three yellow pea and two green pea cultivars, in which plants were inoculated separately with fifteen different *R. leguminosarum* bv. *viciae* commercial inoculants. Significant differences were observed among pea cultivars for the percentage of plant N supplied by bacterial N fixation and also for the number of root nodules formed per plant [10]. In the case of lentils, significant effects among different commercial inoculants were observed on the percentage of total plant N supplied by bacterial N fixation, whereas inoculant effects were not significant for peas [10].

The significant effects of commercial inoculants on N fixation in pea observed by Abi-Ghanem et al. [10] cannot conclusively be due to genetic differences among *R. leguminosarum* bv. *viciae* isolates that constitute the various commercial inoculants. This is because no definitive proof has provided that the commercial inoculants were composed of genetically distinct bacterial isolates. This information is critical in determining if differences among commercial inoculants in their ability to fix N in legume hosts are due to genetic differences among bacterial strains or due to another component or process of the product formulation. Information on genetic diversity among bacterial isolates will also assist in assessing relatedness between indigenous *R. leguminosarum* bv. *viciae* strains and strains present in commercial inoculant preparations. Presently, it is not clearly known how commercial inoculants compete with indigenous *R. leguminosarum* bv. *viciae* strains. Examples have reported on both the inability of commercial inoculants to compete with indigenous strains [11] and the successful competitive ability of commercial inoculants [12].

Applying genetic approaches to rhizobia analysis will better define plant-rhizobia interactions and processes involving nodulation and nitrogen fixation. Genetic diversity among rhizobial symbionts of diverse legume species has been observed based on restriction site polymorphism of 16S rRNA genes and by PCR DNA fingerprinting with repetitive sequences [13]. Genetic polymorphisms have previously been used to improve bacterial taxonomy and characterize novel strains *Mesorhizobium trianshanense* with better N fixing efficiencies [14]. The primary objectives of this study were to collect *R. leguminosarum* isolates from peas in Washington State and examine genetic diversity among these isolates and commercial strains using both 16S rDNA sequence analysis and sequence-related amplified polymorphisms (SRAP). Biological nitrogen fixation was also examined using different pea cultivars and a subset of genetically distinct isolates of *R. leguminosarum* bv. *viciae*.

## 2. Materials and Methods

*2.1. Isolation of R. leguminosarum .* Plant and soil samples were collected from noninoculated pea fields in Whitman County, Washington, USA (Spillman, Shawnee, Oakesdale, Colton, and Colfax). Cultivars were Aragorn for all fields except Colton, where Banner peas were instead planted. Samples were collected at 15–20 cm depth following an W

pattern across the field. Equal amounts of soil were sampled at 15–20 cm depths from five different locations within a field and mixed together. Soil samples were then commercially analyzed to assess soil acidity, nitrate-N, ammonium-N, sulfate-S, phosphorus, potassium, boron, zinc, manganese, copper, and iron contents (Soiltest Farm Consultants, Inc. Moses Lake, WA, USA). Twenty plants were selected based on the presence of rhizobial nodulation. Pea roots were thoroughly washed, and nodules were sterilized with 0.5% sodium hypochlorite for 1 min and then passed quickly through a flame for surface disinfection. Nodules were cut and squeezed to extract liquid containing inoculum, which was applied to an inoculation loop. Inoculum was streaked on yeast extract mannitol agar (YEMA) (Bacto, Inc., Sparks, MD, USA) plates and incubated at 28°C for 3 days, after which a single colony was picked and restreaked on another YEMA plate. A single colony from this was then used to inoculate sterile tubes containing 3 mL of yeast extract mannitol broth (YEMB) (Bacto, Inc., Sparks, MD, USA) and agitated at 250 rpm on an orbital shaker at 28°C for 5 days. An 800 $\mu$L sample of each cultured strain was cryopreserved in 20% glycerol at −80°C. Commercial rhizobial strains used in this study were also stored at −80°C. The commercial strains of *Rhizobium leguminosarum* bv. *viciae* were obtained from cooperating companies. Novozymes Biologicals (Saskatchewan, Canada) provided eight strains (S007A-2, S008B-3, S012 A-3, S016B-4, S019A-1, S030A-4, S068A-1, and S007A-5); Becker Underwood (Iowa, USA) supplied five strains (P2, 082, 212-9, 213-5, ICAR 20), and EMD (Wisconsin, USA) provided three (EMD1, EMD2, and EMD3).

*2.2. DNA Extraction and PCR.* Genomic DNA was extracted from the cultured strains using the Fast DNA SPIN kit (MP Biomedicals Inc., OH, USA) according to the manufacturer's instructions. DNA was quantified with a fluorometer (TD-700; Turner Designs, Inc., Sunnyvale, CA, USA) and diluted to 10 ng/$\mu$L for use in PCR. An 1340 bp product corresponding to nearly the full length of the 16S rRNA gene was amplified by PCR [15] using the forward primer fD1 (5′-AGAGTTTGATCCTGGCTCAG-3′) and reverse primer rD1 (5′-AAGGAGGTGATCCAGCC-3′). PCR reactions were performed using a Veriti thermal cycler (Applied Biosystems, Foster City, CA, USA) in 25 $\mu$L reactions containing 100 ng DNA; 200 $\mu$M each dNTP; 1.5 mM MgCl$_2$; 2.5 units GoTaq Flexi DNA polymerase (Promega Corp., Madison, WI, USA); 5 $\mu$L 5 X GoTaq Flexi Buffer, and 0.2 $\mu$M of both forward and reverse primers. Cycling conditions consisted of a single cycle of 95°C for 2 min, followed by 30 cycles of 94°C for 40 s, 52°C for 40 s, and 72°C for 1 min, and a final extension cycle at 72°C for 7 min. Amplicons were resolved on 1.5% agarose gels in TBE buffer (1 X TBE, 90 mM Tris pH 8.0, 90 mM boric acid, 2 mM EDTA), stained with ethidium bromide and visualized with UV light. Amplicons were purified using a SureClean purification solution (Bioline Inc., MA, USA) as per the manufacturer's instructions, and the purified DNA samples were then commercially sequenced (Elim Biopharmaceuticals, Inc., CA., USA). Sequences were edited, and a consensus sequence for each amplicon was

TABLE 1: Reverse and forward primers used for sequence-related amplified polymorphism (SRAP) analysis.

| Forward primers | Reverse primers |
| --- | --- |
| ME2, 5′-TGAGTCCAAACCGGAGC-3′ | EM1, 5′-GACTGCGTACGAATTAAT-3′ |
| F7, 5′-GTAGCACAAGCCGGAGC-3′ | EM2, 5′-GACTGCGTACGAATTTGC-3′ |
| F9, 5′-GTAGCACAAGCCGGACC-3′ | EM3, 5′-GACTGCGTACGAATTGAC-3′ |
| | EM4, 5′-GACTGCGTACGAATTTGA-3′ |
| | EM5, 5′-GACTGCGTACGAATTAAC-3′ |
| | EM6, 5′-GACTGCGTACGAATTGCA-3′ |
| | R7, 5′-GACACCGTACGAATTTGC-3′ |
| | R10, 5′-GACACCGTACGAATTAAC-3′ |

derived using Geneious software version 5.0 [16]. Consensus sequences were compared with GenBank accessions by a BLAST search. Phylogenetic analyses by the neighbor-joining method and a consensus tree were generated using Molecular Evolutionary Genetics Analysis (MEGA) software version 5.0 [17].

2.3. SRAP Analysis. Ten forward and reverse primer pair combinations were selected based on preliminary tests to generate sequence-related amplified polymorphism (SRAP) [18] among rhizobia isolates. These primer pairs were designated as ME2/EM3, ME2/EM5, F7/EM1, F7/EM3, F7/EM4, F7/EM5, F9/EM2, F9/EM6, F9/R7, and F9/R10 (Table 1). All SRAP reactions were performed using a Veriti thermal cycler (Applied Biosystems, Foster City, CA, USA) in 25 μL reactions containing 100 ng DNA; 200 μM each dNTP; 1.5 mM MgCl$_2$; 2.5 units GoTaqFlexi DNA polymerase (Promega Corp., Madison, WI, USA); 5 μL 5 X GoTaqFlexi Buffer, and 37.5 ng of both forward and reverse primers. The thermal cycling profile for all reactions consisted of a single cycle of 95°C for 2 min followed by 5 cycles of 94°C for 1 min, 35°C for 1 min, 72°C for 1 min, 35 cycles of 94°C for 1 min, 50°C for 1 min, 72°C for 1 min, and a final extension at 72°C for 7 min. PCR products were resolved on 1.4% agarose gels in TBE buffer (1 X TBE, 90 mM Tris pH 8.0, 90 mM boric acid, and 2 mM EDTA) and run at 125 V for 3.5 hr. Gels were stained for 30 min in ethidium bromide, and amplicons were visualized with UV light.

2.4. Data Analysis. Electrophoretic agarose gel images were scored visually for the presence or absence of polymorphic and monomorphic amplicons. All amplicons having molecular weights greater than 150 bp were included in the analysis. Mean genetic distance between all pairwise combinations was calculated using GenAlEx software [19].

Cluster analysis was performed based on a previously described genetic distance method [20]. The unweighted pair group arithmetic average (UPGMA) method was used for analysis and a dendrogram was generated using PAUP software version 4.0 [21]. Bootstrap support for clusters was determined using 1000 permuted data sets [22].

2.5. N Fixation Study. Four strains isolated from four locations and with distinct genetic fingerprints were selected to assess N fixation. The selected isolates were designated as SP1; Oak1; Col1; and Sh2 and were collected from Spillman, Oakesdale, Colton, and Shawnee fields, respectively. The pea varieties selected for examination included three spring food grade green peas (Ariel, Aragorn, and Stirling) and two spring animal food grade yellow peas (Delta and Universal). Seeds were surface-disinfected with 0.5% sodium hypochlorite for three min and washed three times with sterile distilled deionized water. Seeds were germinated for five days on moist sterile filter paper at room temperature (22–24°C). At the day of planting in a greenhouse, seeds were sown, one per pot, in cones (6.4 cm × 36 cm) (SC10 container, Stuewe & Sons, Inc.) containing a mix of sterile sand, perlite, and LECA clay (2:1:1 by volume). Colony density was determined based on spectrophotometry (Bio-Rad, Hercules, CA, USA) using the optical density measured at 640 nm. Each pot was inoculated with one of the selected R. leguminosarum strains using 1 mL of inocula containing 10$^6$ cfu mL$^{-1}$. Control plants were not inoculated with rhizobia. As previously described [23], pots were placed in a plastic tray filled with N-free plant nutrient solution containing 1.0 mM K$_2$SO$_4$, 0.5 mM KH$_2$PO$_4$, 0.25 mM K$_2$HPO$_4$, 0.5 mM MgSO$_4$·7H$_2$O, 2.0 mM CaSO$_4$·2H$_2$O, 25 uM KCl, 13 uM H$_3$BO$_3$, 1.0 uM MnSO$_4$·H$_2$O, 1.0 uM ZnSO$_4$·7H$_2$O, 0.25 uM CuSO$_4$·5H$_2$O, 2.5 uM CoCl$_2$·6H$_2$O, 20 uM FeCl$_3$·6H$_2$O, and 0.25 uM Na$_2$MoO$_4$·2H$_2$O. Pots were irrigated with 50 mL of sterile water on a daily basis for the first 2 weeks and then 3 times per week until harvesting occurred after 10 weeks, with the exception of the second set of the experiment, which was harvested after 8 weeks due to an infestation of mites in the greenhouse.

Upon harvesting, plant roots were washed, and the plants were dried at 60°C for 3 days and then weighed. Dried plants were passed through a Wiley mill to homogenize the sample and then powdered through a roller grinder. Total % nitrogen (N) and % carbon (C) were measured using an elemental combustion system (Costech Instruments, USA) and were then multiplied by the plant total biomass to determine the quantity of N per plant. The quantity of N fixed was estimated by subtracting the quantity of N of the control plants from those treated. Biomass resulting from N fixation was calculated by subtracting the total biomass of control plants from the total biomass of those treated.

TABLE 2: Results of soil analysis from fields where rhizobial isolates were collected.

| Location | | Soil factor | | | | | | | | |
| | pH | Ammonium-N ($mg \cdot kg^{-1}$) | Nitrate-N ($mg \cdot kg^{-1}$) | Phosphorus ($mg \cdot kg^{-1}$) | Potassium ($mg \cdot kg^{-1}$) | Boron ($mg \cdot kg^{-1}$) | Zinc ($mg \cdot kg^{-1}$) | Manganese ($mg \cdot kg^{-1}$) | Copper ($mg \cdot kg^{-1}$) | Iron ($mg \cdot kg^{-1}$) |
| --- | --- | --- | --- | --- | --- | --- | --- | --- | --- | --- |
| Colton | 5.2 | 2.3 | 14.1 | 43 | 537 | 0.24 | 0.8 | 24.7 | 2.9 | 140 |
| Colfax | 5.9 | 1.5 | 6.9 | 17 | 169 | 0.24 | 0.3 | 6.9 | 2.6 | 60 |
| Oakesdale | 5.6 | 2.6 | 6.4 | 16 | 124 | 0.13 | 0.3 | 8.9 | 3.1 | 67 |
| Shawnee | 6.0 | 2.5 | 18.5 | 25 | 274 | 0.42 | 1 | 10.8 | 2.4 | 103 |
| Spillman | 5.8 | 2.8 | 6.5 | 17 | 124 | 0.17 | 0.5 | 5.8 | 2.5 | 73 |

*2.6. Experimental Design and Statistical Analysis.* SAS Proc Mixed procedure [24] was used for statistical analyses. The data were analyzed as a split plot analysis of variance in a completely randomized design. Rhizobia strain, pea variety, and their interactions were considered fixed effects with the three replicate flats considered random. Differences in measured biomass and the amount of N fixed for each experimental unit were calculated and tested for significance at $P \leq 0.05$ using Tukey's multiple comparison method.

## 3. Results

*3.1. Strains Isolation and Identification.* A total of 79 rhizobial strains were isolated from pea root nodules. Comparison of the 16S rRNA gene sequences with GenBank accessions via BLAST search confirmed that the bacteria isolated from pea nodules and commercial isolates were *R. leguminosarum*.

*3.2. Phylogenetic Analysis Based on 16S rRNA Gene.* The genetic diversity of native and commercial strains of *Rhizobium* based on sequence analysis of the 16S rRNA gene was represented by a phylogenetic tree. The tree revealed two clades with bootstraps exceeding 60% which did not demonstrate inter- or intrafield diversity among rhizobial isolates. Isolates from Shawnee appear to be more genetically diverse than those from other Palouse fields since they were found in both clades.

*3.3. Phylogenetic Analysis Based on SRAP.* DNA of 79 native and 16 commercial rhizobial isolates was successfully amplified by the ten SRAP primer pairs, and a total of 102 markers were resolved. The number of markers produced by each primer set ranged from 8 to 14 with an average of 10 markers per primer pair. The most polymorphic markers were present in the ≤10% frequency class with an absence of monomorphisms (Figure 1). The distance-based SRAP marker cluster analysis successfully discriminated the examined rhizobial isolates. A dendrogram revealed 17 clusters with bootstraps exceeding 60% (Figure 2). Nearly half of the isolates fell into a single large cluster (no. 1), but smaller clusters were also detected for isolates from all four field locations. Colton isolates were grouped into 2 lineages (clusters 1 and 2). Spillman isolates were grouped into 4 lineages (clusters 1, 3, 4, and 5). Shawnee isolates were grouped into 5 lineages

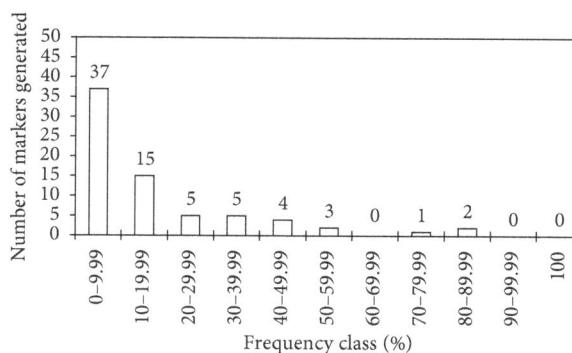

FIGURE 1: Number of sequence-related amplified polymorphism markers generated per frequency class, in percent, over 95 rhizobial DNA samples. Numbers above bars represent the actual numbers of markers.

(clusters 1, 9, 10, and 11). Oakesdale isolates were also grouped into 5 lineages (clusters 1, 6, 7, and 8). Colfax isolates were grouped into 5 lineages (clusters, 12, 13, 14, and 15). None of the isolates from Colfax were in cluster 1, the largest cluster identified in the analysis. Commercial strains were grouped into 3 lineages (clusters 1, 16, and 17). Shawnee field isolates had the highest mean genetic distance, while Spillman field isolates had the lowest mean genetic distance (Table 5).

*3.4. Soil Test Analysis.* Soil test results of the samples were obtained (Table 2). The pH values ranged from 5.2 in Colton to 6 in Shawnee. Ammonium-N was lowest in a Colfax field not previously cultivated for over a decade. Nitrate-N was highest in Colton and Shawnee fields. Phosphorus, potassium, manganese, copper, and iron contents were highest in the Colton field. Boron content was lowest in Oakesdale and Spillman field, and zinc content was highest in Shawnee field.

*3.5. Nitrogen Study.* Pea plant variety ($P = 0.003$) (Table 3) and rhizobial strain ($P = 0.003$) (Table 4) significantly influenced the quantity of N fixed. Ariel plant variety contributed to higher quantity of N fixed than Delta (Table 3). The rhizobial strain (SP1) collected from Spillman field fixed greater quantities of N than Oak1 and Col1 (Table 4). The interaction between plant variety and rhizobial strain did not significantly influence total biomass nor the quantity of N

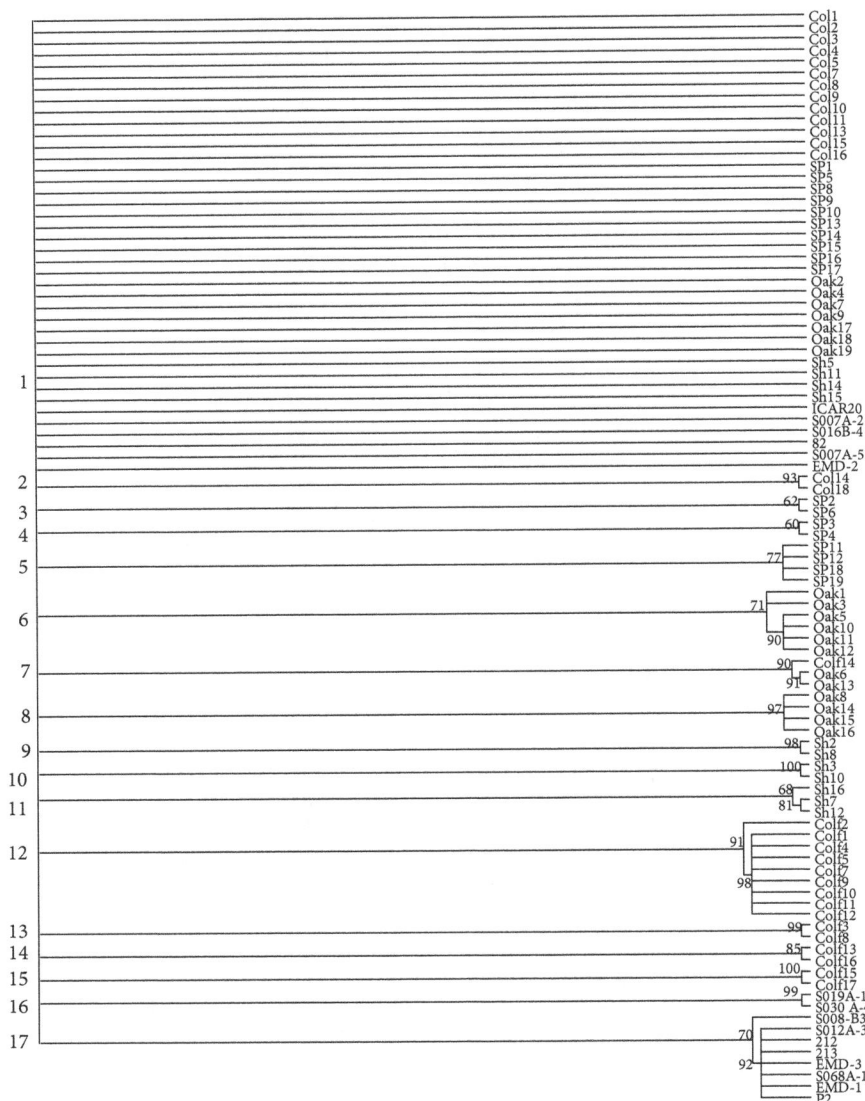

FIGURE 2: Dendrogram representing cluster analysis of 94 strains of *R. leguminosarum* based on 102 sequence-related amplified polymorphisms. Tree was constructed using UPGMA implemented in PAUP program 4 version [21]. Bootstraps values (1,000 replicates) are given at the branch nodes. Branches reproduced in less than 60% of bootstrap replicates were collapsed.

fixed. Pea plant variety ($P = 0.036$) (Table 3) and rhizobial strain ($P = 0.009$) (Table 4) significantly influenced total biomass due to N fixation.

## 4. Discussion

Chromosomal variation was previously reported to be limited in *R. leguminosarum* [25]. However, other studies demonstrated heterogeneity amongst strains within natural populations [26, 27]. The degree of heterogeneity depended on the method used to assess variation and also on the targeted characteristics [28]. In these previous studies, methods based on the characterization of 16S–23S ribosomal DNA internal transcribed spacer PCR-restriction fragment length

polymorphism and plasmid group-specific rep-C PCR amplification were used [3]. In other studies, rhizobial symbiont genetic diversity was determined by mapping restriction site polymorphism of the 16S rRNA genes and by PCR DNA fingerprinting with repetitive sequences [13]. In this study, SRAP markers were used to examine genetic diversity among isolates of *R. leguminosarum* bv. *viciae*. The phylogenetic tree constructed by SRAP analysis demonstrated higher variability among rhizobial isolates than that from 16S rRNA sequences. The distinct nature of Rhizobia from the Colfax field (Figure 2), which had not been cultivated for over a decade, may be attributed to selection pressures associated with the lack of cropping or agrichemical inputs at this location. In contrast to the Spillman field isolates having the lowest mean genetic distance, Shawnee and Colton fields

TABLE 3: Mean values of total biomass, biomass due to nitrogen (N) fixation, quantity (Q) of N per plant, and quantity of N fixed by varieties ($n$ = 24 for the treated plants, and $n$ = 6 for the control plants).

| Varieties | Total biomass (g N) | Biomass due to N fixation (g N) | Q of N/plant (mg N) | Q of N fixed (mg N) |
|---|---|---|---|---|
| Delta | 1.22 | 0.26[a] | 23.78 | 4.96[a] |
| Aragorn | 1.37 | 0.51[ab] | 25.57 | 8.17[ab] |
| Stirling | 1.03 | 0.5[ab] | 18.73 | 8.7[ab] |
| Universal | 1.41 | 0.45[ab] | 28.65 | 11.46[ab] |
| Ariel | 1.6 | 0.93[b] | 31.09 | 19.07[b] |
| $P$ value | 0.146 | 0.036 | 0.12 | 0.043 |
| Delta-Ctr | 0.96 | NA* | 18.7 | NA |
| Aragorn-Ctr | 0.86 | NA | 17.39 | NA |
| Stirling-Ctr | 0.53 | NA | 10.04 | NA |
| Universal-Ctr | 0.96 | NA | 17.19 | NA |
| Ariel-Ctr | 0.67 | NA | 12.02 | NA |
| $P$ value | 0.267 | NA | 0.24 | NA |

Letters that are the same within a column are not significantly different at $P \leq 0.05$.
*Statistical $P$ value is not applicable to this data.

TABLE 4: Mean values of biomass due to nitrogen (N) fixation, quantity (Q) of N per plant, and quantity of N fixed by strains ($n$ = 30).

| Strains | Biomass due to N fixation (g N) | Q of N/plant (mg N) | Q of N fixed (mg N) |
|---|---|---|---|
| Oak1 | 0.14[a] | 15.22[a] | 0.13[a] |
| Col1 | 0.21[a] | 17.13[a] | 2.03[ab] |
| Sh2 | 0.71[ab] | 28.92[ab] | 13.83[bc] |
| SP1 | 1.06[b] | 40.99[b] | 25.9[c] |
| $P$ value | 0.009 | 0.009 | 0.003 |

Letters that are the same within a column are not significantly different at $P \leq 0.05$.

TABLE 5: Mean genetic distance of isolates within locations.

| Isolates | Isolates number | Genetic distance |
|---|---|---|
| Col | 15 | 18.5 |
| SP | 18 | 10.3 |
| Oak | 19 | 16.3 |
| Sh | 11 | 29.5 |
| Colf | 16 | 17.1 |
| Commercial | 16 | 17.9 |

isolates demonstrated higher mean genetic distance, which may reflect a higher diversity of R. leguminosarum isolates within the fields.

Previous studies have demonstrated a significant influence of plant variety on N fixation. Bello et al. [29] found 70% and 25% variability in kg N fixed ha$^{-1}$ among three soybean varieties at two sites. Hafeez et al. [30] found 9%–48% PNF (proportion of plant N supplied by fixation) and 81% variability among Pakistani lentil varieties in a field trial. Abi-Ghanem et al. [10] found significant but relatively low 14.4% and 4.7% PNF variability among five USA lentil and pea varieties, respectively. In this study, host-plant variety significantly influenced the quantity of N fixed, suggesting the possibility of breeding for increased N fixation.

Averaged across all plant varieties, rhizobial strain significantly contributed to increased pea plant total biomass and

quantity of N per plant. The rhizobial strain Sh2 in cluster 7 significantly had higher quantity of N fixed than Oak1. The rhizobial strain SP1 from Spillman field had higher quantity of N fixed than the Oak1 and Col1 clustered in the same cluster 2 (Figure 2), suggesting that the clusters identified in this study are not representative of differences among isolates in their ability to fix N. The higher quantity of N fixed in the Spillman field strain may have been attributed to selection pressures encountered in this field, which had an optimal soil pH of 5.8 and low level of nitrate-N of 6.5 mg·kg$^{-1}$ [31].

The results presented in this study clearly demonstrate genetic differences among isolates of R. leguminosarum bv. viciae both within and between fields. Additional genetic analysis of the isolates examined in this study should allow for increased separation of isolates into distinct lineages. Continuing to examine genetic diversity among rhizobia isolates collected from peas grown in these same field locations will provide opportunities to examine changes in R. leguminosarum bv. viciae population dynamics associated with cultural practices.

## Conflict of Interests

The authors do not have a direct financial relation with the commercial identities mentioned in this paper that might lead to a conflict of interests.

## References

[1] D. M. Sylvia, J. J. Fuhrmann, P. G. Hartel, and D. A. Zuberer, *Principles and Applications of Soil Microbiology*, Prentice Hall, Upper Saddle River, NJ, USA, 2nd edition, 2005.

[2] D. C. Jordan, "Rhizobiaceae," in *Bergey's Manual of Systematic Bacteriology*, N. R. Krieg, Ed., vol. 1, pp. 234–256, Williams & Wilkins, London, UK, 1984.

[3] K. M. Palmer and J. P. W. Young, "Higher diversity of *Rhizobium leguminosarum* biovar viciae populations in arable soils than in grass soils," *Applied and Environmental Microbiology*, vol. 66, no. 6, pp. 2445–2450, 2000.

[4] P. M. Groffman, P. Eagan, W. M. Sullivan, and J. L. Lemunyon, "Grass species and soil type effects on microbial biomass and activity," *Plant and Soil*, vol. 183, no. 1, pp. 61–67, 1996.

[5] I. C. Mendes and P. J. Bottomley, "Distribution of a population of *Rhizobium leguminosarum* bv. *trifolii* among different size classes of soil aggregates," *Applied and Environmental Microbiology*, vol. 64, no. 3, pp. 970–975, 1998.

[6] E. S. Jensen, "The Role of Grain Legume $N_2$ Fixation in the Nitrogen Cycling of Temperate Cropping Systems," RISO National Laboratory, Roskilde, Denamrk, 1997.

[7] M. J. Unkovich, J. S. Pate, and P. Sanford, "Nitrogen fixation by annual legumes in Australian Mediterranean agriculture," *Australian Journal of Agriculture Research*, vol. 48, no. 3, pp. 267–293, 1997.

[8] L. Skot, "Cultivar and rhizobium strain effects on the symbiotic performance of pea (*Pisum sativum*)," *Plant Physiology*, vol. 59, pp. 585–589, 1983.

[9] M. A. Sattar, M. A. Quader, and S. K. A. Danso, "Nodulation, $N_2$ fixation and yield of chickpea as influenced by host cultivar and *Bradyrhizobium* strain differences," *Soil Biology and Biochemistry*, vol. 27, no. 4-5, pp. 725–727, 1995.

[10] R. Abi-Ghanem, L. Carpenter-Boggs, and J. L. Smith, "Cultivar effects on nitrogen fixation in peas and lentils," *Biology and Fertility of Soils*, vol. 47, no. 1, pp. 115–120, 2011.

[11] H. Moawad, S. M. S. Badr El-Din, and R. A. Abdel-Aziz, "Improvement of biological nitrogen fixation in Egyptian winter legumes through better management of Rhizobium," *Plant and Soil*, vol. 204, no. 1, pp. 95–106, 1998.

[12] N. H. Shah, F. Y. Hafeez, M. Arshad, and K. A. Malik, "Response of lentil to *Rhizobium leguminosarum* bv. *viciae* strains at different levels of nitrogen and phosphorus," *Australian Journal of Experimental Agriculture*, vol. 40, no. 1, pp. 93–98, 2000.

[13] G. Laguerre, P. Van Berkum, N. Amarger, and D. Prévost, "Genetic diversity of rhizobial symbionts isolated from legume species within the genera *Astragalus*, *Oxytropis*, and *Onobrychis*," *Applied and Environmental Microbiology*, vol. 63, no. 12, pp. 4748–4758, 1997.

[14] Z. Y. Tan, X. D. Xu, E. N. T. Wang, J. L. Gao, E. Martinez-Romero, and W. X. Chen, "Phylogenetic and genetic relationships of *Mesorhizobium tianshanense* and related rhizobia," *International Journal of Systematic Bacteriology*, vol. 47, no. 3, pp. 874–879, 1997.

[15] W. G. Weisburg, S. M. Barns, D. A. Pelletier, and D. J. Lane, "16S ribosomal DNA amplification for phylogenetic study," *Journal of Bacteriology*, vol. 173, no. 2, pp. 697–703, 1991.

[16] A. J. Drummond, B. Ashton, M. Cheung et al., Geneious v 5.0, 2010.

[17] K. Tamura, D. Peterson, N. Peterson, G. Stecher, M. Nei, and S. Kumar, "MEGA5: molecular evolutionary genetics analysis using maximum likelihood, evolutionary distance, and maximum parsimony methods," *Molecular Biology and Evolution*, vol. 29, pp. 457–472, 2011.

[18] G. Li and C. F. Quiros, "Sequence-related amplified polymorphism (SRAP), a new marker system based on a simple PCR reaction: its application to mapping and gene tagging in *Brassica*," *Theoretical and Applied Genetics*, vol. 103, no. 2-3, pp. 455–461, 2001.

[19] R. Peakall and P. E. Smouse, "GENALEX 6: Genetic analysis in Excel. Population genetic software for teaching and research," *Molecular Ecology Notes*, vol. 6, no. 1, pp. 288–295, 2006.

[20] M. Nei and W. H. Li, "Mathematical model for studying genetic variation in terms of restriction endonucleases," *Proceedings of the National Academy of Sciences of the United States of America*, vol. 76, no. 10, pp. 5269–5273, 1979.

[21] D. L. Swofford, "PAUP: Phylogenetic Analysis Using Parsimony (and Other Methods)," Version 4. Sinauer Associates, Sunderland, Mass, USA, 2003.

[22] J. Felsenstein, "Confidence limits on phylogenies: an approach using the bootstrap," *Evolution*, vol. 39, pp. 783–791, 1985.

[23] T. R. McDermott and M. L. Kahn, "Cloning and mutagenesis of the *Rhizobium meliloti* isocitrate dehydrogenase gene," *Journal of Bacteriology*, vol. 174, no. 14, pp. 4790–4797, 1992.

[24] SAS Institute, "SAS Version 9.2 program and procedures guide," SAS Institute, Cary, NC, USA, 2008.

[25] J. P. W. Young, L. Demetriou, and R. G. Apte, "*Rhizobium* population genetics: enzyme polymorphism in *Rhizobium leguminosarum* from plants and soil in a pea crop," *Applied and Environmental Microbiology*, vol. 53, no. 2, pp. 397–402, 1987.

[26] E. S. P. Bromfield, S. B. Indu, and M. S. Wolynetz, "Influence of location, host cultivar and inoculation on the composition of naturalized populations of *Rhizobium meliloti* on *Medicago sativa* nodules," *Applied and Environmental Microbiology*, vol. 51, pp. 1077–1084, 1986.

[27] N. P. Thurman and E. S. P. Bromfield, "Effect of variation within and between *Medicago* and *Melilotus* species on the composition and dynamics of indigenous populations of *Rhizobium meliloti*," *Soil Biology and Biochemistry*, vol. 20, no. 1, pp. 31–38, 1988.

[28] K. D. Noel and W. J. Brill, "Diversity and dynamics of indigenous *Rhizobium japonicum* populations," *Applied and Environmental Microbiology*, vol. 40, pp. 931–938, 1980.

[29] A. B. Bello, W. A. Ceron-Dias, C. D. Nickell, E. O. Elsheriff, and L. C. Davis, "Influence of cultivar between-row spacing and plantpopulation of fixation of soybeans," *Crop Science*, vol. 20, pp. 751–775, 1980.

[30] F. Y. Hafeez, N. H. Shah, and K. A. Malik, "Field evaluation of lentil cultivars inoculated with *Rhizobium leguminosarum* bv. *viciae* strains for nitrogen fixation using nitrogen-15 isotope dilution," *Biology and Fertility of Soils*, vol. 31, no. 1, pp. 65–69, 2000.

[31] S. P. Harrison, D. G. Y. Jones, and J. P. W. Pung, "*Rhizobium* population genetics: genetic variation within and between populations from diverse locations," *Journal of General Microbiology*, vol. 135, pp. 1061–1069, 1989.

# Permissions

The contributors of this book come from diverse backgrounds, making this book a truly international effort. This book will bring forth new frontiers with its revolutionizing research information and detailed analysis of the nascent developments around the world.

We would like to thank all the contributing authors for lending their expertise to make the book truly unique. They have played a crucial role in the development of this book. Without their invaluable contributions this book wouldn't have been possible. They have made vital efforts to compile up to date information on the varied aspects of this subject to make this book a valuable addition to the collection of many professionals and students.

This book was conceptualized with the vision of imparting up-to-date information and advanced data in this field. To ensure the same, a matchless editorial board was set up. Every individual on the board went through rigorous rounds of assessment to prove their worth. After which they invested a large part of their time researching and compiling the most relevant data for our readers. Conferences and sessions were held from time to time between the editorial board and the contributing authors to present the data in the most comprehensible form. The editorial team has worked tirelessly to provide valuable and valid information to help people across the globe.

Every chapter published in this book has been scrutinized by our experts. Their significance has been extensively debated. The topics covered herein carry significant findings which will fuel the growth of the discipline. They may even be implemented as practical applications or may be referred to as a beginning point for another development. Chapters in this book were first published by Hindawi Publishing Corporation; hereby published with permission under the Creative Commons Attribution License or equivalent.

The editorial board has been involved in producing this book since its inception. They have spent rigorous hours researching and exploring the diverse topics which have resulted in the successful publishing of this book. They have passed on their knowledge of decades through this book. To expedite this challenging task, the publisher supported the team at every step. A small team of assistant editors was also appointed to further simplify the editing procedure and attain best results for the readers.

Our editorial team has been hand-picked from every corner of the world. Their multi-ethnicity adds dynamic inputs to the discussions which result in innovative outcomes. These outcomes are then further discussed with the researchers and contributors who give their valuable feedback and opinion regarding the same. The feedback is then collaborated with the researches and they are edited in a comprehensive manner to aid the understanding of the subject.

Apart from the editorial board, the designing team has also invested a significant amount of their time in understanding the subject and creating the most relevant covers. They scrutinized every image to scout for the most suitable representation of the subject and create an appropriate cover for the book.

The publishing team has been involved in this book since its early stages. They were actively engaged in every process, be it collecting the data, connecting with the contributors or procuring relevant information. The team has been an ardent support to the editorial, designing and production team. Their endless efforts to recruit the best for this project, has resulted in the accomplishment of this book. They are a veteran in the field of academics and their pool of knowledge is as vast as their experience in printing. Their expertise and guidance has proved useful at every step. Their uncompromising quality standards have made this book an exceptional effort. Their encouragement from time to time has been an inspiration for everyone.

The publisher and the editorial board hope that this book will prove to be a valuable piece of knowledge for researchers, students, practitioners and scholars across the globe.

# List of Contributors

**Robert Orangi Nyambati**
Kenya Forestry Research Institute, Maseno Regional Centre, P.O. Box 5199, Kisumu, Kenya

**Peter Asbon Opala**
Department of Soil Science, Maseno University, P.O. Box, Private bag, Maseno, Kenya

**Aderonke Adetutu Okoya**
Institute of Ecology and Environmental Studies, Obafemi Awolowo University, Ile-Ife, Nigeria

**Aderemi Okunola Ogunfowokan and Olabode Idowu Asubiojo**
Department of Chemistry, Obafemi Awolowo University, Ile-Ife, Nigeria

**Nelson Torto**
Department of Chemistry, Rhodes University, Grahamstown 6140, South Africa

**M. B. Scheer**
Research and Development Assistance (APD/DMA), SANEPAR, Curitiba, PR, Brazil
Brazilian Agricultural Research Corporation (Embrapa Florestas), Colombo, PR, Brazil

**G. R. Curcio**
Brazilian Agricultural Research Corporation (Embrapa Florestas), Colombo, PR, Brazil

**C. V. Roderjan**
Federal University of Parana (UFPR), Curitiba, PR, Brazil

**H. Etesami, H. Mirsyedhosseini and H. A. Alikhani**
Department of Soil Science, College of Agriculture & Natural Resources, Tehran University, Karaj 31587-77871, Iran

**Venkat Lakshmi**
Department of Earth and Ocean Sciences, University of South Carolina, Columbia, SC 29208, USA

**K. Rinu, Mukesh Kumar Malviya, Priyanka Sati and Anita Pandey**
Biotechnological Applications, G. B. Pant Institute of Himalayan Environment and Development, Kosi-Katarmal, Almora, Uttarakhand 263 643, India

**S. C. Tiwari**
Department of Botany and Microbiology, HNB Garhwal University, Srinagar, Uttarakhand 246 174, India

**Héctor Estrada-Medina, Juan José María Jiménez-Osornio and Wilian de Jesús Aguilar Cordero**
Departamento de Manejo y Conservaci´on de Recursos Naturales Tropicales (PROTROPICO), Campus de Ciencias Biol´ogicas y Agropecuarias (CCBA), Universidad Aut´onoma de Yucat´an (UADY),
Km 15.5 Carretera M´erida - Xmatkuil, M´erida, Yucat´an 97315, Mexico

**Francisco Bautista**
Centro de Investigaciones en Geografia Ambiental (CIGA), Universidad Nacional Aut´onoma de M´exico (UNAM), Antigua Carretera a Patzcuaro No. 8701, Col. Ex-Hacienda de San Jos´e de La Huerta, Morelia, Michoac´an 58190, Mexico

**José Antonio González-Iturbe**
Facultad de Arquitectura (UADY), Calle 50 S/N x 57 y 59 Ex-Convento de La Mejorada, M´erida, Yucat´an 97000, Mexico

**Tadele Amare**
Centre for Development and Environment, University of Bern, Hallerstrasse 10, CH-3012 Bern, Switzerland
Amhara Regional Agricultural Research Institute, P.O. Box 527, Bahir Dar, Ethiopia

**Christian Hergarten, Hans Hurni and Bettina Wolfgramm**
Centre for Development and Environment, University of Bern, Hallerstrasse 10, CH-3012 Bern, Switzerland

**Birru Yitaferu**
Amhara Regional Agricultural Research Institute, P.O. Box 527, Bahir Dar, Ethiopia

**Yihenew G. Selassie**
Bahir DarUniversity, P.O. Box 79, Bahir Dar, Ethiopia

**Gebreyesus Brhane Tesfahunegn**
College of Agriculture, Aksum University, Shire Campus, P.O. Box 314, Shire, Ethiopia
Center for Development Research, University of Bonn,Walter-Flex-Straße 3, 53113 Bonn, Germany

**Paul L. G. Vlek**
Center for Development Research, University of Bonn,Walter-Flex-Straße 3, 53113 Bonn, Germany

**Xi Zhang, Tingting Liu, Chen Xu, Dechao Duan, Cheng Peng, Shenhai Zhu and Jiyan Shi**
Institute of Environmental Science and Technology, Zhejiang University, Hangzhou, Zhejiang 310058, China

**Feng Li**
Institute of Environmental Science and Technology, Zhejiang University, Hangzhou, Zhejiang 310058, China
College of Materials and Environmental Engineering, Hangzhou Dianzi University, Hangzhou, Zhejiang 310018, China

**Mohamed A. M. Abd Elbasit**
Arid Land Research Center, Tottori University, 1390 Hamasaka, Tottori 680-0001, Japan
Desertification Research Institute, National Center for Research, Khartoum 11111, Sudan

**Hiroshi Yasuda**
Arid Land Research Center, Tottori University, 1390 Hamasaka, Tottori 680-0001, Japan

**Jinbai Huang**
College of Hydraulic and Architecture of North East Agricultural University, Harbin 150030, China

**CSP Ojha**
Department of Civil Engineering, Indian Institute of Technology, Roorkee 247667, India

**Eltayeb O. Adam**
Remote Sensing Authority, National center for Research, Khartoum 11111, Sudan

**R. Rezaei Arshad and Gh. Sayyad**
Department of Soil Science, Faculty of Agrriculture, Shahid Chamran University of Ahvaz, Ahvaz, Iran

**M. Mosaddeghi**
Department of Soil Science, College of Agriculture, Isfahan University of Technology, Isfahan, Iran

**B. Gharabaghi**
School of Engineering, University of Guelph, Guelph, ON, Canada

**Samuel I. Haruna and Nsalambi V. Nkongolo**
Center of Excellence for Geospatial Information Sciences, Department of Agriculture and Environmental Science, Lincoln University, Jefferson City, MO 65102-0029, USA

**Ahmed A. Khalil**
Soils, Water, and Environment Research Institute, Agricultural Research Centre, P.O. Box 175 El-Orman, Giza 12112, Egypt

**Manuel F. Restrepo, Claudia P. Florez, Nelson W. Osorio and Juan D. León**
Universidad Nacional de Colombia, Calle 59A No. 63-20, Oficina 14-225 050034, Medell´ın, Colombia

**Mahmoud Ahmed**
Department of Transportation, Route 9A Project, Suite 1701, 115 Broadway, New York, NY 10006, USA

**Athanase Nduwumuremyi**
Department of Natural Resource Management, Rwanda Agriculture Board (RAB), P.O. Box 5016, Kigali, Rwanda
Department of Agricultural Resource Management, Kenyatta University, P.O. Box 43844-00100, Nairobi, Kenya

**Jayne Njeri Mugwe**
Department of Agricultural Resource Management, Kenyatta University, P.O. Box 43844-00100, Nairobi, Kenya

**Vicky Ruganzu and Athanase Cyamweshi Rusanganwa**
Department of Natural Resource Management, Rwanda Agriculture Board (RAB), P.O. Box 5016, Kigali, Rwanda

**Rita Abi-Ghanem**
Department of Crop and Soil Sciences, Washington State University, Pullman, WA 99164-6420, USA

**Jeffrey L. Smith**
Land Management and Water Conservation Research Unit, USDA-ARS and Washington State University, Pullman, WA 99164-6421, USA

**George J. Vandemark**
Grain Legume Genetics Physiology Research, USDA-ARS and Washington State University, Pullman, WA 99164-6421, USA

www.ingramcontent.com/pod-product-compliance
Lightning Source LLC
Chambersburg PA
CBHW050457200326
41458CB00014B/5214